AF380062

Solid Mechanics and Its Applications

Volume 203

Series Editor

G. M. L. Gladwell
Department of Civil Engineering, University of Waterloo, Waterloo, Canada

For further volumes:
http://www.springer.com/series/6557

Aims and Scope of the Series

The fundamental questions arising in mechanics are: *Why?*, *How?*, and *How much?* The aim of this series is to provide lucid accounts written by authoritative researchers giving vision and insight in answering these questions on the subject of mechanics as it relates to solids.

The scope of the series covers the entire spectrum of solid mechanics. Thus it includes the foundation of mechanics; variational formulations; computational mechanics; statics, kinematics and dynamics of rigid and elastic bodies: vibrations of solids and structures; dynamical systems and chaos; the theories of elasticity, plasticity and viscoelasticity; composite materials; rods, beams, shells and membranes; structural control and stability; soils, rocks and geomechanics; fracture; tribology; experimental mechanics; biomechanics and machine design.

The median level of presentation is the first year graduate student. Some texts are monographs defining the current state of the field; others are accessible to final year undergraduates; but essentially the emphasis is on readability and clarity.

Atul Tiwari

Editor

Nanomechanical Analysis of High Performance Materials

 Springer

Editor
Atul Tiwari
Mechanical Engineering
University of Hawaii
Honolulu, HI
USA

ISSN 0925-0042
ISBN 978-94-007-6918-2   ISBN 978-94-007-6919-9   (eBook)
DOI 10.1007/978-94-007-6919-9
Springer Dordrecht Heidelberg New York London

Library of Congress Control Number: 2013944750

Preface

I was always curious to know the reason behind the failure of a material. What triggers failure in a material at a molecular level was hard to find until the invention of nanoindentation technique. Researchers could gain control over structure and properties of materials as they can select proper ingredients that lead to products with extended service life. The developments of new nanoindentation test methods have helped researchers in analyzing the intrinsic properties of the materials in larger domain. The nanoindentation technique appears fairly straightforward; however, improper knowledge may lead to erroneous results. Since the technique is relatively new, few books are available that can develop a deeper understanding in the younger generation of researchers.

This book is edited for those who are seeking information on how to use the nanoindentation technique and how to analyze the data, so that the relevant information can be extracted. I have invited leaders to contribute their chapters that can be easily understood by the new users. This is a unique book in which both equipment manufacturers and their users have contributed their chapters. New users will learn the techniques directly from the inventors and the manufacturer will gain insight on what else is needed for the wider adaptability of this technique.

New test methods, such as PeakForce QNM from Bruker and substrate independent measurements from Agilent are detailed in this book. Fischer-Cripps describes the technicalities involved in analyzing super hard material, while Hysitron teaches a rapid method for mapping the hardness and modulus of materials. Similarly, Micromaterials has described various factors that should be taken into account while conducting environmental nanomechanical analysis.

This book comprehensively covers a broad range of materials. Chapters are written on polymers, ceramics, hybrids, biomaterials, metal oxide, nanoparticles, minerals, carbon nanotubes, and welded joints. Dedicated chapter introduces the topic and few chapters teach advanced numerical modeling needed to understand the properties of the intricate materials.

Being an editor, I can assure readers that this book will certainly help them in developing a deeper understanding on concepts related to nanomechanical analysis of materials. The book has been written for a broad readership and will help young researchers as well as senior learners.

Atul Tiwari, Ph.D.

Contents

Part III   Contributions from Users

Contributors

Phillip Agee Agilent Technologies Inc., Chandler, AZ, USA

Tunik Alla Paton Electric Welding Institute, National Academy of Sciences of Ukraine, 11, Bozhenko Str., Kyiv 150, Kiev 03680, Ukraine

Ishchenko Anatoliy Paton Electric Welding Institute, National Academy of Sciences of Ukraine, 11, Bozhenko Str., Kyiv 150, Kiev 03680, Ukraine

M. R. Ayatollahi Fatigue and Fracture Research Laboratory, School of Mechanical Engineering, Iran University of Science and Technology, Tehran, Iran

Annalisa Bandini Department of Civil, Chemical, Environmental and Materials Engineering, University of Bologna, via Terracini 28, 40131 Bologna, Italy

Ben D. Beake Micro Materials Ltd.,, Willow House, Yale Business Village, Ellice Way, Wrexham LL13 7YL, UK

C. Begau Lehrstuhl Werkstoffmechanik, Interdisciplinary Centre for Advanced Materials Simulation, Ruhr-Universität, 44780 Bochum, Germany

Gerard A. Bell Micro Materials Ltd.,, Willow House, Yale Business Village, Ellice Way, Wrexham LL13 7YL, UK

E. Bemporad Department of Mechanical and Industrial Engineering, University of Rome "ROMA TRE", Via della Vasca Navale 79, 00146 Rome, Italy

Paolo Berry Department of Civil, Chemical, Environmental and Materials Engineering, University of Bologna, via Terracini, 28, 40131 Bologna, Italy

Parag Bhargava Metallurgical Engineering and Materials Science, IIT Bombay, Mumbai, India

F. Carassiti Department of Mechanical and Industrial Engineering, University of Rome "ROMA TRE", Via della Vasca Navale 79, 00146 Rome, Italy

A. Charitidis School of Chemical Engineering, National Technical University of Athens, 9, Heroon Polytechneiou st., Zografos, Athens 157 80, Greece

Jian Chen Jiangsu Key Laboratory of Advanced Metallic Materials, School of Materials Science and Engineering, Southeast University, Nanjing 211189, China

Liming Dai Department of Macromolecular Science and Engineering, Case Western Reserve University, Cleveland, OH 44106, USA

Hanshan Dong School of Metallurgy and Materials, University of Birmingham, Birmingham B15 2TT, UK

A. Dragatogiannis School of Chemical Engineering, National Technical University of Athens, 9, Heroon Polytechneiou st., Zografos, Athens 157 80, Greece

Feng Du Department of Macromolecular Science and Engineering, Case Western Reserve University, Cleveland, OH 44106, USA

P. S. Engels Lehrstuhl Werkstoffmechanik, Interdisciplinary Centre for Advanced Materials Simulation, Ruhr-Universität, 44780 Bochum, Germany

A. C. Fischer-Cripps Fischer-Cripps Laboratories Pty Ltd,, Sydney, Australia

S. Gupta Lehrstuhl Werkstoffmechanik, Interdisciplinary Centre for Advanced Materials Simulation, Ruhr-Universität, 44780 Bochum, Germany

Ude D. Hangen Hysitron, Inc. Technologiezentrum am Europaplatz, Dennewartstrasse, Aachen, Germany

A. Hartmaier Lehrstuhl Werkstoffmechanik, Interdisciplinary Centre for Advanced Materials Simulation, Ruhr-Universität, 44780 Bochum, Germany

Ajay Kumar Jena Metallurgical Engineering and Materials Science, IIT Bombay, Mumbai, India

Johnson Joseph Department of Mechanical Engineering, University of Kentucky, Lexington, KY 40506, USA

Khokhlova Julia Paton Electric Welding Institute, National Academy of Sciences of Ukraine, 11, Bozhenko Str., Kyiv 150, Kiev 03680, Ukraine

A. Karimzadeh Fatigue and Fracture Research Laboratory, School of Mechanical Engineering, Iran University of Science and Technology, Tehran, Iran

Elias P. Koumoulos School of Chemical Engineering, National Technical University of Athens, 9, Heroon Polytechneiou st., Zografos, Athens 157 80, Greece

Y. Charles Lu Department of Mechanical Engineering, University of Kentucky, Lexington, KY 40506, USA

A. Ma Lehrstuhl Werkstoffmechanik, Interdisciplinary Centre for Advanced Materials Simulation, Ruhr-Universität, 44780 Bochum, Germany

Khokhlov Maksym Paton Electric Welding Institute, National Academy of Sciences of Ukraine, 11, Bozhenko Str., Kyiv 150, Kiev 03680, Ukraine

Mohsen Mohseni Department of Polymer Engineering and Color Technology, Amirkabir University of Technology, Tehran, 15875-4413, Iran

J. Němeček Faculty of Civil Engineering Prague, Czech Technical University, Prague, Czech Republic

A. H. W. Ngan Department of Mechanical Engineering, University of Hong Kong, Pokfulam Road, Hong Kong, People's Republic of China

Sharanabasappa B. Patil Metallurgical Engineering and Materials Science, IIT Bombay, Mumbai, India

Bede Pittenger Bruker Nano Surfaces Division, Robin Hill Road, Santa Barbra, CA 93117, USA

Bahram Ramezanzadeh Department of Polymer Engineering and Color Technology, Amirkabir University of Technology, Tehran, 15875-4413, Iran

B. Schmaling Lehrstuhl Werkstoffmechanik, Interdisciplinary Centre for Advanced Materials Simulation, Ruhr-Universität, 44780 Bochum, Germany

N. Schwarzer Saxonian Institute of Surface Mechanics, Tankow 1, 18569 Ummanz/Rügen, Germany

M. Sebastiani Department of Mechanical and Industrial Engineering, University of Rome "ROMA TRE", Via della Vasca Navale 79, 00146 Rome, Italy

Douglas D. Stauffer Hysitron, Inc. Technologiezentrum am Europaplatz, Dennewartstrasse, Aachen, Germany

Chanmin Su Bruker Nano Surfaces Division, Robin Hill Road, Santa Barbra, CA 93117, USA

S. A. Syed Asif Hysitron, Inc. Technologiezentrum am Europaplatz, Dennewartstrasse, Aachen, Germany

Atul Tiwari Department of Mechanical Engineering, University of Hawaii at Manoa, Honolulu, HI 96822, USA

Hossein Yahyaei Department of Polymer Engineering and Color Technology, Amirkabir University of Technology, Tehran, 15875-4413, Iran

Hossein Yari Department of Polymer Engineering and Color Technology, Amirkabir University of Technology, Tehran, 15875-4413, Iran

Qiuhong Zhang University of Dayton Research Institute, University of Dayton, 300 College Park, Dayton, OH 45469, USA

Part I
Introduction

Nanotribological Characterization of Polymeric Nanocoatings: From Fundamental to Application

Mohsen Mohseni, Hossein Yahyaei, Hossein Yari
and Bahram Ramezanzadeh

Abstract Polymers are chemical compounds or mixture of compounds consisting of repeating structural units created through a process known polymerization. These are important groups of materials made up of long chain carbon, covalently bonded together. Polymerization is a process in which monomeric molecules react together chemically to form macromolecules. Polymers are now finding increasing use in engineering applications due to unique properties. Mechanical strength of polymers is of prime importance in engineering applications. Polymers in their service life are exposed to different mechanical and thermal stresses. Durability of polymer strongly depends on the resistance of these materials against environmental condition. In order to assess the strength of material, good knowledge on mechanic of materials is imperative. In this manner, this section aims at introducing mechanical properties of polymers.

1 Stress and Strain: Back to Fundamentals

In order to characterize the mechanical strength of polymers, knowledge about definition and types of stress and strain is necessary. Assume a piece of material with length l_0 and cross-sectional area A_0. Applying an axially external force along the length of the specimen, stress field is developed. As a result of this external force (F); stress τ normal to the cross-section is created according to Eq. 1 (Nielsen 1962):

$$\tau = \frac{F}{A_0} \tag{1}$$

This is a simple term that is possible for definition of stress, but in general, it is necessary to resolve the stresses on a specimen into nine components. Stress

M. Mohseni (✉) · H. Yahyaei · H. Yari · B. Ramezanzadeh
Department of Polymer Engineering and Color Technology, Amirkabir University of Technology, 158754413 Tehran, Iran
e-mail: mmohseni@aut.ac.ir

A. Tiwari (ed.), *Nanomechanical Analysis of High Performance Materials*,
Solid Mechanics and Its Applications 203, DOI: 10.1007/978-94-007-6919-9_1,
© Springer Science+Business Media Dordrecht 2014

state is described in terms of a stress tensor, τ_{ij} according to Eq. 2 (Kinloch and Young 1983):

$$\tau_{ij} = \begin{bmatrix} \tau_{11} & \tau_{12} & \tau_{13} \\ \tau_{21} & \tau_{22} & \tau_{23} \\ \tau_{31} & \tau_{32} & \tau_{33} \end{bmatrix} \tag{2}$$

The first subscript gives the normal to the plane on which the stress acts and the second subscript defines the direction of the stress. The components τ_{11}, τ_{22} and τ_{33} are known as direct or normal stresses because the applied force is perpendicular to plane and parallel with the normal vector of plane. Accordingly, this stress is named tensile stress and gets a positive sign if the force and normal vector of plane are in the same direction and is called compress stress with negative sign if it has an opposite direction. In other component the force is perpendicular on normal vector of plane and can shear the specimen and is named shear stress.

Response of specimen against development of stress is named strain. In general, it is defined as the difference in length of specimen in one dimension per unit of length. Strain is defined according to Eq. 3 (Nielsen 1962):

$$\varepsilon = \frac{\Delta l}{l_0} = \frac{l - l_0}{l_0} \tag{3}$$

where Δl is change in length and l_0 is the initial length. The complete analysis of the strain in a specimen requires careful consideration of the relative displacements in the body under stress.

1.1 Stress–Strain Behavior

When the stress is directly proportional to strain, the material is said to obey Hooke's law. The slope of the straight line portion of curves in Fig. 1 is equal to

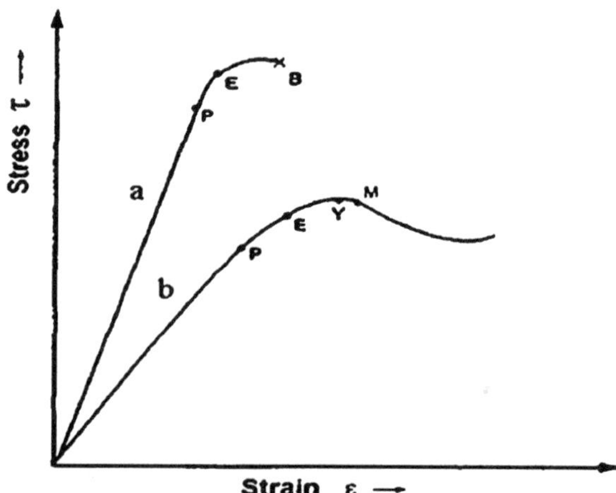

Fig. 1 A typical stress–strain curve of materials. **a** Metals. **b** Polymers

the modulus of elasticity. The maximum stress point on the curve, up to which stress and strain remain proportional, is called the proportional limit (Nat 1980).

In fact, with increasing the stress, deformation in specimen increases. Strain is proportional to stress increase up to a certain point (point p in Fig. 1) at which point the kind of deformation of material against stress changes.

Most materials return to their original size and shape, even if the external load exceeds the proportional limit. The elastic limit represented by the point E in Fig. 1 is the maximum load which may be applied without leaving any permanent deformation of the material. If the material is loaded beyond its elastic limit, it does not return to its original size and shape, and is said to have been permanently deformed. On continued loading, a point is reached at which the material starts yielding. This point (point Y in Fig. 1b) is known as the yield point, where an increase in strain occurs without an increase in stress. The point B in Fig. 1a represents break of the material. In polymers before breaking may occur strain softening (cold drawing) appears. In this region subsequent deformation can occur without further increase in stress. This deformation is named plastic deformation which will be reviewed in the next part. After cold drawing, strain hardening step occurs. In this region, polymer starts to be stronger before breaking commences.

1.2 Tensile Modulus

According to Eqs. 2 and 3 one can obtain the modulus of elasticity, E, which is known as Young's Modulus:

$$E = \frac{\tau}{\varepsilon} \tag{4}$$

This equation expresses the required tension stress for a determined strain. The modulus may be thought of as stiffness, or a material's resistance to elastic deformation. The greater the modulus, the stiffer is the material, or the smaller is the elastic strain.

1.3 Shear Properties

Figure 2 shows the influence of a shear force F_t acting on the area A of a rectangular sample which causes the displacement Δu. The valid expressions are defined by shear strain (Eq. 5):

$$\gamma = \frac{\Delta u}{l} \tag{5}$$

And shear stress is defined as in Eq. 6:

$$\tau = \frac{F_t}{A} \tag{6}$$

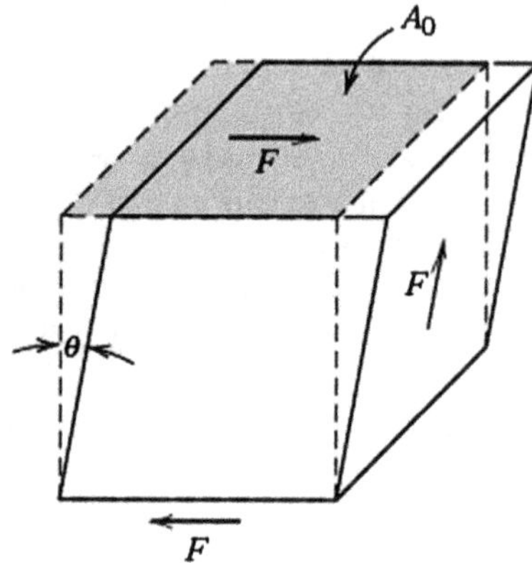

Fig. 2 Shear force on a specimen

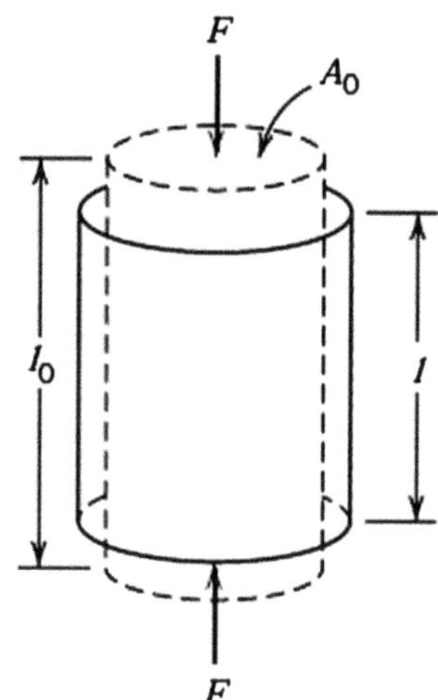

Fig. 3 Compression test

1.4 Shear Modulus

The ratio of shear stress to shear strain represents the shear modulus G (Eq. 7).

$$G = \frac{\tau}{\gamma} \tag{7}$$

1.5 Compressive Properties

The isotropic compression due to the pressure acting on all sides of the parallel cylinder is shown in Fig. 3 (Eq. 8):

$$\kappa = \frac{\Delta V}{V_0} \tag{8}$$

where ΔV is the reduction of volume due to deformation of the specimen with the original volume V_0.

2 Deformation of Polymers

2.1 Elastic Deformation

The elastic behavior of polymers reflects the deformation of the structure on a molecular level. In the high-modulus polymers, the deformation takes place essentially through bending and stretching of the aligned polymer backbone bonds which requires high forces. On an atomic scale, macroscopic elastic strain is manifested as small changes in the interatomic spacing and the stretching of interatomic bonds. As a consequence, the magnitude of the modulus of elasticity is a measure of the resistance to separation of adjacent atoms/ions/molecules, that is, the interatomic bonding forces.

Differences in modulus values between metals, ceramics, and polymers are a direct consequence of the different types of atomic bonding that exist for the three materials types. The mechanism of elastic deformation in semi-crystalline polymers in response to tensile stresses is the elongation of the chain molecules from their stable conformations, in the direction of the applied stress, by the bending and stretching of the strong covalent bonds. In addition, there may be some slight displacement of adjacent molecules, which is resisted by relatively weak secondary or Van der Waals bonds.

2.2 Viscoelasticity

A distinctive feature of the mechanical behavior of polymers is their response to an applied stress or strain depends upon the rate or time period of loading. In elastic materials the stress is proportional to strain whereas viscose materials, such as liquids tend to obey Newton's law whereby the stress is proportional to strain-rate. The behavior of many polymers can be thought of as being somewhere between elastic solid and viscose liquid. Owing to this, dual behavior and displaying elastic and viscose behaviors in the same time make polymers viscoelastic.

Since polymers are viscoelastic, the mechanical behavior of polymers depends upon testing rate as well as temperature, therefore let us review time related properties of polymers in this section.

2.3 Creep

Under the action of a constant load a polymeric material experiences a time dependent increase in strain called creep. Creep is therefore the result of increasing strain over time under constant load.

2.4 Relaxation

Relaxation is the stress reduction which occurs in a polymer when it is subjected to a constant strain. This data is of significance in the design of parts which undergo long-term deformation.

2.5 Plastic Flow

Ductile polymers tend to have a fairly well defined yield point as shown in Fig. 1. Some semi-crystalline polymers can be cold-drawn to extension ratios in excess during which a stable neck extends along the specimen (Bowden 1973).

From molecular point of view, during the initial stage of deformation, the chains in the polymer slip past each other and align in the loading direction. This causes the lamellar ribbons simply to slide past one another as the tie chains within the amorphous regions become extended. Continued deformation in the second stage occurs by tilting the lamellae so that the chain folds become aligned with the tensile axis. Next, crystalline block segments separate from the lamellae, which segments remain attached to one another by tie chains. In the final stage, the blocks and tie chains become oriented in the direction of the tensile axis. Thus, appreciable tensile deformation of semi-crystalline polymers produces a highly oriented structure. During deformation the spherulites experience shape changes for moderate levels of elongation. However, for large deformations, the spherulitic structure is virtually destroyed.

2.6 Glass Transition Temperature

Glass transition temperature (T_g) is a thermomechanical property of polymers and is very important in mechanical studying. At temperatures above the glass transition temperature, at least at slow to moderate rates of deformation, the amorphous polymer is soft and flexible and is either a rubber or a very viscous liquid. Mechanical properties show profound changes in the region of the glass transition, whilst below this temperature polymer is hard, rigid and glassy. For this reason, T_g can be considered the most important material characteristic of a polymer as far as mechanical properties are concerned. Many physical and mechanical properties change rapidly with temperature in the glass transition region (Nielsen 1962).

Thermodynamic theory based on Gibbs and Di Marzio's propose, considers that the glass transition temperature of a given polymer corresponds to the value at which a given chain exhibits only one unique conformation. From the calculation of the variation of entropy of a chain as a function of the temperature, T_g is defined as the temperature at which the conformational entropy is equal to zero.

2.7 Physical Damaging of Polymers

Polymer resists different stress up to a certain amount after which it deforms irreversibly, depending on its structure and condition of damaging (such as rate of stressing, temperature) in two forms.

2.8 Ductile Damage

As mentioned earlier after passing from yield stress, polymer deforms plastically. This damage may occur in two ways.

2.8.1 Shear Yielding

One important mechanism which can lead to plastic deformation in polymers is shear yielding (Bowden 1973). It takes place essentially at constant volume and leads to permanent change in specimen shape. There is obviously a translation of molecules past each other during shear yielding but on annealing a deformed glassy polymer above its T_g it often completely recovers its original shape. The molecules are well anchored at entanglements in the structure and are not broken to any motions when they have sufficient mobility above T_g (Bowden and Young 1974).

2.8.2 Crazing

A craze is initiated when an applied tensile stress causes microvoids to nucleate at point of high stress concentrations in the polymer created by scratches, flaws, cracks, dust particles, molecular heterogeneities.

The microvoids develop in the plane perpendicular to the maximum principal stress. The resulting localized yielded region therefore consists of interpenetrating system of voids and polymer fibrils and is known as craze. This typically results in slow cavity expansion, followed by crack extension through the craze, and finally crack propagation proceeds by a craze.

Shear yielding occurs essentially at constant volume whereas crazing occurs with an increase in volume.

2.9 Fracture Damage

In most practical situations a fracture originates from local concentrations of stress at flaws, scratches or notches. When a polymer solid is deformed the molecules slide past each other and tend to uncoil, breaking secondary bonds. Molecular fracture through the scission of primary bonds will take place if, for any reason, the flow of molecules past each other is restricted due to the nature of the polymer structure.

3 Scratch and Mar Resistance of the Polymeric Materials

The lifetime of polymers is often dependent on their mechanical properties. For example, as a result of abrasion, loss of optical properties can occur. Scratch and mar are two types of frequent damages occurred when the surface of polymeric materials is exposed to any mechanical force and/or stress. In these kinds of damages the dimensions of failures vary from very small, i.e. micron size scratches, to very large up to a millimeter width. Mar is referred to those mechanical damages with small size, a few microns in width and depth, produced typically by automatic car wash brushes, hand washing and sand abrading particles (Courter and Mar 1997; Jardret et al. 2000; Hara et al. 2000; Barletta et al. 2010; Bautista et al. 2011). In fact, the cause of mar is a slight mechanical stress which is not easily visible with human eyes. However, sharp objects such as tree branches and keys induce normal forces. The scratch groove will be created when a sharp object moves across the polymer surface. The schematic illustration of the process named scratch is shown in Fig. 4.

Scratch involves different issues including friction, wear and lubrication. The friction is occurred when an object moves onto on another one. Depending on the types of the materials, the friction may follow three main forms including (1) the friction force is proportional to the normal load, (2) the friction force is independent of the apparent area of the contact zone and (3) the friction force is independent of the sliding velocity. Polymers are viscoelastic materials with indentation softness which follows only the third friction form. Friction can be classified into two categories. In the first one the friction is due to adhesion (F_{adh}) and in the second the friction is a result of deformation (F_{def}). The former is related to the molecular attractive forces that operate at the asperities and the latter is attributed to energy dissipation by the surface deformation when friction occurs (Hutchings 1992; Cartledge et al. 1996; Yang and Wu 1997; Persson 2000; Schwarzentruber 2002; Tahmassebi et al. 2010; Ramezanzadeh et al. 2011a, b).

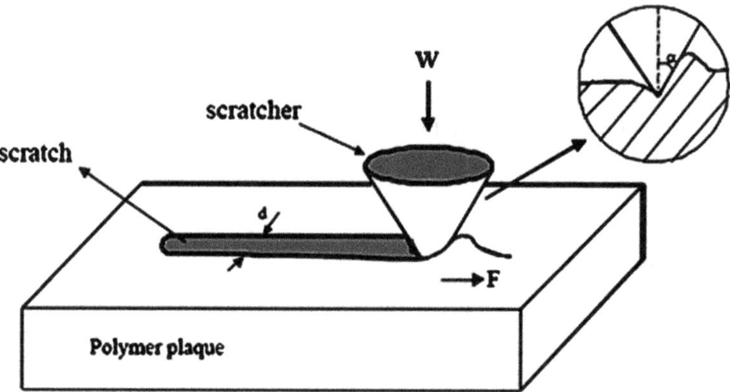

Fig. 4 The schematic illustration of scratch formation on schematic of polymeric materials surfaces

Depending on the cause of scratch, stress intensity, scratch velocity and viscoelastic properties of the polymer, a wide range of surface deformations including elastic (no damage), visco-plastic (damage) and fracture (severe damage) can be observed. Furthermore, the tribological properties of polymeric materials, depending on the scratching factors, are shown in four types of failure damages including detachment, delamination, cracking and/or spalling of the materials. It seems that different parameters including polymer physical and mechanical properties, substrate and the interfacial interaction play significant roles on the polymer tribological properties (Hutchings 1992; Cartledge et al. 1996; Briscoe et al. 1996a, b; Yang and Wu 1997; Persson 2000; Bertrand-Lambotte et al. 2002; Tahmassebi et al. 2010).

Different parameters like scratching indenter tip morphology (tip radiance and stiffness), tip velocity and polymer viscoelastic properties can affect the polymer response against applied stresses. Figure 5 shows that stress applied to the polymer surface during scratch test can be divided into two forms of tangential and vertical vectors. Tangential forces are responsible for compression and stretching of the polymer in front and behind of the scratching tip respectively. Tensile stresses produced behind such a tip can lead to cracks in the polymer and/or aid in scratch formation. Consequently, the tensile stress/strain behavior of the polymer surface can be used to predict scratch behavior. Based on the stress applied, both ductile and brittle behaviors can be seen. At low intensities of the stress, the polymer behaves more ductile and therefore scratch with plastic morphology will be formed. For ductile polymers, the plastic deformation correlates with the evolution of the ratio of elastic modulus/hardness (E/H). On the other hand, fracture deformation occurs when the stress applied to the polymer is high enough that polymer cannot tolerate

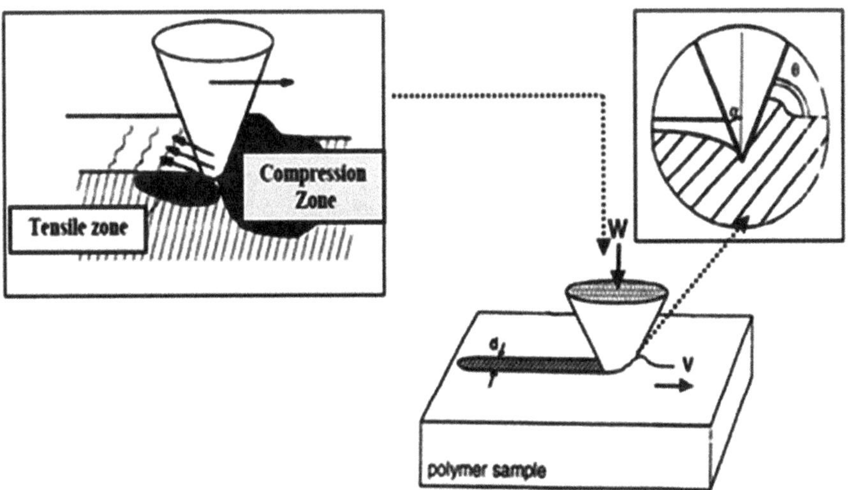

Fig. 5 Schematic illustration of the applied stress by scratch indenters on the polymer surface deformation

it. In this case, polymer show more brittleness and the scratch may be linked to the tensile behavior of the polymer (Cartledge et al. 1996; Briscoe et al. 1996a, b; Yang and Wu 1997).

4 Polymer Damage Classification from a Tribological View Point

There are three main classifications of polymer damage including (a) damage without material exchange, (b) damage with loss of material, i.e. wear and (c) damage with material pick up. In the first type, only surface geometry and/ or topography of the polymer surface without material exchange will be influenced. The severity of this kind of damage depends on polymer characteristics like Young's modulus, hardness and toughness. In fact, the hardness and toughness of the polymeric materials are crucial parameters which could influence the polymer surface topography changes and surface cracking (Cartledge et al. 1996; Briscoe et al. 1996a, b; Yang and Wu 1997; Persson 2000; Bertrand-Lambotte et al. 2002; Ramezanzadeh 2012a, b). Wear is another type of damage including polymer material loss both by the polymer detachment and/or gradual removal of the polymeric material. Damage with material pick up is another cause of polymeric materials tribology as a result of surface indentation and/or scratch (Briscoe et al. 1996a, b; Bertrand-Lambotte et al. 2002; Ramezanzadeh 2012a, b).

5 Polymer Surface Damage Classification from Morphological View Point

Depending on stress severity performed on the polymer surface, scratching condition as well as the viscoelastic properties of the polymer, scratches with plastic or fracture morphology can be created on the polymer surface. The fracture type scratches are caused by more severe contact damage from larger asperities. They have irregular shapes with sharp edges and mostly followed by material pick up from the polymer surface. The scratch width in this case varies from very small (under 25 μm) to very large (over 1 mm) (Hutchings 1992; Yang and Wu 1997; Persson 2000; Schwarzentruber 2002; Tahmassebi et al. 2010). The electron microscope images of such morphology are shown in Fig. 6.

The materials removal during fracture types of scratch depends on both scratching conditions and polymer properties. It has been shown that fracture type scratches have two main morphologies; (a) polymer material removal with polymeric particulates residue inside scratch and (b) polymer material pick up without polymeric particles residue. The visual observations of these two different fracture types of scratches are shown in Fig. 7.

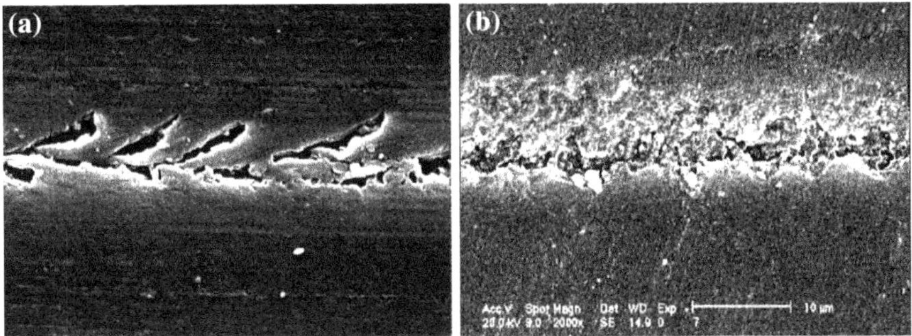

Fig. 6 SEM micrographs of the scratches with fracture morphology created at different stresses. **a** Surface rupture. **b** Material pike up (Tahmassebi et al. 2010)

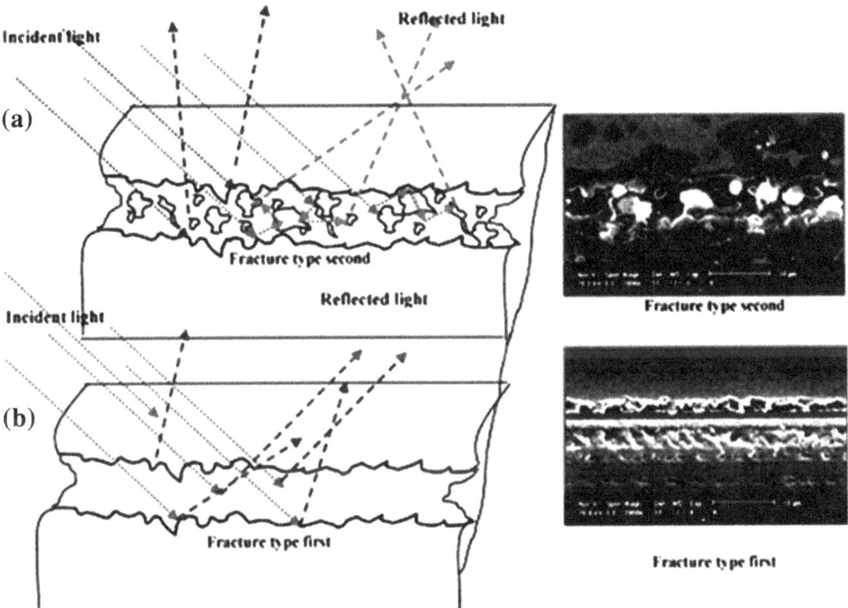

Fig. 7 Two types scratches with fracture morphology. **a** With polymeric particulates residue inside scratch. **b** Without polymeric particulates residue inside of the scratch (Ramezanzadeh et al. 2011a, b)

It can be seen from Fig. 4 that both fracture type's scratches can scatter visible light intensively, leading to significant reduction in polymer gloss. However, it will be shown that scratches with detached pieces of polymeric particles can scatter visible light even more intensively than the other type. Another characteristic of the fracture type scratch is its low self-healing ability in exposure to environmental condition. In fact, the fracture type deformation is irreversible and cannot

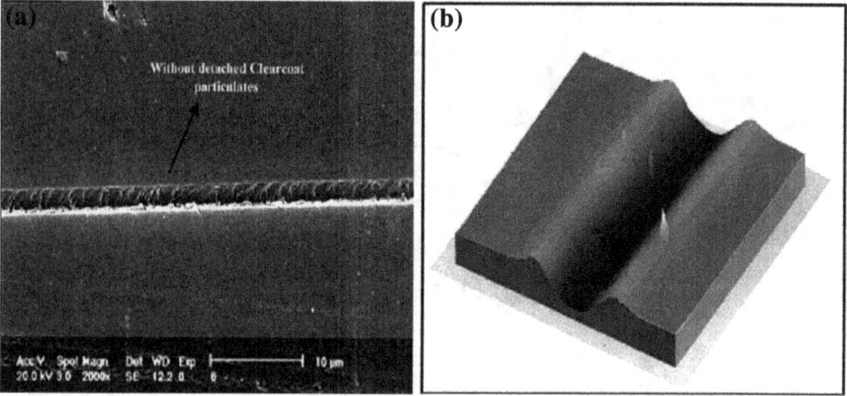

Fig. 8 Plastic type morphology of the scratch. **a** SEM micrograph. **b** Schematic illustration (Tahmassebi et al. 2010; Ramezanzadeh et al. 2011a, b)

be healed in anyways due to the polymer chains tearing (Schwarzentruber 2002; Tahmassebi et al. 2010; Ramezanzadeh et al. 2011a, b).

Unlike fracture type scratch, the plastic type has smooth regular surface with lower capability of light scattering. In addition, this type of deformation has inherent ability to self heal with time or with increased temperature. The visual performance of a plastic type scratch is shown in Fig. 8.

Because of the smooth surface, this kind of scratch is not visible easily by human eyes.

6 Effective Parameters Influencing Polymer Scratch Resistance

In order to enhance the polymers resistance against wear and scratch, there are two main strategies; (1) increasing surface slippage and hardness and (2) enhancing cohesive forces within polymer matrix that modify the viscoelastic properties (Courter 1997; Jardret et al. 1998; Hara et al. 2000). In the first approach, the force needed by the scratching objects to produce scratch on the polymer surface can be significantly increased by the enhancement in polymer surface slippage and hardness. However, it has been shown that greater hardness does not necessarily guarantee the polymer against scratch. There are problems with highly increased polymer surface hardness. For examples, when the applied forces are greater than the critical force, it leads to fracture type scratches creation. Increasing polymeric materials surface hardness can also result in an increase in polymer brittleness and therefore reduction of other properties like flexibility. For the hard polymers, the scratch resistance is often related to the adhesive strength of the interface with their substrate. Failure of these polymers is commonly identified as a delamination process.

In the second approach, enhancing the cohesive forces within a polymer could prevent sharp objects from polymer chains tearing. Therefore, changing polymer viscoelastic properties is another way of controlling scratch resistance (Courter 1997; Jardret et al. 1998; Hara et al. 2000; Jardret and Ryntz 2005). In order to investigate polymers scratch resistance precisely, it is necessary to evaluate their viscoelastic–viscoplastic properties as well as the difference between the tensile and compressive behaviors. Polymers are known as viscoelastic materials with high ability of dissipating energy as heat during deformation. The amount of energy dissipated per unit distance during this process can be defined as friction. The frictional force, $F_{viscoelastic}$, is shown by Eq. 9:

$$F_{viscoelastic} = 0.17 \ \beta W^{4/3} R^{-2/3} \left(1 - \upsilon^2\right)^{1/3} E^{-1/3} \tag{9}$$

where υ, E, β and W are respectively Poisson's ratio, real part of Young's modulus, fraction of the total energy that is dissipated and projected area to scratching tip (Cartledge et al. 1996).

Different parameters like cross-linking density, polymer chains molecular weight, glass transition temperature (Tg) and elastic modulus are effective parameters influencing the polymer viscoelastic properties. The polymer scratch resistance can be under control of such parameters changes. In this regard, two main strategies are found effective; (1) producing polymers with low enough Tg showing reflow behavior and (2) producing extraordinary high cross-linking density and/or molecular weight polymers. In the first approach, although the scratching objects could penetrate into the polymer matrix easily, the polymer deformation would be mostly in form of plastic which can be easily healed by temperature (Tahmassebi 2010; Ramezanzadeh et al. 2011a, b, 2012a, b). Moreover, the plastic type scratches have fewer effects on the polymer appearance changes. The only negative point mentioned for this strategy of scratch resistance improvement of the polymers is low polymeric materials resistance against sharp objects penetration. In fact, polymers with low Tg cannot tolerate high level of stresses. As a result, the number of scratches produced in long period of application time will be considerably increased. Producing polymers with the mentioned properties can significantly increase its scratch resistance. The stress needed for the scratching objects to diffuse into the polymer matrix will be considerably increased when the polymer cross-linking density and/or molecular weight is high. Increasing cross-linking density and/or molecular weight results in higher polymer ability to stress storing and stress relaxation. The cross-links behave like a spring which can store stress when the polymer is under stress and relax it when leaving stress. In this way, the polymer resistance against scratch is high enough that no damage occurs. However, the high polymer elasticity (elastic modulus) cannot necessarily lead to its protection against scratch at all scratching conditions. In fact, when the stress applied to the polymer exceeds a critical stress that polymer can tolerate the polymer chains start tearing resulting into fracture type scratch. Moreover, the fracture type scratch influences the polymer appearance severely in a negative way. Based on these explanations, polymers with the low and high cross-linking density and/or elastic modulus have advantages and disadvantages from the scratch resistance view point. In our recent findings, it has been shown that in order to design a polymer with appropriate scratch resistance, its toughness should be enhanced. In fact, the tough polymeric materials have high capability of stress damping together with high scratch resistance (Courter 1997; Hara et al. 2000; Jardret and Morel 2003; Jardret and Ryntz 2005; Groenewolt 2008). The scratches on tough polymers are both in forms of plastic (with self-healing ability) and fracture deformations. There are lots of methods to enhance the polymers toughness which will be discussed later.

7 Methods and Instrumentation

In order to evaluate any phenomenon, it is needed to characterize it quantitatively. As mentioned in the previous section, scratch affects the visual properties of polymeric coatings. Below, we review testing methods that evaluate changes in visual performance of surfaces of polymeric coatings. After that, we proceed with explaining the tests that mimic a single scratch on the surface and measure important parameters on scratch resistance such as: elastic modulus of surface, surface hardness, depth and width of scratch, pile up and depth recovery.

7.1 Amtec Laboratory Car Wash

The automotive and paint industry are already interested in this method. The method allows a realistic simulation of the strain caused by an automatic car wash. The clearcoat to be tested is moved back and forth 10 times under a rotating car wash brush. The brush is sprayed with washing water during the cleaning procedure. A defined amount of quartz powder is added to the washing water as a replacement for street dirt. A gloss measurement in 20° geometry is used to evaluate the scratch resistance. The initial gloss and the gloss after the cleaning procedure are measured. The percentage of residual gloss with regard to the initial gloss is a measure for scratch resistance. High values indicate good scratch resistance (Osterhold and Wagner 2002).

7.2 Crockmeter

An automatic crockmeter is used for this test. This instrument is equipped with an electrical motor, so that a uniform stroke rate of 60 double strokes per min is reached. The sample is fixed on a flat pedestal. The sample is exposed to linear rubbing caused by a "rubbing finger". A special testing material is attached to the bottom side of the "rubbing finger". The "rubbing finger" is 16 mm in diameter and its downward force is 9 N. Ten double strokes are carried out over a length of 100 mm. For this method, the percentage of residual gloss is also used as a measure for scratching (Osterhold and Wagner 2002).

7.3 The IST Method

In this modified sanding process, using a spherical capped grinder (calotte) the measure is the amount of abraded material per wear-inducing work. The calotte sanding machine works with a rotating ball, 3 cm in diameter that rotates on the coated surface (Fig. 9). Dirt and dust are simulated by covering the ball with an abrasive suspension during the sanding process. After a defined exposure time (sanding time), the diameter of the sanded calotte is measured using a light microscope. The wear volume is calculated from the measured diameter. By adjusting the exposure time, the process can be set up so that the depth of the calotte is only a few micrometers. The wear value (V_W) can be calculated from the sanding distance (S), downward force (F_N), and wear volume (V_{Kal}) (Eq. 10):

$$V_W = \frac{V_{Kal}}{SF_N} \tag{10}$$

7.4 Taber

The Taber test is used to measure abrasion resistance of surfaces. In this test, the abrasion is produced by the contact of a test sample, turning on a vertical axis, with the sliding rotation of two abrading wheels. The wheels are driven by the sample in opposite directions about a horizontal axis displaced from the axis of the sample. One abrading wheel rubs the specimen outward toward the periphery and the other, inward toward the center. The resulting abrasion marks form a pattern of crossed scratches over an area of approximately 30 square centimeters. Wheel pressure against the specimen is usually included in the standard specification. Ordinarily, one of three standard weights, i.e., 250, 500 or 1,000 g, is used

Fig. 9 Schematic diagram of IST test

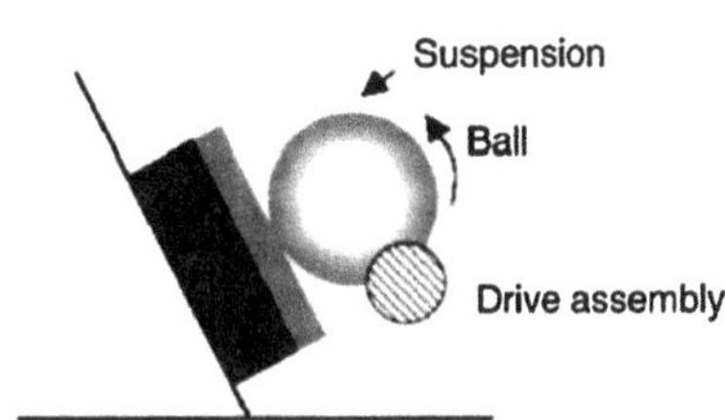

for each wheel. In order to evaluate coating abrasion resistance after abrading, two methods can be utilized:

1. Weight loss: in this approach, the weights of specimen before and after abrasion are to be compared.
2. Optical method: In this method optical property of coatings such as gloss, transmission and haze are measured and compared before and after abrasion (Sun et al. 2002).

These tests investigate scratch of surface globally in a micron scale. Since mar and scratch occur on the surface of coatings, it is important to study mechanical properties of coatings at the top layer of surface. In this regard, nanoindentation and nanoscratch are very helpful in order to investigate surface properties.

8 Nanoindentation

Figure 10 shows a schematic representation of load versus indentation time and depth respectively from the indentation data, the contact depth, h_c is calculated from the indenter load, P, total penetration, h, and contact stiffness, S (dP/dh) measured at the beginning of unloading [Eq. 11 (Oliver and Pharr 1992)]:

$$h_c = h_{max} - \varepsilon \frac{p_{max}}{S} \tag{11}$$

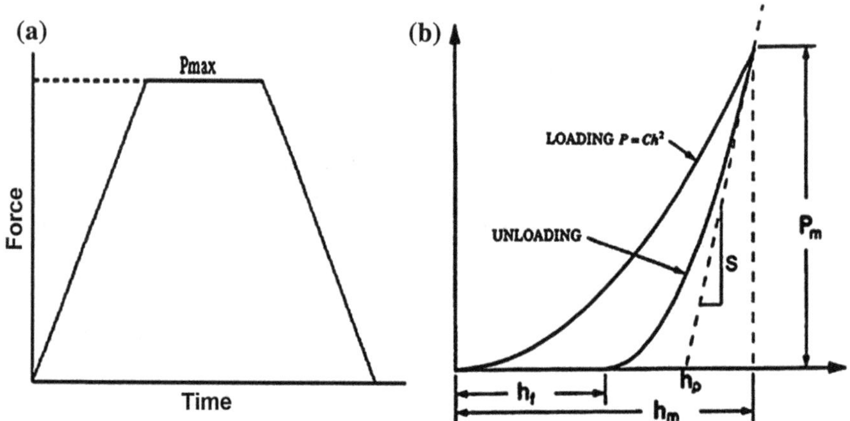

Fig. 10 Schematic representation of load versus time (**a**) and load versus indentation depth (**b**) for an indentation experiment. The quantities shown are p_{max}: the maximum indentation load, h_{max}; maximum indentation depth at maximum load, h_f final indentation depth at zero load; and hc: extrapolated indentation depth, S; contact stiffness (*unloading slope curve*)

where ε is a constant (≈ 0.75). The contact stiffness, S, is obtained from a regression function fitted to the unloading curve. The contact area A is calculated from contact depth ($A = f(h_c)$), with the calibration function according to Eq. 12 (Oliver and Pharr 1992):

$$A = 24.5 h_c^2 \tag{12}$$

The composite modulus E_r is determined from the contact stiffness S and contact area [Eq. 13 (Oliver and Pharr 1992)]

$$E_r = \frac{\sqrt{\pi}}{2\beta} \frac{S}{\sqrt{A}} \tag{13}$$

β is the correction factor for the indenter shape ($\beta \approx 1.07$; see reference (Oliver and Pharr 1992).

The Elastic modulus of the specimen is then calculated from the composite modulus using Eq. 14 (Oliver and Pharr 1992):

$$\frac{1}{E_r} = \frac{1 - v^2}{E} + \frac{1 - v_i^2}{E_i} \tag{14}$$

This equation reflects the fact that the total elastic compliance at the contact consists of the compliance of the specimen (no subscript) and the indenter (subscript i); v is Poisson's ratio. Hardness is defined as the extent of the material resistance to local plastic deformation. Thus, the hardness, H, can be measured from the maximum applied load, P_{max}; divided by the contact area, A; according to Eq. 15 (Oliver and Pharr 1992):

$$H = \frac{P_{max}}{A} \tag{15}$$

Generally speaking, a polymer film may respond to an indentation force in one of three ways: viscoelastic deformation, viscoplastic deformation, or fracture. Nanoindentation measurements can only distinguish between recoverable and unrecoverable deformation. For deformations that recover in the time scale of the measurement, polymers are generally behaving elastically. For those that do not recover in the time scale of the measurement, they are taken to be plastic deformations. Usually, a parameter called the "plasticity index", Ψ, is used to describe the relative plastic/elastic character of a material as Eq. 16 (Binyang et al. 2001).

$$\Psi = \frac{w_p}{w_e + w_p} \tag{16}$$

where w_p is the area encompassed between the loading and unloading curves and is equal to the plastic work done during indentation. w_e is the area under the unloading curve and is equal to the viscoelastic recovery as illustrated in Fig. 11.

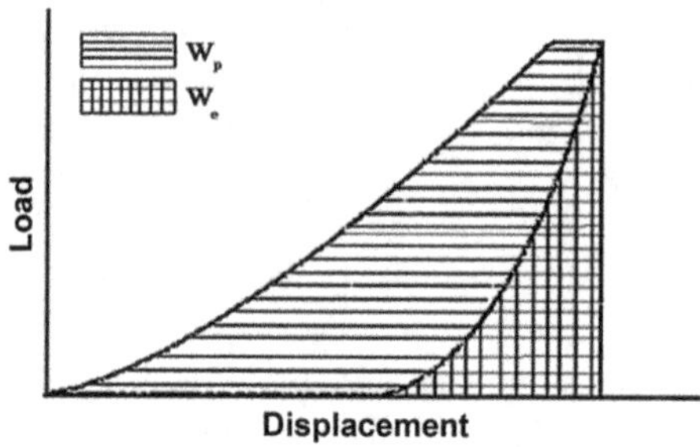

Fig. 11 The schematic diagram of plasticity index (Binyang et al. 2001)

9 Nanoscratch

In this test, a scratching tip in a constant load or a progressive variable normal force moves on the surface of coating with a determined velocity. After creating scratch on the sample a monitoring devices such as AFM could scan the scratch region and prepare picture from scratched domain. The scratch dimensions, i.e. scratch depth, pile-up height of created scratches and percent elastic recovery could be derived from AFM image.

In this method, in addition to pictorial analysis, applied force is plotted versus displacement.

10 Case Studies

It is well known that tribological properties of polymers can be generally improved with the incorporation of reinforcing and/or lubricating substances. While, the lubricants mainly decrease the surface energy and friction coefficient, reinforcing fillers increase the strength of polymeric materials. Examples of conventional lubricating fillers include graphite, poly (tetrafluoroethylene) (PTFE), molybdenum disulfide (MoS_2), and some synthetic oils; while common reinforcing fillers include Silica, titania, alumina and so on.

10.1 Lubricating Fillers

In the severe applications, employing high performance materials would not be sufficient to resist intensive tribological actions. Therefore, the tribological properties of these materials have to be improved via other mechanisms. To make it clear, the following example is given. Polyetheretherketone (PEEK) is intrinsically a high performance engineering thermoplastic polymer with excellent mechanical and long-term properties. Utilization of PEEK in bearing and sliding parts causes a high friction coefficient as a result of direct contact between PEEK coating and other metals in contact. In a study done at Nottingham University (Hou et al. 2008), researchers attempted to reduce the friction coefficient of the PEEK using incorporating inorganic fullerene-like tungsten disulfide (IF-WS2) nanoparticles as novel lubricant in the PEEK matrix. Their results revealed that significant improvement in the tribological properties of the PEEK coatings was resulted. Up to 70 % (from 0.4 to 0.15) decrease in friction coefficient as a consequence of incorporation of 2.5 wt % of IF nanoparticles was attributed to the lower shearing strength of the lubricating nanoparticles. This modification led also to as much as 60 % increase in hardness of the coating.

10.2 Reinforcing Fillers

It is well understood that addition of a low loading of rigid nanoparticles into polymers significantly enhances their mechanical properties, especially stiffness and scratch resistance. These improvements are mainly attributed to the much more extensive interfacial areas of nanoparticles compared to their macro- and micro-scale counterparts. In order to reach the highest level of improvement, homogenous dispersion of nano-sized particles in the matrix and good interfacial interaction between nanoparticles and matrix are required. A few examples of using various reinforcing fillers into the polymeric matrices for the purpose of promoting tribological properties of polymers are briefly presented and discussed here.

In a research on UV cured coatings done by Salleh and co-workers, it was demonstrated that increasing in silica content of the coating improved its scratch resistance (Salleh et al. 2009). This modification also led to an increase in hardness. The hardness increase was because of increase in modulus as well as the presence of the hard silica nanoparticles. It was claimed that nanoparticles via accelerating the cure reaction and restricting the segmental motion of the polymeric chains (as a result of nanoparticle incorporation which promotes hydrogen bonding and viscosity) enhanced the gel content and cross-linking density of UV cured systems.

In recent works (Tahmassebi et al. 2010; Ramezanzadeh et al. 2011a, b, 2012a, b), hydrophobic nano silica particles were incorporated into an acrylic/melamine resin, which is used as an automotive clearcoat, to improve its abrasion and scratch resistance. Scratch resistance is one of the primary requirements of an automotive clearcoat to prevent or decrease gloss losses caused by various abrasive factors to which the coating may be exposed. Figure 12 shows optical microscope images of surfaces of different clearcoats containing various loadings of nanosilica scratched by a simulated carwash (Yari et al. 2012).

The optical images clearly indicate that as nano silica increases, the abrasion resistance enhances. Sample containing higher loadings of silica have the best performance against carwash brushes so that the sample loaded with 3.75 wt % nanosilica seems remained intact during carwash simulation test. In addition, it was also found that as the silica content increased the morphology of damages

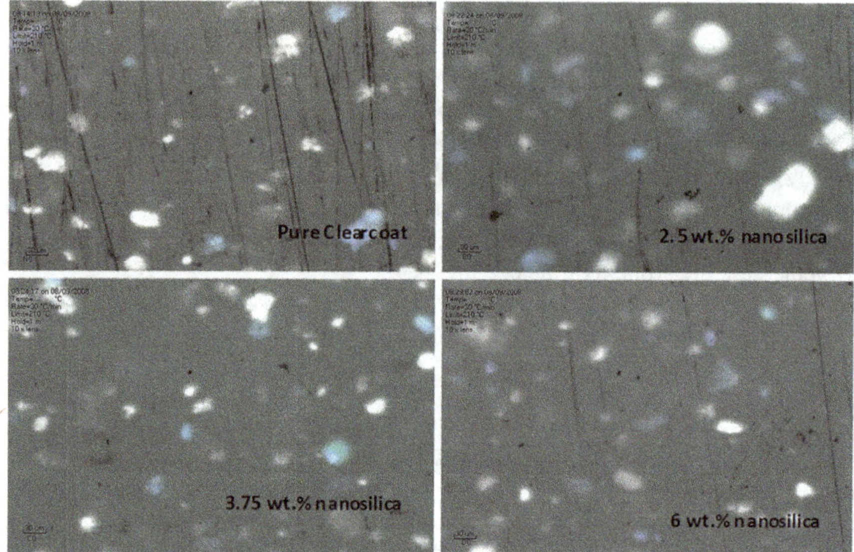

Fig. 12 Optical images of clearcoat surfaces after carwash simulation test

altered from fracture-type to plastic-type (Ramezanzadeh et al. 2011a, b). This was attributed to the ball-bearing effect of silica particles which ease the motion of polymeric chain and plastic deformation (Ramezanzadeh et al. 2011a, b; Yari et al. 2012). The plastic scratches created on such nanocomposites have been shown in previous parts (Fig. 8). Beside enhanced scratch resistance, it was proved that nanosilica particles function as a UV absorber in the clearcoat. This resulted in a considerable improvement in weathering performance and durability of the automotive clearcoats which considered as important as the scratch resistance in automotive industry (Yari et al. 2012).

Other mechanisms have been also proposed for enhanced hardness and scratch resistance of coatings containing nanosilica particles. In couple of studies (Messori et al. 2003; Amerio 2005), it was revealed that the increase in hardness of nanocomposites results from segregation of nanoparticles from bulk of the matrix toward its surface. The XPS results proved that silica nanoparticles segregated onto the outer surface, making the surface completely inorganic.

Polycarbonate (PC) is an engineering transparent thermoplastic material with good dimensional and thermal stability. In some optical applications such as glazing, eye wear and compact disks the tribological properties like scratching or wear performance of PC are very crucial. In a research paper (Carrión et al. 2008), it was attempted to modify the tribological properties through introducing nano-sized organoclay sheets into PC system. It was concluded that nanoclay had a plasticizing effect in the matrix which lowered glass transition temperature, easing movement of polymer chain segments. This modification led to a slight increase in hardness, but a significant decrease in friction coefficient (up to 88 %). These variations enhanced the wear performance of nanocomposite nearly two orders of magnitude with respect to the neat polymer. Such improvement in tribological performance was assigned to the presence of nanoclay and its uniform dispersion in the PC matrix.

In another study targeted in improving the tribological properties of PC films, it was demonstrated that addition of nano-SiO_2 particles increased the hardness, modulus, elastic recovery and scratch resistance of the PC films. The increased scratch resistance was thought to be due to the rolling effect of the nanoparticles which decreased the friction coefficient of the surface (Wang et al. 2010).

Polyhedral oligomeric silsesquioxanes (POSS) are a novel class of organic–inorganic compounds, which have been utilized to modify various properties especially mechanical ones of polymeric matrices (Cordes et al. 2010). In numerous papers they have been utilized to reinforce the matrix structure of a wide range of thermosets and/or thermoplastics. These hybrid materials have been used for improving the surface properties of a typical methacrylic coating (Amerio et al. 2008). The final properties of UV cured films demonstrated that introducing 5 and 10 % of POSS content in the organic resin enhanced the scratch resistance around 5 and 8 times, respectively. This level of improvement was beholden to presence of POSS cages in the matrix which resulted in an increase in T_g, cross-linking density and modulus of the coating.

As shown in previous examples, nano-filler embedded polymers showed enhanced scratch and wear resistance provided that the fillers have been well dispersed in the polymeric matrix. However, obtaining appropriate dispersion requires surface modification of the particles and/or using different dispersing techniques (Rostami et al. 2010; Rostami 2012). In-situ generating inorganic phase inside organic matrix using sol–gel technique has attracted lots of attentions in recent years. Various Organic/inorganic precursors can be used to produce in situ inorganic network in the matrix. These precursors, either as network former, such as tetraethyl orthosilicate (TEOS) or network modifier such as methacryloxy propyl trimethoxysilane (MEMO) and glycidoxy propyl trimethoxysilane (GPTS), can be introduced to the main polymeric film former to obtain a so-called hybrid nanocomposite films. As a result of various hydrolysis and self-condensation reactions, a three-dimension cross-linked network is formed (Ramezanzadeh and Mohseni 2012; Ramezanzadeh 2012a, b). The hydrolyzed precursors are able to react with the functional groups of polymeric matrix such as polyol and other curing cross-linkers such as amino or isocyanate compounds. In this way, a hybrid nanocomposite containing nano-sized organic/inorganic phases can be obtained (Fig. 13).

In an attempt to modify the abrasion resistance of PC substrates, a novel sol–gel based UV cured hybrid coatings were prepared by combining different silanes with various urethane acrylate monomers (Yahyaei et al. 2011; Yahyaei and Mohseni 2013). A schematic presentation of coating structure is shown in Fig. 14. The tribological aspects of the PC substrate coated with different hybrid coatings were evaluated in both micro- and nano-scale by the aid of precise characterizing techniques.

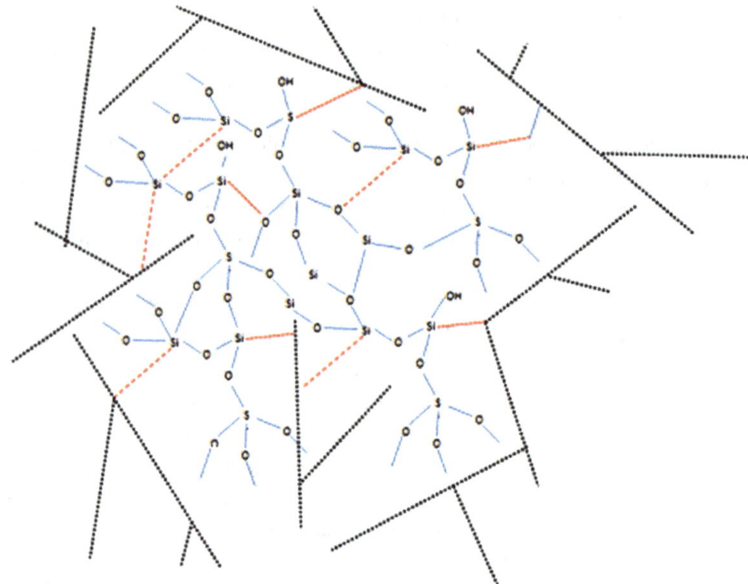

Fig. 13 Schematic illustration of a sol–gel based automotive clearcoat containing organic/inorganic precursors

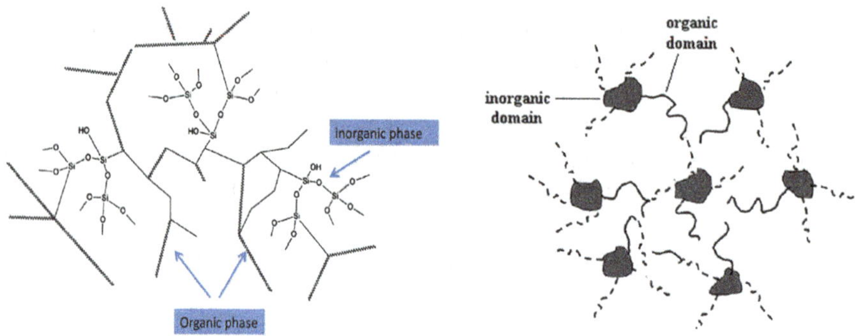

Fig. 14 A pictorial representation of hybrid materials

Abrasion test results demonstrated that abrasion index (AI) values varied noticeably. While abrasion index of blank PC was 0.28, those of coated with sol–gel containing films had declined to values less than 0.1. Lower AI means a higher abrasion resistance. Increased abrasion resistance in presence of the inorganic phase was attributed to enhanced elastic modulus and hardness of the hybrid films. However, it was concluded that in order to observe a greater abrasion resistance a medium content of inorganic and organic is needed to meet a balance of hardness and flexibility. As seen in Fig. 15, films possessing medium hardness have the best abrasion resistance. This balance allows less deformation and higher ability of the deformed area to recover and heal after removing the force.

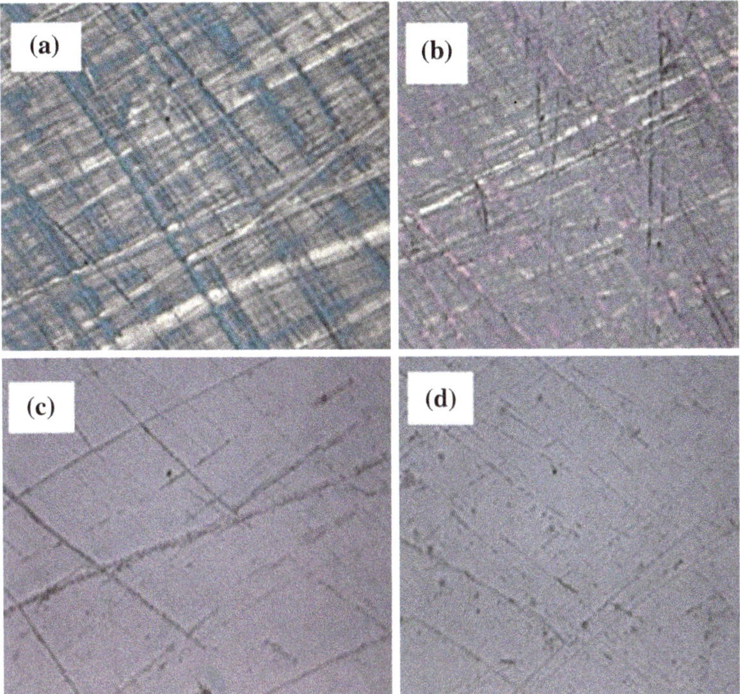

Fig. 15 Cross polarized optical micrographs of taber abraded tracks. **a** Sample with low hardness. **b** Blank PC. **c** Sample with high hardness. **d** Sample with medium hardness

11 Concluding Remarks

Polymers have different usages depending on their structures and their properties. In all applications, mechanical characteristics and the way polymers respond to an internal and/or external force is crucial. In addition, the effect of temperature and the testing method are also important. In some applications it is also needed to appreciate the mechanical properties at the bulk as well as the very top layer of the material close to the interface of material and air. In this chapter we have attempted to describe the fundamental of polymer properties and the methods by which they can be evaluated. Strategies to enhance the surface and bulk mechanical properties of polymers together with some case studies have been discussed.

References

Amerio E, Sangermano M, Malucelli G, Priola A, Voit B (2005) Preparation and characterization of hybrid nanocomposite coatings by photopolymerization and sol–gel process. Polymer 46:11241

Amerio E, Sangermano M, Colucci G, Malucelli G, Messori M, Taurino R, Fabbri P (2008) UV curing of Organic-Inorganic hybrid coatings containing Polyhedral Oligomeric Silsesquioxane blocks. Macromol Mater Eng 293:700–707

Barletta M, Bellisario D, Rubino G, Ucciardello N (2010) Scratch and wear resistance of transparent topcoats on carbon laminates. Prog Org Coat 67:209

Bautista Y, Gómez MP, Ribes C, Sanz V (2011) Correlation between the wear resistance, and the scratch resistance, for nanocomposite coatings. Prog Org Coat 70(4):178

Bertrand-Lambotte P, Loubet JL, Verpy C, Pavan S (2002) Understanding of automotive clearcoats scratch resistance. Thin Solid Films 420–421:281

Binyang D, Ophelia KC, Qingling Z, Tianbai H (2001) Study of elastic modulus and yield strength of polymer thin films using atomic force microscopy. Langmuir 17:3286

Bowden PB (1973) In: Haward RN (ed) The physics of glassy polymers. Applied Science Publisher Ltd., London

Bowden PB, Young RJ (1974) Deformation mechanisms in crystalline polymers. J Mater Sci 9:2034

Briscoe BJ, Evans PD, Pelillo E, Sinha SK (1996a) Scratching maps for polymers. Wear 200:137

Briscoe BJ, Pelillo E, Sinha SK (1996b) Scratch hardness and deformation maps for polycarbonate and polyethylene. Polym Eng Sci 36(24):2996

Carrión FJ, Ao Arribas, Bermu'dez MD, Guillamon A (2008) Physical and tribological properties of a new polycarbonate-organoclay nanocomposite. Eur Polymer J 44:968–977

Cartledge HCY, Baillie C, Mai YW (1996) Friction and wear mechanisms of a thermoplastic composite GF/PA6 subjected to different thermal histories. Wear 194:178

Cordes DB, Lickiss PD, Rataboul F (2010) Recent developments in the chemistry of cubic Polyhedral Oligosilsesquioxanes. Chem Rev 110:2081–2173

Courter JL (1997) Mar resistance of automotive clearcoat: I. Relationship to coating mechanical properties. J Coat Technol 69(866):57

Groenewolt M (2008) Highly scratch resistant coatings for automotive applications. Prog Org Coat 61:106

Hara Y, Mori T, Fujitani T (2000) Relationship between viscoelasticity and scratch morphology of coating films. Prog Org Coat 40:39

Hou X, Shan CX, Choy KL (2008) Microstructures and tribological properties of PEEK-based nanocomposite coatings incorporating inorganic fullerene-like nanoparticles. Surf Coat Technol 202:2287

Hutchings IM (1992) Tribology-Friction and wear of engineering materials. CRC Press, Boca Raton

Jardret V, Morel P (2003) Viscoelastic effects on the scratch resistance of polymers: relationship between mechanical properties and scratch properties at various temperatures. Prog Org Coat 48:322

Jardret V, Ryntz R (2005) Visco-Elastic Visco-Plastic analysis of scratch resistance of organic coatings. J Coat Technol Res 2(8):591

Jardret V, Zahouani H, Loubet JL, Mathia TG (1998) Understanding and quantification of elastic and plastic deformation during a scratch test. Wear 218:8

Jardret V, Lucas BN, Oliver W (2000) Scratch durability of automotive clear coatings: a quantitative, reliable and robust methodology. J Coat Technol 72(907):79

Kinloch AJ, Young RJ (1983) Fracture behavior of polymers. Applied Science Publishers Ltd., London

Messori M, Toselli M, Pilati F, Fabbri E, Fabbri P, Busoli S, Pasquali L, Nannarone S (2003) Flame retarding poly(methyl methacrylate) with nanostructured organic–inorganic hybrids coatings. Polymer 44:4463

Nat DS (1980) A text book of materials and metallurgy. Katson Publishing House, Ludhiana

Nielsen LE (1962) Mechanical properties of polymers and composites. Dekker M. INC

Oliver WC, Pharr GM (1992) An improved technique for determining hardness and elastic modulus using load and displacement sensing indentation experiments. J Mater Res 7:1564

Osterhold M, Wagner G (2002) Methods for characterizing the mar resistance. Prog Org Coat 45:365

Persson BNJ (2000) Sliding Friction–Physical principles and applications, 2nd edn. Springer, Berlin

Ramezanzadeh B, Mohseni M (2012) Preparation of sol–gel based nano-structured hybrid coatings: effects of combined precursor's mixtures on coatings morphological and mechanical properties. J Sol-Gel Sci Technol 64:232–244

Ramezanzadeh B, Moradian S, Khosravi A, Tahmasebi N (2011a) A new approach to investigate scratch morphology and appearance of an automotive coating containing nano-SiO2 and polysiloxane additives. Prog Org Coat 72(3):541

Ramezanzadeh B, Moradian S, Tahmasebi N, Khosravi A (2011b) Studying the role of polysiloxane additives and nano-SiO2 on the mechanical properties of a typical acrylic/melamine clearcoat. Prog Org Coat 72:621

Ramezanzadeh B, Mohseni M, Karbasi A (2012a) Preparation of sol–gel-based nanostructured hybrid coatings, part 1: morphological and mechanical studies. J Mater Sci 47:440–454

Ramezanzadeh B, Moradian S, Khosravi A, Tahmassebi N (2012b) Effect of polysiloxane additives on the scratch resistance of an acrylic melamine automotive clearcoat. J Coat Technol Res 9(2):203

Rostami M, Ranjbar Z, Mohseni M (2010) Investigating the interfacial interaction of different aminosilane treated nano silicas with a polyurethane coating. Appl Surf Sci 257:899–904

Rostami M, Mohseni M, Ranjbar Z (2012) An attempt to quantitatively predict the interfacial adhesion of differently surface treated nanosilicas in a polyurethane coating matrix using tensile strength and DMTA analysis. Int J Adhes Adhes 34:24–31

Salleh NGN, Yhaya MF, Hassan A, Bakar AA, Mokhtar M (2009) Development of Scratchand Abrasion-Resistant coating materials based on nanoparticles, cured by radiation. Int J Polym Mater 58:422

Schwarzentruber P (2002) Scratch resistance and weatherfastness of UV-curable clearcoats. Macromol Symp 187:531

Suna J, Mukamal H, Liu Z, Shen W (2002) Analysis of the Taber test in characterization of automotive side windows. Tribo Lett 13:49

Tahmassebi N, Moradian S, Ramezanzadeh B, Khosravi A, Behdad S (2010) Effect of addition of hydrophobic nano silica on viscoelastic properties and scratch resistance of an acrylic/melamine automotive clearcoat. Tribo Intern 43:685

Wang ZZ, Gu P, Zhang Z (2010) Indentation and scratch behavior of nano-SiO2/polycarbonate composite coating at the micro/nano-scale. Wear 269:21–25

Yahyaei H, Mohseni M (2013) Use of nanoindentation and nanoscratch experiments to reveal the mechanical behavior of sol–gel prepared nanocomposite films on polycarbonate. Tribol Int 57:147–155

Yahyaei H, Mohseni M, Bastani S (2011) Using Taguchi experimental design to reveal the impact of parameters affecting the abrasion resistance of sol–gel based UV curable nanocomposite films on polycarbonate. J Sol-Gel Sci Technol 59:95–105

Yang ACM, Wu TW (1997) Wear and friction in glassy polymers: micro-scratch on blends of polystyrene and poly (2,6-dimethyl-1,4-phenylene oxide). J Polym Sci B: Polym Phys 35:1295

Yari H, Moradian S, Tahmasebi N, Arefmanesh M (2012) The effect of weathering on tribological properties of an acrylic melamine automotive nanocomposite. Tribol Lett 46:123

Part II
Contributions from Manufacturers

Mechanical Property Mapping at the Nanoscale Using PeakForce QNM Scanning Probe Technique

Bede Pittenger, Natalia Erina and Chanmin Su

Abstract Development of PeakForce QNM® a new, powerful scanning probe microscopy (SPM) method for high resolution, nanoscale quantitative mapping of mechanical properties is described. Material properties such as elastic modulus, dissipation, adhesion, and deformation are mapped simultaneously with topography at real imaging speeds with nanoscale resolution. PeakForce QNM has several distinct advantages over other SPM based methods for nanomechanical characterization including ease of use, unambiguous and quantitative material information, non-destructive to both tip and sample, and fast acquisition times. This chapter discusses the theory and operating principles of PeakForce QNM and applications to measure mechanical properties of a variety of materials ranging from polymer blends and films to single crystals and even cement paste.

1 Introduction

Since its invention in the early 1980's, the field of scanning probe microscopy (SPM) and its most popular member, atomic force microscopy (AFM), have become a main nanoscale characterization tool. The heart of the AFM, in contrast to other electron or optical based microscopies, is a mechanical interaction between a very sharp silicon cantilever/tip assembly and the surface. Because the AFM tip mechanically interacts with or touches the surface, it provides unique potential for characterizing mechanical properties of materials and surfaces.

The development of TappingMode™ imaging in 1993 (Zhong et al. 1993) at Digital Instruments (now Bruker) was a key step forward in the functionality of SPM. In TappingMode the probe is vibrated near or at the resonant frequency of

B. Pittenger (✉) · N. Erina · C. Su
Bruker Nano Surfaces Division, Robin Hill Road, Santa Barbara, CA 93117, USA
e-mail: bede.pittenger@bruker-nano.com

A. Tiwari (ed.), *Nanomechanical Analysis of High Performance Materials*, Solid Mechanics and Its Applications 203, DOI: 10.1007/978-94-007-6919-9_2, © Springer Science+Business Media Dordrecht 2014

the cantilever while it raster scans across the sample. The tip only contacts the surface for a small percentage of the time, keeping the tapping force low and the lateral forces negligible. Since the probe is oscillating, it experiences both attractive and repulsive forces depending on its position in the cycle in a way that is analogous to force curves. Consequently, TappingMode™ has the ability to generate high-quality data for a wide range of samples, making it the dominant imaging mode for most SPM applications over the last two decades.

The data types obtained from TappingMode™ SPM are primarily topography and phase. The phase of the TappingMode cantilever vibration relative to the drive is a useful indication of different mechanical properties. Unfortunately, the phase signal is a mixture of material properties, depending both on dissipative and conservative forces (Cleveland et al. 1998; Tamayo and Garcia 1997). Since elasticity, hardness, adhesion and energy dissipation all contribute to phase shift, using this single data channel to solve for multiple unknown variables can hardly be quantitative. Additionally, the phase signal depends on imaging parameters such as drive amplitude, drive frequency, and set-point. This makes it difficult and sometimes impossible to interpret the source of the contrast, leaving the user to conclude that there are differences in the sample without further knowledge of contributing physical factors. Though qualitatively useful, the phase signal is a complicated convolution of multiple material properties, and therefore is prone to artifacts during imaging and is unable to be related directly to material properties.

There are two other TappingMode-based modes that have recently gained popularity: TappingMode while observing simultaneously a separate harmonic (integer multiple of the drive frequency) of the tapping drive (Sahin 2007), and tapping mode while observing simultaneously a higher cantilever eigen-mode (bimodal

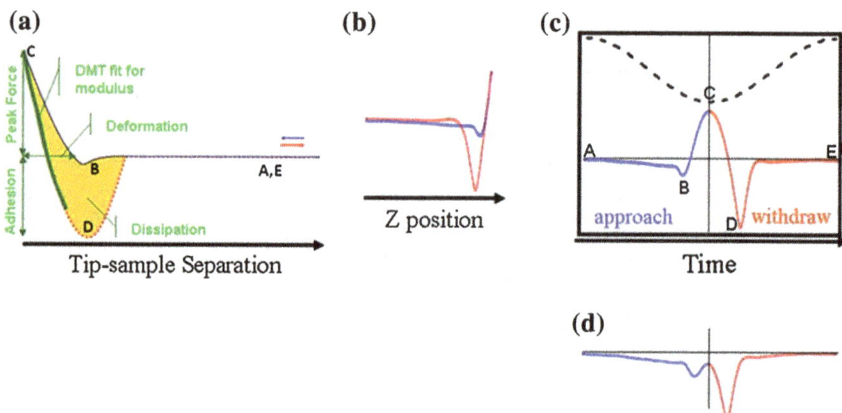

Fig. 1 Force curves and information that can be obtained from them: **a** Force versus tip sample separation including (*B*) jump-to-contact (*C*) PeakForce (*D*) Adhesion; **b** A "traditional" force curve measuring force versus z piezo position; **c** Force as a function of time including (*B*) jump-to-contact (*C*) PeakForce (*D*) Adhesion; **d** Force versus time with small PeakForce

AFM imaging) (Proksch 2006; Rodriguez and Garcia 2004). Single harmonic imaging depends on either special cantilevers or a lucky coincidence of an overtone with a harmonic, while bimodal AFM imaging adds a second frequency (usually at an overtone or higher eigen-mode of cantilever vibration) to the vibration driving the cantilever. Both of these techniques provide contrast that is analogous to phase contrast in that they do not fully separate the mechanical properties.

Perhaps the simplest measured interaction between the AFM tip and sample is a single point measurement where the tip is lowered into the surface, and the force exerted on the tip is measured as a function of the tip-sample distance. This kind of single point measurement is typically referred to as a "force curve" and lies at the heart of Peakforce Tapping and Peakforce Quantitative Nanomechanical Mapping (PeakForce QNM or PF-QNM) measurements recently developed by Bruker (Pittenger et al. 2010). In PeakForce tapping, the maximum force (or PeakForce, see Fig. 1) on the tip is controlled during the acquisition of the force curves. PeakForce QNM combines the control provided by PeakForce Tapping with the ability to extract quantitative material properties from the acquired force curves. As described in detail below, there are numerous advantages of PF-QNM over other nanomechanical measurement methods presently available. These advantages include unambiguous and quantitative measurement, ease of use, non-destructive to tip and sample, and high resolution property mapping. The rest of this chapter describes the PeakForce QNM capability and theoretical background, followed by applications to measure mechanical properties of a variety of materials.

2 Basis of QNM Measurements: Force Curves

Force curves are the basis of the PeakForce QNM technology. A typical force curve plots the force exerted on the tip as a function of tip-sample separation (Fig. 1a). Note that the x-axis here is tip-sample *separation*, as opposed to z-piezo modulation, another common form of plotting force curves.

When the tip is far from the surface (point A) there is little or no force on the tip. As the tip approaches the surface, the cantilever is pulled down toward the surface by attractive forces (usually van der Waals, electrostatics, or capillary forces) as represented by the negative force (below the horizontal axis). At point B, the attractive forces overcome the cantilever stiffness and the tip is pulled to the surface. The tip then stays on the surface and the force increases until the Z position of the modulation reaches its bottom-most position at point C. This is where the peakforce occurs. The peakforce (force at point C) during the interaction period is controlled by the trigger value (traditional force curves) or the Force Setpoint (in PF-QNM). The probe then starts to withdraw and the force decreases until it reaches a minimum at point D. The adhesion is given by the force at this point. The point where the tip comes off the surface is called the pull-off point. This often coincides with the minimum force. Once the tip has come off the surface, only long range forces affect the tip, so the force is very small or zero when the vibration is at its apex (point E).

As mentioned previously, traditional AFM force curves are collected as a function of z piezo position, which then have to be translated to tip-sample separation. The corresponding force versus z piezo position is shown in Fig. 1b. The tip-sample separation is different from the Z position of the modulation since the cantilever bends.

It can also be useful to plot the force curve as a function of time, as shown in Fig. 1c. Here, the different points A-E are again labeled for a PF-QNM force curve on a silicon sample. The top (dashed) line represents the Z-position of the sinusoidal modulation used in PFQNM as it goes through one period plotted as a function of time. The lower line (solid) represents the measured force on the probe during the approach of the tip to the sample (blue segment), while the red segment represents the force while the tip moves away from the sample. Since the modulation frequency is about 2 kHz in the current implementation, the time from point A to point E is about 0.5 ms.

As the system scans the tip across the sample, the feedback loop of the system maintains the instantaneous force at point C—the peakforce—at a constant value by adjusting the extension of the Z piezo. Figure 1d illustrates an interesting and unique challenge for peakforce control when imaging is done at a *low* peakforce. Here the controlled peakforce at point C is actually attractive. This can occur when the peakforce setpoint is small and the attractive forces are relatively large. Looking at the measured force in this plot, one might infer that the force at C is not the maximum force. In fact, the addition of the long range attractive forces cause the stress beneath the center of the probe tip to be compressive (and greater) at point C even though the measured force on the probe is less than that at point A. The small peak in the attractive background is caused by the repulsive force at the very apex point of the tip. The total interaction force is integrated over all of the tip atoms. While the tip apex atoms feel a repulsive force, the neighbor atoms, which consist of far more volume, can still be feeling an attractive force. This leads to a net negative force overall. Even when the peakforce is negative, Peakforce Tapping can recognize the local maximum and maintain control of the imaging process.

Once acquired, the force curves can be analyzed to determine a variety of properties of the material beneath the tip. These material properties include elastic modulus, tip-sample adhesion, energy dissipation, and maximum deformation. The analysis of force curves to obtain these material properties is described in detail below.

One serious limitation to the utility of force curves has been the time needed to acquire a set of force curves over the entire surface since these force curves can only provide data at one point on the sample surface at a time.

The earliest mechanical property mapping with AFM was performed by force volume (Radmacher 1994), which is still often used to acquire quantitative nanomechanical data. Force volume collects force curves triggered by the same maximum repulsive force while scanning back and forth over the surface. Collecting a force volume image usually takes several hours because individual force curves generally take about a second to collect, and a map needs thousands of force curves to be useful. Additionally, force volume imaging employs linear z ramping and requires discrete triggers at each ramps. In addition to slowing down the method, this results in a fairly aggressive tip-sample interaction where the tip hits the sample at full velocity, and turnaround at high speed results in ringing or hysteresis. This speed limitation was greatly improved by pulsed-force mode, which modulates the Z piezo at about 1 kHz, allowing property mapping in much shorter time. Pulsed-force mode (Rosa-Zeiser et al. 1997) is primarily used as a property mapping method with a trigger force of a few nanonewtons or more. Below one nanonewton, parasitic motion of the cantilever can dominate and cause feedback instability. As described below in the operational advantages section, Peakforce QNM is able to overcome these disadvantages.

Finally, extracting force curves from the high speed mode of TappingMode is impractical since resonant behavior of the probe also acts as a filter, preventing the ability to reconstruct the force curves with sufficient precision to extract quantitative mechanical information (Legleiter et al. 2006; Stark et al. 2002). In 2008,

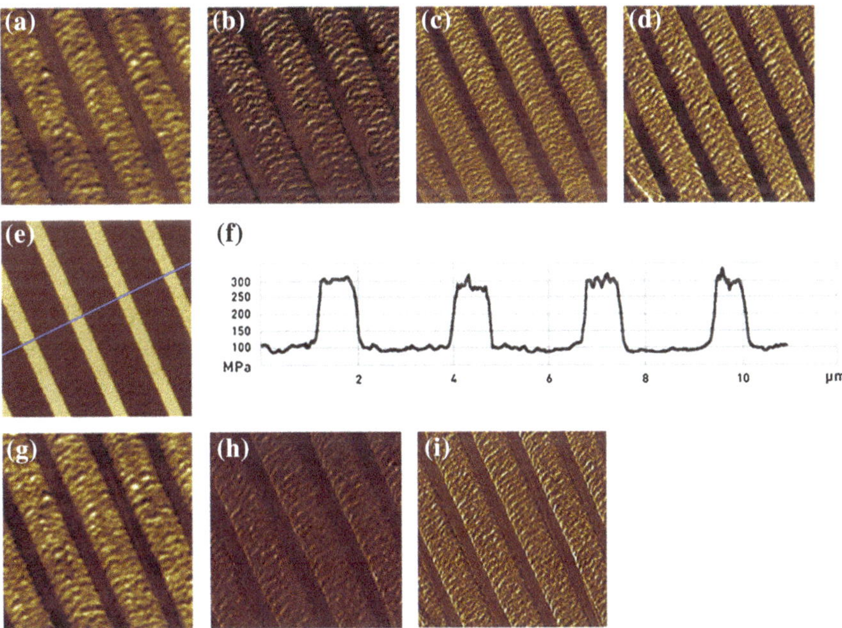

Fig. 2 Multilayer polymer optical film comparing results obtained with PeakForce QNM (**a** height, **b** adhesion, **c** dissipation, **d** deformation, **e** and **f** DMT modulus) and TappingMode Phase Imaging (**g** height, **j** and **i** phase) (10 μm scan size). The phase image in (**h**) was collected with an amplitude setpoint of 80 % of the free amplitude, while the amplitude setpoint for the phase image in (**i**) was 40 % (collected in two different scans). The plot in (**f**) is the modulus along the blue line in (**e**) from left to right. Note that the tapping phase result (**h**) is nearly identical to the PeakForce QNM Adhesion image (**b**)

Veeco (now Bruker) released HarmoniX® as a solution to this problem. HarmoniX adds a second sensor with a much higher bandwidth by offsetting the tip and measuring the torsional signal (Sahin 2005, 2007; Sahin et al. 2007). This technique has been successful in resolving material components in complex polymeric systems. The downsides of this approach are that (1) it requires special probes (2) the operation of the technique can be complicated (especially in fluid), and (3) interpretation of the results is sometimes difficult.

PeakForce QNM is a new mode developed by Bruker that provides the capabilities of HarmoniX without the complexity of operation and interpretation. In peakforce QNM, the force curves are acquired at high speed and with high precision low force control. The force curves are then analyzed to obtain the properties of the sample (adhesion, modulus, deformation, and dissipation) and the information is sent to one of the image data channels while imaging continues at usual imaging speeds. A representative set of PeakForce QNM maps of a multilayer polymer optical film is shown in the Fig. 2a–f where the various channels are shown for PeakForce QNM height (Fig. 2a), adhesion (Fig. 2b), dissipation (Fig. 2c), deformation (Fig. 2d), and DMT modulus (Fig. 2e, f). The result is images that contain maps of material properties (false colored with a user selectable color table) collected in real time.

Since the system can acquire up to eight channels at once, it is possible to map all of the currently calculated properties in a single pass. Offline analysis functions can calculate statistics of the mechanical properties of different regions and sections through the data to show the spatial distribution of the properties.

3 Operational Advantages of Peakforce QNM

PeakForce QNM operates with Peakforce Tapping where the maximum force (or Peakforce, point C in Fig. 1c) on the tip is controlled during the acquisition of the force curves. This is in contrast to conventional TappingMode, where the system keeps the cantilever vibration amplitude constant as the tip is raster scanned on the surface and does not involve force curve acquisition. The combination of the control of Peakforce Tapping with the ability to extract quantitative material properties from the acquired force curves provides a number of operational advantages for Peakforce QNM.

3.1 User Friendly Technique

Peakforce Tapping does not require any cantilever tuning like conventional TappingMode. Additionally, Peakforce Tapping enables Bruker's ScanAsyst™ feature, which automatically adjusts the scanning parameters in real-time to optimize the image and protect the probe and sample. This requires minimal intervention from the user, and enables users from a wide variety of experience levels to benefit from Peakforce Tapping (and PeakForce QNM). No special AFM cantilevers are required.

It is often interesting or necessary to image samples at real world conditions such as under fluid or at temperatures above or below ambient. PeakForce QNM works well in these environments, especially as it is not necessary to re-tune the cantilever when the temperature is changed or when changing from air to fluid operation. This is in contrast to conventional TappingMode™, where a temperature or fluid environment change changes the cantilever environment, and thus its resonant frequency and Q, thereby requiring a retune. With Peakforce Tapping, the system is not being driven at the cantilever resonance, so it is not sensitive to changes in probe resonant frequency and Q. By operating at a frequency far below the resonance, Peakforce Tapping removes the complex resonant dynamics and replaces it with simple and stable feedback on the peakforce. Furthermore, Peakforce Tapping can achieve equal or better force control than TappingMode imaging in all the environments by using broad range of cantilevers, making high quality imaging much easier to achieve with this control mode. Finally, the lack of a need for special probes means that PeakForce QNM can be used with other techniques that do require special probes such as Nanoscale Thermal Analysis with VITA.

3.2 High-Resolution Mapping of Mechanical Properties

Scanning speeds and number of pixels in an image are similar to TappingMode. Analysis of force curve data is done on the fly, providing a map of multiple mechanical properties that has the same resolution as the height image. Sample deformation depths are limited to a few nanometers, minimizing the loss of resolution that can occur with larger tip-sample contact areas.

3.3 Non-Destructive to Tips and Samples

In scanning probe microscopy, there are two primary causes of tip and sample damage. Any lateral force that the tip exerts on the sample can cause the sample to tear (the tip plows through the sample). Likewise, lateral forces from a hard sample can cause the end of the tip to fracture and break off. Normal forces can also cause damage to both tip and sample. Even if there is not enough normal force to damage the sample, there can still be enough to deform the sample, increasing the contact area (and the effective probe size) and reducing the resolution of the scan.

Peakforce Tapping provides direct control of the maximum normal force of the sample, while eliminating lateral forces. This preserves both the tip and sample. By controlling the maximum force exerted on the tip during the force curve acquisition of Peakforce QNM, the tip and sample are protected from damage while allowing the tip-sample contact area to be minimized.

3.4 High Precision Force Control of Interaction Forces with Peakforce Tapping

Peakforce Tapping modulates the Z piezo in a similar fashion to force volume and pulse force mode. However, it can operate with interaction forces orders of magnitude lower, i.e., piconewtons. Such high-precision force control is enabled by data pattern analysis within each interaction period. When the relative Z position between the probe and sample is modulated, various parasitic cantilever motions can occur. These motions include cantilever oscillation excited by the pull-off, as well as deflections triggered by harmonics of the piezo motion or viscous forces in air or fluid. The parasitic deflection (defined as the deflection signal variation when the tip is not interacting with the sample) limits the ability of pulsed force mode to operate with very low forces. Poor force control happens to be the most important factor in achieving high-resolution imaging and property measurements (see image of calcite crystal in Fig. 5). For example, if the tip end has 1 square nanometer area, 1 nanonewton force will lead to 1 gigapascal stress at the tip end, which is enough to break a silicon tip. To lower the stress below the fracture stress of silicon, the control force needs to be no more than a few hundred piconewtons. For materials softer than silicon, the required controlling force is even lower.

During Peakforce Tapping operation, the parasitic deflection signal and its data pattern are analyzed by comparing the measured cantilever deflection during each modulation cycle with the cantilever deflection observed with the cantilever slightly out of contact. The signature of the interaction, in the shape of a "heart beat" signal, is extracted from the parasitic deflections. The "heart-beat" signal is the interaction force curve plotted in the time domain (see Fig. 1c). The feedback loop can choose any point in the force curve to control tip-sample interactions instantaneously. For Peakforce Tapping, the peak point in the repulsive interaction is automatically chosen as the control parameter, similar to the triggered level in force volume mapping.

In comparison, the feedback in all TappingMode techniques uses the near or at-resonance amplitude as a control parameter. In a normal tapping control, the peak interaction force varies from a fraction of a nanonewton to tens of nanonewtons, depending on the operating amplitude, cantilever spring constant, and set point. Such interaction force is well controlled when cantilever oscillation is in a steady-state over a single point. However, when the tip is scanning on a sample surface with varying material properties or topography the force is not constant. This occurs because the amplitude of oscillation is influenced by sample properties such as dissipation as well as peakforce. For example, when scanning from an area with high dissipation to one of low dissipation on a flat sample, the piezo will extend in an attempt to keep the amplitude constant, causing the peakforce to increase. Changes in peakforce can also occur when scanning rough surfaces, where the amplitude error occurring at the sharp edges can correspond to interaction force one order of magnitude higher than that of a steady-state. Amplitude error incurred force is a leading cause of tip damage. However, even with low amplitude error, damage can occur because the feedback is not directly controlling interaction force. In contrast, PeakForce Tapping directly controls the PeakForce on the sample. This protects the tip and sample while maintaining excellent surface tracking.

3.5 Broad Range of Measurements

A major advantage of Peakforce Tapping is its broad operating force range due to its ability to use a wide range of cantilever probes in all environments. Using cantilevers with a spring constant between 0.3 and 300 N/m, the peakforce imaging control is able to achieve force control from piconewtons to micronewtons. At the high force end, it coincides with traditional mechanical mapping techniques, namely force volume and pulse force mode, yet can generate quantitative data by fitting incoming force distance data various models in real-time. In the low force regime, it matches the interaction force achievable by extremely light tapping in TappingMode SPM, but with much improved stability and ease of use in all environments. Piconewton force control is normally only possible in fluid imaging but now can also be applied in ambient imaging, achieving improved imaging quality and tip protection over even best practice TappingMode. With the proper choice of tip radius and cantilever spring constant; this force range can facilitate quantitative characterization of nearly all materials, with modulus ranges from hydrogels to metals and semiconductors.

3.6 Unambiguous and Quantitative Data Over a Wide Range of Materials

Analysis of the entire force curve for each tap allows different properties to be independently measured. Since a wide selection of probes is available, it is possible to cover a very broad range of modulus or adhesion parameters while maintaining excellent signal-to-noise ratios.

Figure 2 compares PeakForce QNM (Fig. 2a–f) for a multilayer polymer sample with TappingMode phase images of the same area (Fig. 2g–i). As mentioned above, phase imaging while qualitatively useful, is prone to imaging artifacts and complicated interpretation. It is often assumed that the phase contrast is primarily caused by variations in sample modulus. The images are topography (Fig. 2g), tapping phase image collected at an amplitude setpoint of 80 % (Fig. 2h) and an amplitude setpoint of 40 % (Fig. 2i). Comparing Fig. 2e and (Fig. 2h or i), it is clear that this is not true in this case. By tapping harder (reducing the amplitude setpoint), one would expect to deform the sample more, increasing the contribution of the modulus to the phase. However, in (Fig. 2i) the relative contrast does not change significantly. The PeakForce QNM data (Fig. 2b–e) shows that the phase signal is dominated by the adhesion independent of tapping setpoint for this tip-sample interaction. It is easy to see that one must be very careful in interpreting phase results, even for qualitative use. The PeakForce QNM modulus channel (Fig. 2e), on the other hand, has unambiguous contrast that can be quantified, as shown in the section plot (Fig. 2f). The narrow strips have a modulus of about 300 MPa, while the wide ones have a modulus of about 100 MPa. The ability of PeakForce QNM to measure elastic moduli compares favorably with instrumented indentation as measured on a variety of polymer samples (Dokukin and Sokolov 2012; Young et al. 2011).

4 Extracting Quantitative Material Properties with PeakForce QNM

The foundation of material property mapping with PeakForce QNM is the ability of the system to acquire and analyze the individual force curves from each tap that occurs during the imaging process. To separate the contributions from different material properties and provide quantitative mechanical property information such as adhesion, modulus, dissipation, and deformation, it is necessary to measure the instantaneous force on the tip rather than a time-average of the force or dissipation over time, as is done in TappingMode Phase Imaging. This requires a force sensor that has a significantly higher bandwidth than the frequency of the periodic interactions. In Peakforce Tapping, the vibration frequency is intentionally chosen to be at a significantly lower (usually several decades lower) frequency than the cantilever resonant frequency, for example a few hundred Hz to several kHz. The force measurement bandwidth of a cantilever is approximately equal to the resonant frequency of the fundamental bending mode used for force detection. As a result, a properly chosen Peakforce Tapping cantilever can respond to changes in instantaneous interaction force with an immediate deflection change.

As mentioned above, the force curve is usually converted to a force versus separation plot (see Fig. 1a) for fitting and further analysis. The separation, which is the negative of the deformation (sometimes called indentation depth), is obtained by adding the Z position of the piezo modulation to the cantilever deflection. A constant can be added to the separation to make it zero at the point of contact if the point of contact can be determined, but this is not required for many analyses. This process is the same as removing frame compliance in indentation measurements. These force-separation curves are analogous to the load-indentation curves commonly used in nanoindentation.

The curves are then analyzed to obtain the properties of the sample (adhesion, modulus, deformation, and dissipation) and the information is sent to one of the image data channels while imaging continues at usual imaging speeds. The result is a set of images that contain maps of material properties along with the usual height image.

Figure 1a illustrates how common mechanical properties are extracted from calibrated force curves. Analysis of the force curves with other models is also possible by capturing raw force curve data with either the "High-Speed Data Capture"

(HSDC) function or the "Peakforce Capture" (PFC) capture mode. HSDC acquires all of the curves from a ~32 s period (providing about 64,000 raw force curves) from any time during scanning and can later be correlated with the analyzed data in the image. Peakforce Capture acquires a single curve for each pixel in an image. Depending on the scan rate and number of samples, there may be some averaging of the force curve data. Either method allows users to recalculate properties using different models or parameters and apply their own models to the raw data to study more unusual materials and properties.

4.1 Elastic Modulus

To obtain the Young's Modulus, the retract curve is fit (see the green line in Fig. 1a) using the Derjaguin–Muller–Toporov (DMT) model (Maugis 2000) show in Eq. 1:

$$F - F_{adh} = \frac{4}{3}E * \sqrt{R(d - d_0)^3} \tag{1}$$

F-F_{adh} is the force on the cantilever relative to the adhesion force, R is the tip end radius, and d-d_0 is the deformation of the sample. The result of the fit is the reduced modulus E^*. If the Poisson's ratio is known, the software can use that information to calculate the Young's Modulus of the sample (Es). This is related to the sample modulus by the Eq. 2:

$$E^* = \left[\frac{1 - v_s^2}{E_s} + \frac{1 - v_{tip}^2}{E_{tip}} \right]^{-1} \tag{2}$$

We assume that the tip modulus E_{tip} is infinite, and calculate the sample modulus using the sample Poisson's Ratio (which must be entered by the user into the NanoScope® "Cantilever Parameters.") The Poisson's ratio generally ranges between about 0.2 and 0.5 (perfectly incompressible) giving a difference between the reduced modulus and the sample modulus between 4 and 25 %. Since the Poisson's ratio is not generally accurately known, many publications report only the reduced modulus. Entering zero for the parameter will cause the system to return the reduced modulus. Models for force curve fitting are also incorporated into the offline analysis including the Hertz and a conical Sneddon model.

PeakForce QNM provides quantitative modulus results over the range of 700 kPa to 70 GPa provided the appropriate probe is selected and calibrated and provided that the DMT model is appropriate. While calibration is still a several step process, it has been made significantly easier than HarmoniX calibration by eliminating several steps and by automating the calculation of several parameters. An experienced user can complete a calibration in less than ten minutes.

Figure 3 demonstrates that this works for a wide range of materials from polydimethylsiloxanes to Silica. The data in Fig. 3 was collected with a set of probes that were selected to have the most accuracy over a range of modulus. Data has also been collected on a range of polymers from LDPE (modulus 200 MPa) to PMMA (modulus 4.8 GPa) (Young et al. 2011). Figure 4 lists the probe types used and gives the approximate modulus range for each probe type. The data in Fig. 4 was acquired on homogeneous samples and the system was calibrated by independently measuring the tip radius, spring constant (the "absolute method").

If the DMT model is not appropriate, the modulus map will still return the fit result, but it will be only qualitative. Some cases where the DMT model are not appropriate include those where the tip-sample geometry is not approximated by a hard sphere (the tip) contacting an elastic plane, cases where mechanisms of deformation other than elastic deformation are active during the retracting part of the curve (at these time scales), and cases where the sample is confined vertically or laterally by surrounding material (close enough to effect the strain in the deformed region). If this is suspected, High Speed Data Capture or Peakforce Capture can acquire the individual force curves across a section of interest to examine the force curves directly and potentially apply more advanced fitting models. Additionally, the analysis software supplies a Matlab portal to simplify the reading of force curves or data into Matlab for customized analysis.

PeakForce QNM can be quite repeatable if care is taken in the calibration process. Recent experiments on homogeneous samples where the absolute method (discussed below) was used measure samples in a range between 1 and 400 MPa ten times each (with different probes) resulted in a relative standard deviation of less than 25 % for all samples. If the goal is to discriminate between components in a multi-component system where the modulus of one component is known, the

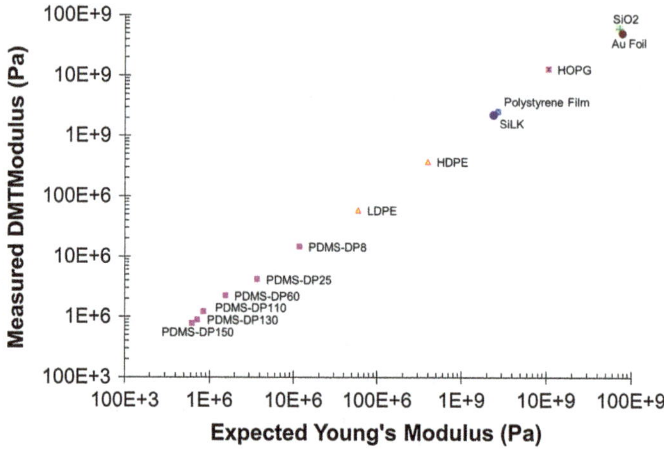

Fig. 3 Plot of measured modulus versus expected Young's modulus (from the literature, except LDPE and PS film which are from SPM nanoindentation). Multiple probes were used with different spring constants to cover the entire range. Each probe was individually calibrated using the absolute method

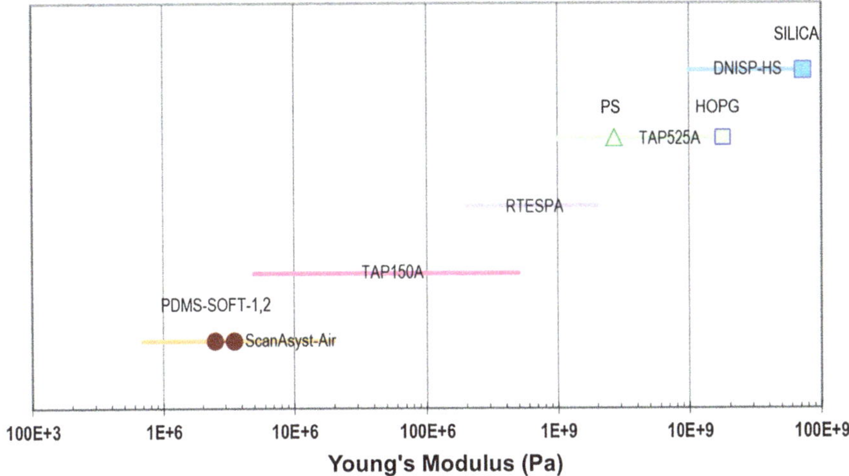

Fig. 4 Modulus ranges covered by various probes. The modulus of the reference sample for each range is also indicated

modulus noise level is more illuminating. For the measurements in the study, the relative standard deviation in a single image was never more than 6 %.

4.2 Adhesion

The second mechanical property acquired in the mapping is the adhesion force, illustrated by the minimum force in Fig. 1a. The source of the adhesion force can be any attractive force between the tip and sample. In air, van der Waals, electrostatics, and forces due to the formation of a capillary meniscus can all contribute with the relative strengths of the contributions depending on such parameters as Hammaker constants, surface charges, and hydrophilicity. For example, if either sample or the probe surface is hydrophilic, a capillary meniscus will typically form, leading to higher adhesion that extends nanometers beyond the surface. For polymers in which the long molecules serve as a meniscus, the adhesion can extend tens of nanometers beyond the surface. The adhesion typically increases with increasing probe end radius. Simple models based on surface energy arguments predict the adhesion to be proportional to the tip end radius, but recent investigations show that this is an oversimplification for real SPM tips (Israelachvili 1992; Thoreson et al. 2006). The area below the zero force reference (the horizontal line in the force curve) and above the withdrawing curve is referred to as "the work of adhesion." The energy dissipation is dominated by work of adhesion if the peakforce set point is chosen such that the non-elastic deformation area (the hysteresis above the zero force reference) in the loading–unloading curve is negligible compared to the work of adhesion (Rosa-Zeiser et al. 1997).

Adhesion force becomes a more meaningful and important quantity if the tip is functionalized. In this case, the amount of adhesion reflects the chemical interaction between specific molecules on the tip and sample. The adhesion map in this case carries the chemical information.

4.3 Dissipation

Energy dissipation is given by the Force times the velocity integrated over one period of the vibration (represented by the gold area in the Fig. 1a) as in Eq. 3:

$$W = \int \vec{F} \cdot d\vec{Z} = \int_0^T \vec{F} \cdot \vec{v}dt \tag{3}$$

where W represents energy dissipated in a cycle of interaction. F is the interaction force vector and dZ is the displacement vector. Because the velocity reverses its direction in each half cycle, the integration is zero if the loading and unloading curves coincide. For pure elastic deformation there is no hysteresis between the repulsive parts of the loading–unloading curve, corresponding to very low dissipation. In this case the work of adhesion becomes the dominant contributor to energy dissipation. Energy dissipated is presented in electron volts as the mechanical energy lost per tapping cycle.

4.4 Deformation

The fourth property is the maximum deformation, defined as the penetration of the tip into the surface at the peakforce, after subtracting cantilever compliance. As the load on the sample under the tip increases, the deformation also increases, reaching a maximum at the peakforce. Because the measured deformation is based on the approach curve, it includes both elastic and plastic contributions. With known tip shape and contact area, this parameter can also be converted to the hardness (although this is usually only applied in cases where the dominant deformation mechanism is plastic deformation). Maximum sample deformation is calculated from the difference in separation from a point where the force is near zero (controlled by parameter "Deformation Fit Region") to the peakforce point along the approach curve (see Fig. 1a). There may be some error in this measurement due to the fact that the tip first contacts the surface at the jump-to-contact point (Fig. 1a, point B) rather than at the zero crossing (actual deformations are usually larger than the reported value).

5 Practical Applications

5.1 Atomic Resolution of Calcite Crystal

The ability of Peakforce QNM to achieve atomic resolution while simultaneously providing maps of properties derived from the individual force interactions in real-time is shown in the images shown in Fig. 5. Both topography (a) and modulus (b) channels are shown of a single step in a calcite crystal under fluid. Calcite is known to be rhombohedral and belongs to the R3(bar)c space group (Rode et al. 2009). The cleavage planes are (1 0 1 bar 4) planes and the height image shows the topography of the protruding oxygen atoms with a difference in height between the plane on the upper right and the upper plane on the lower left of about 300 pm as expected. Unexpectedly, the DMTModulus image (b) shows that alternate rows of atoms have significantly increased contact stiffness (blue circles mark the positions of the oxygen atoms in the height image). The stiffer rows switch with the less stiff rows when crossing over the step. More work must be done to understand the source of this variation in modulus, but it seems likely that it arises due to the interaction of the partially ordered water near the surface and the calcite lattice itself. While the concept of modulus breaks down at the level of individual atoms, we can still use the map to learn about the qualitative stiffness of different positions on the surface. Additionally, force curves can be collected at each atomic site using HSDC or Peakforce Capture to study the interaction in more detail.

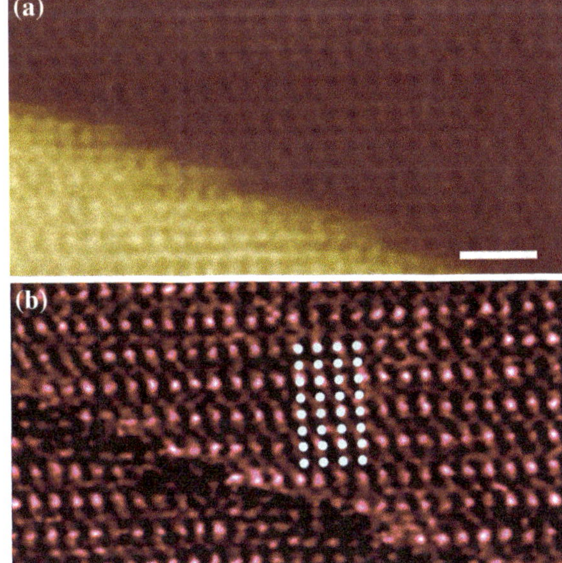

Fig. 5 **a** height and **b** modulus images of atomic resolution of a calcite step under fluid. **a** shows the topography of the protruding oxygen atoms **b** DMT modulus image shows that alternate rows of atoms have significantly increased contact stiffness (blue circles mark the positions of the oxygen atoms in the height image). The stiffer rows switch when crossing over the step. Data collected in collaboration with Dr. Daniel Ebeling, University of Maryland

5.2 Thermal Transitions in a Polymer Blend

Peakforce QNM is used to image a blend of syndiotactic polypropylene (sPP) and polyethylene oxide (PEO) in Fig. 6, where the blend was exposed to 2 different heating–cooling cycles. The initial state of the blend (at 25C) is observed in (a). In the first heating cycle, the temperature was initially raised (60 °C) so that only the PEO-matrix melted (b). The temperature was then allowed to rapidly dropped, causing quick crystallization of the PEO (c). One can see that the PEO topography underwent minor changes compared to the initial morphology

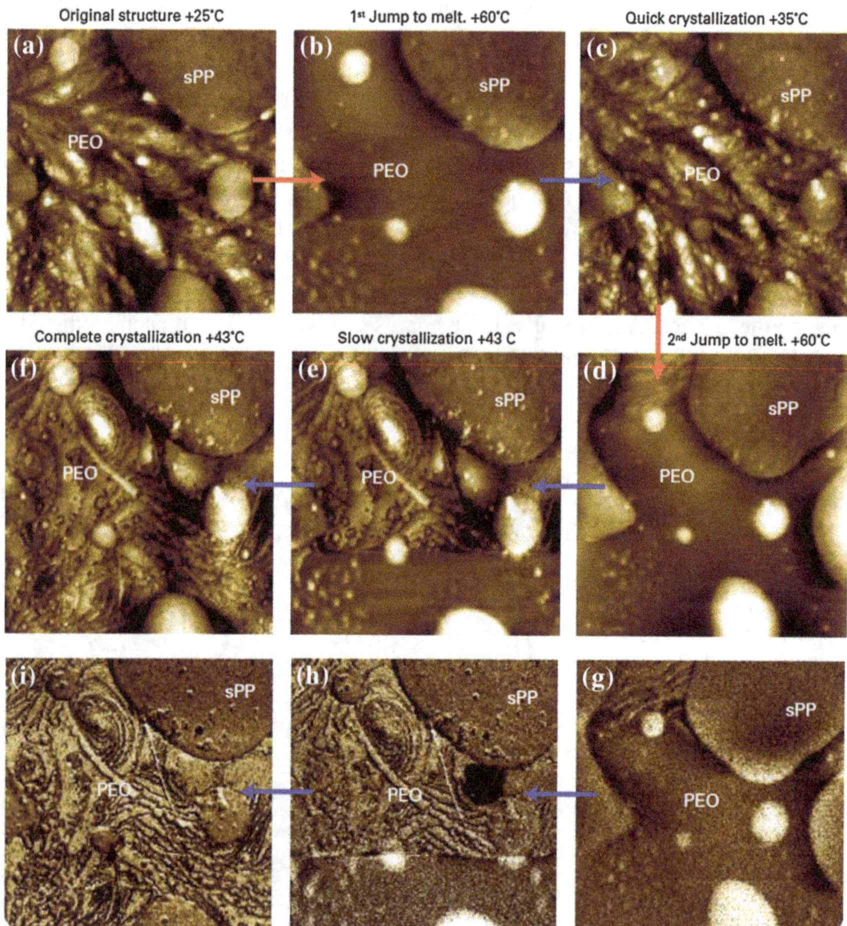

Fig. 6 Cycle of heating and cooling of polymer blend of syndiotactic polypropylene and polyethylene oxide. Images **a–f** shows height of the surface during the process, while **g–i** shows the modulus of frames **d–f** respectively

(a). This demonstrates the so-called "memory" effect of fast cooling, where the original crystallization nuclei are still active even though polymer appears to be totally melted.

During the second heating–cooling cycle, the temperature was lowered more gradually. Figure 6e shows abrupt transition from melt to solid state 1/3 of the way from the frame bottom (the system was scanning up the frame at the time). It's worth noting that, in this case, the PEO morphology becomes completely different from the original morphology in (a), indicating reorientation of the lamellae from an edge-up state to a flat-up state (f). The images in (g–i) are modulus maps corresponding to the second crystallization cycle. Based on topography in the upper portion of (e) the PEO appears to be completely crystallized, but looking at the corresponding modulus map in (h), it is clear that this is not the case—there is a soft (dark) area just to the right of the center of the image disappears when the sample has fully crystallized (i).

As described above, the ease of conducting this experiment at a variety of temperatures is a feature of PeakForce QNM. If this experiment were done in TappingMode™, it would be necessary to adjust drive amplitude and frequency several times during each heating–cooling cycle (generally this is done just before collecting an image if the temperature has changed by more than about 10 °C). As long as the laser reflection stays on the photodetector, the imaging can proceed continuously and without adjustment during any experiment involving sample heating and cooling (note: for quantitative results, some recalibration will be required if the laser spot moves significantly).

5.3 High Resolution Features in Bottle-Brush Molecules

Large polymer macromolecules have been an interesting but challenging sample for SPM since the 1990s (Magonov et al. 2005). One example of this class of molecule is the Poly (butyl acrylate) (PBA) bottle-brush molecule (Sheiko et al. 2008). This molecule has a long backbone with many short, flexible side chains. The conformation and physical properties of these molecules are controlled by a competition between steric repulsion of densely grafted side chains (brushes) with and attractive forces between the brushes and the substrate. They can be either flexible or stiff, depending on the grafting density and the length of side chains. Molecules can switch their conformation in response to alterations in surrounding environment: surface pressure, temperature, humidity, pH, ionic strength, and other external stimuli. Molecular brushes are a very informative model system for experimental studies of polymer properties.

Figure 7 shows a set of images collected on PBA using PeakForce QNM. In the height image Fig. 7a the backbone is clearly visible both as a long isolated molecule and as a folded set of molecules (a molecular ensemble). In the modulus map (Fig. 7b) the very soft (dark) backbones are surrounded by an area that is slightly stiffer, presumably where the short side chains are present. Figure 7e is a histogram showing the relative frequency of various modulus values in the image. The peak at 62 MPa is from the background (mica covered by low molecular weight amorphous polymer). The region given by the red square in Fig. 7b shows two peaks in the modulus of the molecular ensemble—one at about 25 MPa corresponding to the backbone of the chains, and one at about 32 MPa corresponding to the surrounding region filled with short brushes. These numbers are not expected to be quantitative since the molecules are small in comparison to the tip and the deformation of the sample. However, even though the DMT model is not appropriate in this case, the qualitative interpretation that darker regions are softer than the brighter regions leads to the speculation that the polymer backbones are being partially supported by the short side chains.

In the adhesion image (Fig. 7c); the background appears dark with very little adhesion as indicated in the adhesion histogram (Fig. 7f) by the peak at 0.47 nN. The histogram also has three other peaks: the backbones at about 0.63 nN, the brushes of single molecules at about 0.71 nN, and the brushes of molecular

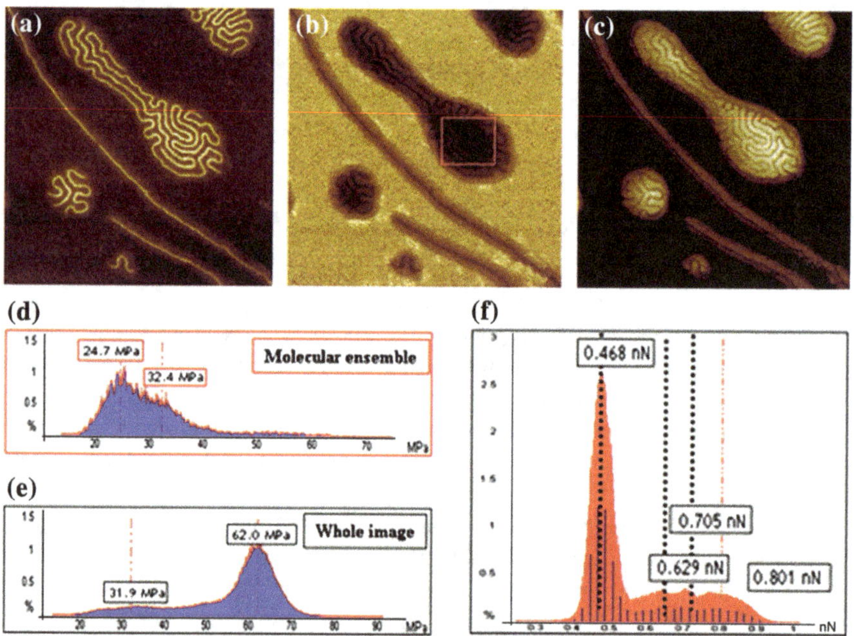

Fig. 7 Poly(butyl acrylate) brush-like macromolecules and molecular ensembles on a mica substrate. **a** Height; **b** modulus; **c** adhesion; **d** histogram of modulus map; **e** histogram of area within red box in modulus map; **f** histogram of adhesion map. Sample was imaged with a MultiMode 8 using PeakForce QNM with a scan size of 500 nm. Sample courtesy of Sergei Sheiko (University of North Carolina, Chapel Hill) and Krzysztof Matyjaszewski (Carnegie Mellon University, Pittsburg)

ensembles at about 0.8 nN. The greater adhesion for the molecular ensembles is likely due to the greater number of brushes available to bind with the tip in those regions.

5.4 Discriminating Domains in Cement Paste

PeakForce QNM was used to characterize the nanomechanical properties of hardened cement paste. This material is significantly harder than the other applications

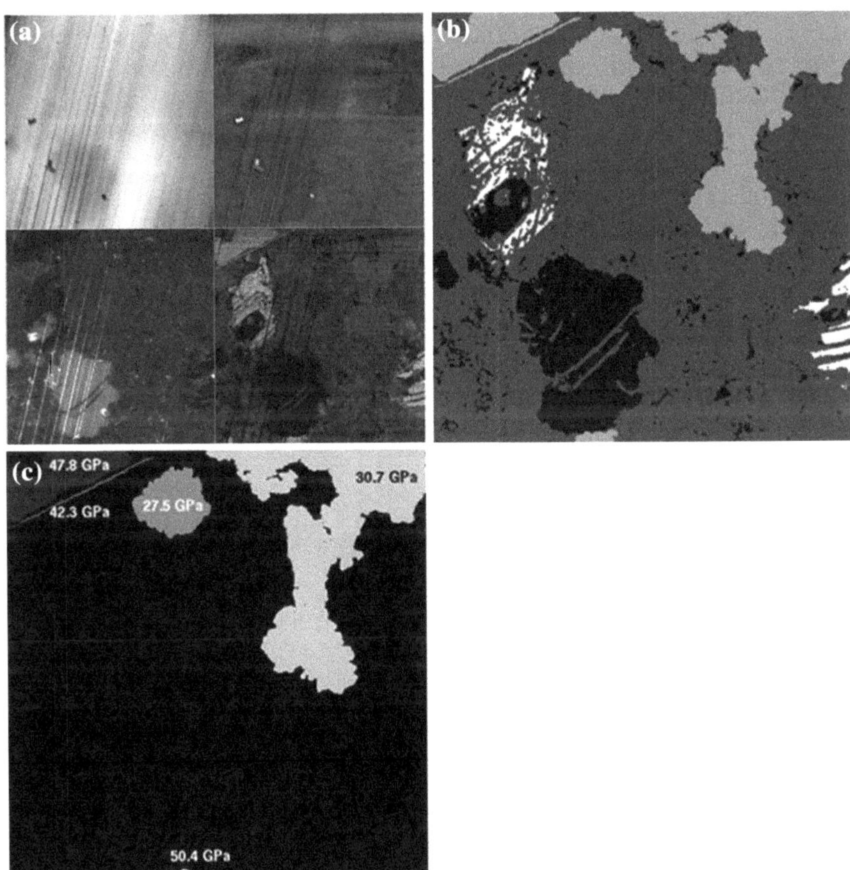

Fig. 8 **a** PeakForce QNM maps of cement paste showing from top left, clockwise topography, adhesion, DMT modulus, and deformation. **b** DMT modulus map thresholded to identify the chemically different domains observed in the EDS images. **c** DMT modulus map with further thresholding to discriminate between various calcium hydroxide domains due to orientation. Reprinted with permission from (Trtik et al. 2012)

described above, with modulus of up to ~100 GPa. In addition, nanomechanical characterization of cementitious materials is typically done with nanoindentation but it suffers from challenges of surface roughness, volume effects from nanoindentor tip interaction volumes, and the need for a statistically large number of samples to get meaningful representation of the mechanical properties. Peakforce QNM is able to circumvent many of these challenges due to its high resolution, use of a significantly smaller tip, and fast acquisition speed where a large number of data points can be collected quickly.

The cement paste was initially characterized by SEM (Trtik et al. 2012). SEM show excellent resolution but poor contrast between domains and no easy way to identify components. EDS provides chemical information on the different components of the sample, but the information is at low spatial resolution. By combining mechanical data collected by PeakForce QNM (Fig. 8) with elemental data from EDS (not shown), a complete picture of the various domains emerges. Figure 8a shows the various channels obtained from Peakforce QNM including topography (top left), adhesion (top right), deformation (bottom left), and DMT-Modulus (bottom right). Figure 8b shows the PeakForce QNM modulus map thresholded to identify the chemically different regions observed in the EDS image. The light gray domains are calcium hydroxide, while the white domain is unhydrated. The dark gray background is other hydrates and the black domain is the epoxy filled pores. Figure 8c further thresholds the modulus map to show that some of the calcium hydroxide regions have different modulus from the others, suggesting that the crystal orientation of the CH is different for the different regions. The Peakforce QNM is the only method that is able to differentiate the various differently oriented domains of calcium hydroxide at this level of resolution.

6 Concluding Remarks

Peakforce QNM is a novel method that provides spatial maps of nanomechanical properties of surfaces including modulus, adhesion, deformation, and dissipation. This technique provides several important advantages over other available SPM methods including ease of use, unambiguous and quantitative material information, non-destructive to both tip and sample, and high speed scanning. The basis of the Peakforce QNM method is force distance curves, which are analyzed directly with appropriate models, so that there is no ambiguity regarding the source of image contrast as often occurs in other techniques. Mechanical property maps are quantitative, low noise, and can span a wide range of property values. Applications of Peakforce QNM to measure properties of materials from single crystals to polymer materials to hardened cement paste are described. These capabilities of PeakForce QNM will provide researchers with critical material property information to enable better understanding of materials at the nanoscale.

References

Cleveland JP, Anczykowski B, Schmid AE, Elings VB (1998) Energy dissipation in tapping-mode atomic force microscopy. Appl Phys Lett 72(20):2613–2615

Dokukin ME, Sokolov I (2012) Quantitative mapping of the elastic modulus of soft materials with HarmoniX and PeakForce QNM AFM modes. Langmuir 28(46):16060–16071. doi:10.1021/la302706b

Israelachvili N (1992) Intermolecular and surface forces Academic Press, New York

Legleiter J, Park M, Cusick B, Kowalewski T (2006) Scanning probe acceleration microscopy (SPAM) in fluids: Mapping mechanical properties of surfaces at the nanoscale. Proc Nat Acad Sci United States of Am 103(13):4813–4818. doi:10.1073/pnas.0505628103

Magonov S, Erina N (2005) Modern Trends in Atomic Force Microscopy of Polymers. Bruker application note AN84

Maugis D (2000) Contact, adhesion and rupture of elastic solids. Springer-Verlag, Berlin

Pittenger B, Erina N, Su C (2010) Quantitative Mechanical Property Mapping at the Nanoscale with Peak Force QNM. Bruker application note AN128

Proksch R (2006) Multifrequency, repulsive-mode amplitude-modulated atomic force microscopy. Appl Phys Lett 89(11):113121–113123

Radmacher M, Cleveland JP, Fritz M, Hansma HG and Hansma PK (1994) Mapping interaction forces with the atomic force microscope. Biophys J 66:2159–2165

Rode S, Oyabu N, Kobayashi K, Yamada H, Kühnle A (2009) True atomic-resolution Imaging of (1014) calcite in aqueous solution by frequency modulation atomic force microscopy. Langmuir 25(5):2850–2853. doi:10.1021/la803448v

Rodriguez TR, Garcia R (2004) Compositional mapping of surfaces in atomic force microscopy by excitation of the second normal mode of the microcantilever. Appl Phys Lett 84(3):449–451

Rosa-Zeiser A, Weilandt E, Hild S, Marti O (1997) The simultaneous measurement of elastic, electrostatic and adhesive properties by scanning force microscopy: pulsed-force mode operation. Meas Sci Technol 8(11):1333

Sahin O (2005) Harmonic force microscope: a new tool for biomolecular identification and material characterization based on nanomechanical Measurements. Stanford University

Sahin O (2007) Harnessing bifurcations in tapping-mode atomic force microscopy to calibrate time-varying tip-sample force measurements. Rev Sci Instrum 78(10):103707–103704

Sahin O, Magonov S, Su C, Quate CF, Solgaard O (2007) An atomic force microscope tip designed to measure time-varying nanomechanical forces. Nat Nanotechnol 2(8):507–514

Sheiko SS, Sumerlin BS, Matyjaszewski K (2008) Cylindrical molecular brushes: Synthesis, characterization, and properties. Progress in Polymer Sci 33(7):759–785. doi:http://dx.doi.org/10.1016/j.progpolymsci.2008.05.001

Stark M, Stark RW, Heckl WM, Guckenberger R (2002) Inverting dynamic force microscopy: From signals to time-resolved interaction forces. Proc Nat Acad Sci 99(13):8473–8478. doi:10.1073/pnas.122040599

Tamayo J, Garcia R (1997) Effects of elastic and inelastic interactions on phase contrast images in tapping-mode scanning force microscopy. Appl Phys Lett 71(16):2394–2396

Thoreson EJ, Martin J, Burnham NA (2006) The role of few-asperity contacts in adhesion. J Colloid Int Sci 298(1):94–101. doi:http://dx.doi.org/10.1016/j.jcis.2005.11.054

Trtik P, Kaufmann J, Volz U (2012) On the use of peak-force tapping atomic force microscopy for quantification of the local elastic modulus in hardened cement paste. Cem Concr Res 42(1):215–221. doi:http://dx.doi.org/10.1016/j.cemconres.2011.08.009

Veeco Application note #84

Veeco (2010) Quantitative mechanical property mapping at the nanoscale with PeakForce QNM

Young TJ, Monclus MA, Burnett TL, Broughton WR, Ogin SL, Smith PA (2011) The use of the PeakForce TM quantitative nanomechanical mapping AFM-based method for high-resolution Young's modulus measurement of polymers. Meas Sci Technol 22(12):125703

Zhong Q, Inniss D, Kjoller K, Elings VB (1993) Fractured polymer/silica fiber surface studied by tapping mode atomic force microscopy. Surf Sci Lett 290(1–2):L688–L692. doi:http://dx.doi.org/10.1016/0167-2584(93)90906-Y

Measurement of Hardness of Very Hard Materials

A. C. Fischer-Cripps

Abstract In the last twenty years or so, the development of hard thin coatings has progressed to the state that hardness testing based on an instrumented technique has become very popular since for this application, the depth of penetration has to be kept within a small percentage of the overall coating thickness and the resulting impressions are too small for an accurate traditional optical measurement. Despite the well-known methods of analyzing instrumented indentation data, considerable problems arise when this test is applied to very hard materials. The underlying boundary conditions for instrumented indentation analysis are often ignored by practitioners who are sometimes accepting of the results at face value, since often they provide a very pleasing and desirable estimation of hardness of their samples. This chapter reviews the essential features of instrumented indentation analysis and points out the significance of those issues that can affect the computed values of both hardness and elastic modulus. In particular, the significance of the geometry factor ε, the indenter area function, and the mean pressure elastic limit. These interrelated factors can conspire to increase the computed value of hardness by up to a factor of 2 if not properly taken into account. This chapter educates and informs the reader so that results of hardness for very hard materials may be properly interpreted when either viewed in the literature or obtained experimentally so as to avoid incorrect conclusions and results.

1 Introduction

Historically, hardness measurements were performed on metallic materials. Familiar terms such as Brinell, Vickers, Knoop, and Rockwell are often associated with testing of metals in an engineering context. Modern instrumented methods of the measurement of hardness are influenced by these early measurements. For example, the

A. C. Fischer-Cripps (✉)
Fischer-Cripps Laboratories Pty Ltd, Sydney, Australia
e-mail: tony.cripps@ibisonline.com.au

A. Tiwari (ed.), *Nanomechanical Analysis of High Performance Materials,*
Solid Mechanics and Its Applications 203, DOI: 10.1007/978-94-007-6919-9_3,

face angle of a modern Berkovich indenter used in nanoindentation testing is made so that the ratio of the contact area to indentation depth is the same as that of a traditional four-sided Vickers indenter (Smith and Sandland 1922). The 136° face angle of a Vickers indenter, was made so that the indentation strain (a/R where a is the contact radius and R is the indenter radius) would be equivalent to that of a spherical (Wahlberg 1901), Brinell indenter at $a/R = 0.2$—a strain at which a fully developed plastic zone would be formed in a typical metal. The Brinell hardness number is favored by some engineers because of the existence of an empirical relationship between it and the ultimate tensile strength of the specimen material.

Nearly all the traditional methods mentioned above are based on an optical measurement of the size of a residual impression made in the specimen surface after the loading of an indenter placed in contact. That is, we measure how hard something is by touching it, not with our fingers, but with a carefully controlled force using a carefully shaped probe. A hard material will leave a smaller residual imprint in the surface compared to a softer surface. Hardness then, is really a measure of plastic yield—a circumstance first described by Hertz (1881, 1882) .

While it is sometimes useful to have a comparison with past measures of hardness, we will see in this chapter that such a historical connection brings with it several problems when the hardness of very hard materials is desired to be measured.

The testing of metals remains an extremely important application of hardness measurements, but it is the hardness of relatively thin coatings that has been given much more attention in the last 20 years as the application of hard coatings to engineering products like cutting tools and hard-wearing surfaces has significant economic benefits. Because hardness impressions made in a thin coatings are almost impossible to accurately measure using optical techniques, modern testing methods employ an instrumented indentation approach whereby the load is applied to the indenter, but the depth of penetration is measured (often to sub-nm resolution) and the area of contact determined from the known geometry of the indenter. The ultimate aim of the modern hard coatings researcher is the production of a coating with hardness equal to or exceeding that of diamond—the hardest known material.

2 Contact Mechanics

Hertz's original analysis of the mechanics of elastic contacts focused on those between glass lenses and has subsequently been applied to the contact between a spherical indenter and a flat semi-infinite surface. Hertz measured the area of contact using impressions made in lamp black, a carbon film. For instrumented hard coatings hardness testing, it is best to use a sharp-tipped pyramidal indenter so as to induce plasticity in the material at the smallest possible load. The intention is to attain a depth of penetration low enough so that the readings are not influenced by the properties of the substrate. In this case, it is usual to treat the pyramidal indenter as an equivalent cone that has the same area to depth ratio as the original indenter to take advantage of symmetry of the problem in the mathematical analysis. The equations of contact for a conical indenter are similar to the spherical case and the most common analytical solution (for elastic contact) is given by Sneddon (1948), where radius of circle of contact is related to the indenter load by Eq. 1:

$$P = \frac{\pi a}{2} E^* a \cot \alpha \tag{1}$$

In this formula, a is the radius of the circle of contact, α is the cone semi-angle, and E^* is the combined elastic modulus of the indenter and the specimen given by Eq. 2:

$$\frac{1}{E^*} = \frac{1 - v^2}{E} + \frac{1 - v_i^2}{E_i} \tag{2}$$

where the subscript i refers to the properties of the indenter. The displacement profile of the deformed surface within the area of contact with respect to the specimen free surface is a function of the radial distance r from the axis of symmetry and is given by Eq. 3:

$$h = \left(\frac{\pi}{2} - \frac{r}{a}\right) a \cot \alpha \quad r \leq a \tag{3}$$

The quantity $a \cot \alpha$ is the depth of penetration h_c measured at the circle of contact. Substituting Eq. 1 into 3 with $r = 0$, we obtain Eq. 4:

$$P = \frac{2E^* \tan \alpha}{\pi} h_t^2 \tag{4}$$

where h_t is the total depth of penetration of the tip of the indenter beneath the original specimen free surface.

In general, contact between an indenter and a specimen may result in both elastic and plastic deformations. For a spherical indenter (and for conical indenters with a rounded tip) the contact is initially elastic. As the load is increased, the mean contact pressure also increases as does the level of shear stress in the indentation stress field. Eventually plastic deformation occurs at the location of greatest shear (about $0.5a$ below the contact surface). If the mean contact pressure is plotted against the ratio a/R (where R is the indenter radius), then it is observed that there is a linear region followed by a plateau at which the mean contact pressure shows no increase with increasing indenter load. When this happens, the plastic zone is said to be "fully developed". The indentation hardness is defined as the mean contact pressure for the condition of a fully developed plastic zone and is computed from the load P_{max} divided by the projected area of the contact A (Eq. 5):

$$H = \frac{P_{max}}{A} \tag{5}$$

For an ideal conical indenter, the contact has the property of geometrical similarity and the mean contact pressure is independent of load since the plastic zone is fully developed from the moment of first contact. In practice, conical and pyramidal indenters are not perfectly sharp so there is usually some initial elastic response before the formation of a fully developed plastic zone, even for very soft materials. This behavior places a limit on the measurement of hardness for very thin films and near surface regions of materials since the load needs to be large enough to induce full plasticity, yet no so large so as to cause an undesired depth of penetration into the sample.

In a typical hardness test, the loading part of the cycle consists of both elastic and plastic deformations. During unloading, the contact is usually entirely elastic, and so the equations of contact above can be used even in the presence of the plastic zone under the indenter, because it is only the elastic strains that relax with the plastic zone somewhat frozen in place.

The most well known method of analyzing indentation data is that of Oliver and Pharr (1992). In this method, the pyramidal indenter is represented by an equivalent cone and Sneddon's equations above used on the elastic unloading part of the indentation process. Making use of Eq. 2 at $r = a$, it is relatively straightforward to calculate the contact depth h_c from Eq. 6:

$$h_c = h_t - \left[\frac{2(\pi - 2)}{\pi}\right]\frac{P_{\max}}{dP/dh} \tag{6}$$

where $P_{\max}$ (the maximum load) and dP/dh (the contact stiffness) are experimentally measured quantities. The square-bracketed term in Eq. 6 evaluates to 0.727 but it is common practice to use a value of 0.75 since this takes into account the upward curvature of the residual impression during unloading.

It is possible to also determine the elastic modulus of the specimen material. The elastic modulus is found from the contact stiffness. The derivative of Eq. 4 with respect to h is:

$$\frac{dP}{dh} = 2\frac{2E^* \tan\alpha}{\pi}h \tag{7}$$

With some rearrangement, and substitution involving Eqs. 1 and 3, it can be shown that:

$$E^* = \frac{dP}{dh}\frac{1}{2}\frac{\sqrt{\pi}}{\sqrt{A}} \tag{8}$$

where A is the area of contact at full load as determined from h_c. Note that the method given above does not require any direct measurement of the size of the contact area. In conventional hardness tests, it is the size of the residual impression that is measured. In instrumented, or depth-sensing, nanoindentation tests, it is the size of the contact under full load that is computed.

3 Instrument Corrections

The application of contact mechanics to indentation tests is now fairly routine, and it is possible to measure elastic moduli and hardness for a wide range of materials from soft biological materials to metals and ceramics. The scale over which these properties can be measured depends very much on the nature of the indenter. If we wish to measure the hardness of a specimen, we do not want plastic deformation to occur in the indenter, and so for this reason, the majority of indenters used in indentation testing area made from diamond. Diamond is the hardest known material with a hardness of about 110 GPa. We might well ask, what happens in the case where the specimen is as hard as diamond. Can we still measure E and H reliably?

To answer this question, it is perhaps best to perform an experiment on a material whose properties are known and then determining if the results of the above analysis provide realistic results. The number of materials with hardness approaching that of diamond is very limited and the reported measurements are not always supported. A very good test of the method would be to perform an indentation test on a specimen of diamond, using a diamond indenter.

Performing such testing requires significantly more attention to the various corrections made to the raw experimental data than is usually the case. There are three main corrections to consider—the initial penetration, the instrument compliance, and the indenter area function.

The initial penetration correction is applied as a constant initial indentation depth into the sample made at the initial contact force. This is required because it is at the initial contact force that the depth sensor is zeroed. But by necessity, the initial contact force results in the indenter penetrating the specimen surface, and it is this initial penetration that must be determined and added to subsequent depth readings so that the depth datum is at the original free surface. While some instruments expect the user to determine the contact point graphically by eye, the most subjective method is to fit the Hertzian elastic equations of contact to this initial data (usually assumed to be elastic by virtue of the tip rounding). The method of fitting is described in detail in (Fischer-Cripps et al. 2001). It is essentially a least squares power law fit to the initial contact data to yield a value of the initial penetration h_i. Thus, the corrected depth h', for subsequent depth readings, is (Eq. 9):

$$h' = h + h_i \qquad (9)$$

The depth measuring system of a typical instrumented indentation hardness tester registers the depth of penetration of the indenter into the specimen and also any displacements of the load frame arising from reaction forces during the application of load to the indenter. These displacements are proportional to the load. Thus, the unloading stiffness dP/dh has contributions from both the elastic responses of the specimen and the instrument. The contribution from the instrument includes the compliance of the loading frame, the indenter shaft, and the specimen mount. If not corrected for, the compliance of the indenter material is included in the composite modulus E^*.

The value of the instrument compliance C_f (typical units μm/mN) can be estimated by an analysis of the area function data (see below). Once obtained, a correction may be made to the indentation depths h' (already corrected for initial contact) to give a further corrected depth h'' according to (Eq. 10):

$$h'' = h' - C_f P \qquad (10)$$

It is the application of the elastic equations of contact to the unloading data that ultimately provides the value of the contact depth h_c, and hence the area of contact. These equations (Eqs. 6, 7 and 8 above) assume that the indenter itself has the ideal shape of an atomically sharp three sided (in the case of a Berkovich indenter) or four-sided (for a Vickers indenter) pyramid. Of course in practice, such an indenter cannot be manufacturer and so there is by practical necessity a finite tip radius. Further, the indenter itself may contain irregularities in its surface arising from the polishing and some error in the nominal face angle. To account for these departures from the idea shape, a table or an equation called the area function is usually measured for each individual indenter used in indentation testing. The data for the area function is usually obtained by a reverse analysis of the equations of contact whereby the known value of elastic modulus for a standard specimen is used as an input and the tip geometry calculated accordingly. For sharp-tipped indenters, it is the tip radius that determines the minimum depth whereby a fully developed plastic zone may be induced in the specimen material and so provide a reliable measurement of hardness.

The area correction is not applied as a correction to the depth readings, but instead is incorporated as a correction to the contact area as a ratio A/A_i where for a value of contact depth h_c, A is the measured contact area (Fig. 1), and A_i is the ideal contact area so that the hardness and elastic modulus are given by Eqs. 11 and 12.

$$H = \frac{P}{A}\left[\frac{A_i}{A}\right] \tag{11}$$

$$E^* = \frac{dP}{dh}\frac{\sqrt{\pi}}{2\beta\sqrt{A}}\sqrt{\frac{A_i}{A}} \tag{12}$$

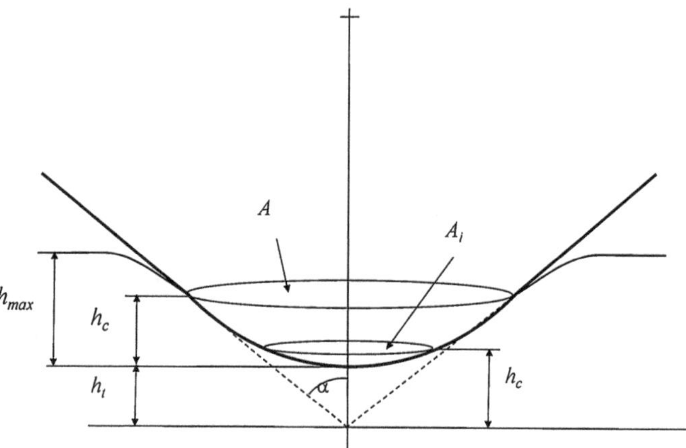

Fig. 1 Schematic of the significance of the area function for real indenters used in indentation testing. The ideal area A_i is that which should be obtained for a given depth of penetration if the indenter geometry were ideal, while the actual area A is that actually obtained with a real indenter with a finite tip radius

The above corrections apply to nearly any indentation test data taken with a conventional depth sensing nanoindentation test instrument. The important point to note is that the greater the elastic modulus of the specimen, the greater the significance of the compliance correction because in these cases, the frame compliance makes a greater contribution to the overall contact stiffness compared to that obtained on a less stiff material.

The act of making an indentation into a very hard material may cause the indenter tip to become blunt during the actual test and so the area function usually has to be measured after testing on the very hard specimen material—preferably after several tests, so that the tip radius may become stabilized.

4 Measurement of Hardness of Very Hard Materials

The sharpness of an indenter tip is essentially its capacity to produce a fully formed plastic zone in the specimen material since this is the primary condition for measurement of hardness. For elastic contact with a conical indenter, the mean contact pressure is given by Eq. 13:

$$p_m = \frac{E^*}{2}\cot\alpha \qquad (13)$$

and is independent of load due to the geometrical similarity of this type of indenter geometry. The significance of this is that in real materials—where indentation plasticity occurs, the mean contact pressure is limited to the hardness value H. However, if the combination of E^* and the angle α are such that the mean contact pressure given by Eq. 13 falls below the specified hardness value H, then the contact is entirely elastic (Caw 1969). This then places a limit on the combination of E^* and indenter angle that may be used for the measurement of hardness of very hard materials.

For a measurement of hardness, we require a fully formed plastic zone. In order to obtain a fully developed plastic zone, we require the limiting value of mean contact pressure p_m for an elastic contact as given by Eq. 13 to be equal to or greater than H. For the case of fused silica, we can take representative values of $E^* = 69.7 \ \alpha = 70.3°$ and to obtain a limiting value of $p_m = 12.5$ GPa. Since the hardness of fused silica is about 9.5 GPa, we can conclude that a perfectly formed diamond Berkovich indenter will reliably measure the hardness of this material. Much the same calculation can be made for other materials. For example, sapphire has a hardness of about 28 GPa (unannealed state) and an elastic modulus of about 450 GPa. Equation 13 shows that the critical value of mean contact pressure and the critical angle becomes larger with increasing values of E^*. This means that the usefulness of a particular indenter in measuring hardness (i.e. the effective sharpness) depends upon the material being tested. With sapphire, an indenter is conceptually "sharper" when used with sapphire than with fused silica due to the modifying effect of the much larger value of elastic modulus of sapphire in Eq. 13. The higher value of hardness for sapphire compared to fused silica tends to increase the plastic depth but it is the far greater elastic modulus that dominates in this case and causes it to be reduced.

When a sample of diamond is tested, we find that due to the very high elastic modulus of diamond ($\approx$1,140 GPa) a reasonably used indenter can still be expected to provide realistic values of hardness because the limiting mean contact pressure becomes 105 GPa which is similar to the expected hardness for this material of about 110 GPa.

Figure 2 shows the indentation curve obtained on a sample of industrial diamond indented with a Berkovich indenter to a load of 200 mN. This results in a depth of penetration of 396 nm. Note that the response is nearly completely elastic, but since the unloading curve is offset from the loading curve, plasticity is in evidence. After the data is corrected for initial penetration, instrument compliance and area function, the results of the unloading analysis yield a value of elastic modulus of 1,198 GPa and a hardness of 108.5 GPa. On the Vickers scale, this corresponds to HV = 10,262 kgf/mm^2. The elastic limit for this material is expected to be about 105 GPa and this is why the contact appears so elastic in character—we are at the limit of obtaining a reading of hardness with the 65.3° face angle of a Berkovich indenter.

For so-called ultra-hard nanocomposite coatings (Fischer-Cripps et al. 2012), we find that for an elastic modulus of about 450 GPa, the limiting contact pressure is of the order of about 65 GPa and so a standard Berkovich or Vickers indenter cannot be expected to provide a mean contact pressure greater than this value, and any values of hardness quoted in excess of this should be treated with extreme caution.

It is important to note that the limiting value of mean contact pressure depends upon the indenter angle and the combined, or reduced, elastic modulus (Eq. 2) and so the figures mentioned above are all with respect to a diamond indenter. In order

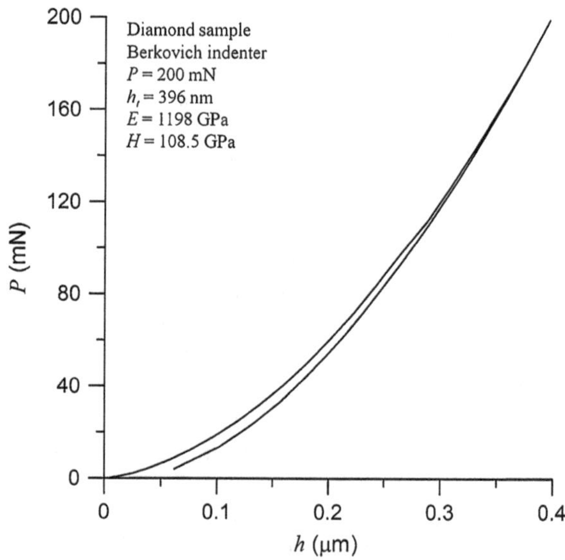

Fig. 2 Load-displacement curve for indentation into diamond with a diamond Berkovich indenter at 200 mN

to determine the limiting value of mean contact pressure, one has to first determine the combined elastic modulus of the indenter and specimen material, and then the equivalent cone angle (same area to depth ratio as the actual indenter—which more often than not, is of pyramidal geometry).

Although it appears that a Berkovich indenter may be used for measurement of hardness of nearly all materials, some headroom may be attained by use of a cube corner indenter—especially in situations where the limiting value of mean contact pressure is about the same as that of the expected hardness of the material. For fused silica, a diamond cube corner indenter (with an effective cone angle $\alpha = 42.3°$) has a limiting value of mean contact pressure of 37.6 GPa. For sapphire ($E = 450$ GPa; $E^* = 340$ GPa), the limiting mean contact pressure becomes about 187 GPa, while for diamond on diamond we reach a potential measurement of hardness of about 300 GPa with a cube corner indenter.

Researchers claiming hardness measurements in excess of the limiting value of mean contact pressure for a particular indenter and specimen material pair with a Vickers or Berkovich indenter would do well to verify their readings with a cube-corner indenter.

5 Concluding Remarks

The instrumented indentation test method is very versatile and is conceptually very simple. Corrections to instrumented test data, such as the initial penetration, instrument compliance, and the indenter area function, are not usually considered in large scale hardness testing where optical measurements are usually taken. For small scale instrumented indentation testing, such issues become important, and when applied to the measurement of very hard, and very stiff specimens it is essential that they be correctly applied. An important consideration in the testing of very hard materials is the elastic limit imposed by the combined elastic modulus of the indenter and specimen E^* and the angle of the indenter. Injudicious choice of indenter angle can result in measurements of hardness that are too low, and without knowledge of these matters, such a mistake can be easily overlooked.

It is hoped that by drawing attention to the particular items that require close attention in the indentation testing of very hard materials, this chapter might educate and inform the reader so that when such results are obtained, or found in the literature, a proper interpretation may be made so as to avoid incorrect conclusions and results.

References

Caw W (1969) The elastic behaviour of a sharp obtuse wedge impressed on a plane. J Phys E: Sci Instrum 2(2):73–78

Fischer-Cripps AC, Bull SJ, Schwarzer N (2012) Critical review of claims for ultra hardness in nanocomposite coatings. Phil Mag 92(13):1601–1630

Fischer-Cripps A, Bendeli A, Bell T, Field J, Jamting A (2001) Methods of correction for analysis of depth-sensing indentaton test data for spherical indenters. J Mater Res 16(8):2244–2250

Hertz H (1881) On the contact of elastic solids. J Reine Angew Math, 92. Translated and reprinted in English in Hertz's Miscellaneous Papers, 1896 Ch. 5. Macmillan & Co., London, pp 156–171

Hertz H (1882) On hardness. Berh. Ver. Beforderung Gewerbe Fleisses, 61. Translated and reprinted in English in Hert's Miscllenaneous Papers, 1896 Ch. 6. Macmillan & Co., London, p 410

Oliver W, Pharr G (1992) An improved technique for determining hardness and elastic modulus using load and displacement sensing indentation experiments. J Mat Res 7(4):1564–1583

Smith RL, Sandland G (1922) An accurate method of determining the hardness of metals with particular reference to those of high degree of hardness. Proc Inst Mech Eng 1:623–641

Sneddon IN (1948) Boussinesq's problem for a rigid cone. Proc Cambridge Philos Soc 44:492–507

Wahlberg A (1901) Brinell's method of determining hardness. J Iron Steel Inst London 59:243–298

Environmental Nanomechanical Testing of Polymers and Nanocomposites

Jian Chen, Ben D. Beake, Hanshan Dong and Gerard A. Bell

Abstract The ever-increasing popularity of nanomechanical testing is being accompanied by the development of more and more novel test techniques and adaptation of existing techniques to work in increasingly environmentally challenging test conditions. Considerable progress has been made and reliable mechanical properties of materials can now be obtained at a range of temperature and surrounding media, greatly aiding development for operation under these environmental conditions. In this chapter several of these developments are reviewed, focussing on their use in the non-ambient nanomechanical testing of polymers and nanocomposites.

1 Introduction

Applications of nanomechanical testing are continually increasing, including for example, characterizing the mechanical properties of advanced materials and systems such as micro/nano-mechanical devices (Beake et al. 2009), hard coatings for engineering tools, thin films in semiconductor (Beake and Lau 2005), advanced polymers (Kranenburg et al. 2009; Chen et al. 2010; Beake et al. 2002a, b) and even biomedical tissues (Chen and Lu 2012; Mencik et al. 2009) etc. The successes can be attributed to three main advantages. Firstly, the technique has a high spatial resolution and depth

J. Chen
Jiangsu Key Laboratory of Advanced Metallic Materials, School of Materials Science
and Engineering, Southeast University, Nanjing 211189, China
e-mail: j.chen@seu.edu.cn

B. D. Beake (✉) · G. A. Bell
Micro Materials Ltd, Willow House, Yale Business Village, Ellice Way,
Wrexham LL13 7YL, UK
e-mail: ben@micromaterials.co.uk

H. Dong
School of Metallurgy and Materials, University of Birmingham, Birmingham B15 2TT, UK

A. Tiwari (ed.), *Nanomechanical Analysis of High Performance Materials*,
Solid Mechanics and Its Applications 203, DOI: 10.1007/978-94-007-6919-9_4,
© Springer Science+Business Media Dordrecht 2014

sensitivity. The obtained information is therefore highly localized within a particular area of interest with less influence from its surroundings, for example, in quantitative assessment of the properties of inclusions in metallic alloys. With the spatial resolution and repositioning accuracy down to submicron scale and better, highly detailed mechanical property mapping can be performed to optimize the materials or system under study, explaining why one early name of such instruments was a mechanical properties microprobe. Moreover, the depth sensitivity is sub-nanometer. Intrinsic mechanical properties of (ever thinner) thin films can be obtained by limiting the penetration depth to avoid/minimise the influence of the substrate or sublayers. Secondly, the load and penetration depth are continuously recorded allowing investigation of basic mechanical properties (hardness and elastic modulus) (Fischer-Cripps 2006), time-dependent behavior (Kranenburg et al. 2009; Oyen 2007; Chen et al. 2010), phase transformations (Chinh et al. 2004), stress–strain behavior, fracture (Beake 2005; Casellas et al. 2007) and fatigue behavior (Beake and Smith 2004). Finally, it is routine to schedule large arrays of test experiments to be run automatically over extended periods. Many properties such as hardness and elastic modulus can be calculated without the necessity for post-test imaging, leading to high efficiency and throughput. The current generation of nanomechanical test instrumentation now can perform additional test techniques, such nanoscratch and nano-impact testing, alongside nanoindentation to provide enhanced capability in a single test instrument or platform. The information with the different tests is quite complementary, e.g. nanoindentation to obtain the hardness and elastic modulus of elastic–plastic materials, nanoscratch tests to evaluate their tribological performance and nano-impact to probe their dynamic high strain rate properties. Even compression or bending experiments can be performed using the positioning accuracy and precise load control of the instrument to investigate elastic deformation, yield and fracture of suspended beams or micropillars.

Theoretical analyses of the relationship between contact load and displacement for flat-punch, spherical, or conical indenters into a linear-elasticsolid were derived by Boussinesq (Johnson 1985), Hertz (Johnson 1985), and Sneddon (1965). Based on Sneddon's work, Oliver and Pharr (Oliver and Pharr 1992) developed a simple method to obtain the elastic modulus and hardness using pyramidal indenters where the unloading behavior is assumed to be entirely elastic. In this popular method it is assumed that no plastic deformation or continuing creep deformation occurs during unloading. The results thus are well-suited for elastic–plastic materials. The slope of the unloading curve at any point is called the contact stiffness. In this analysis, the reduced modulus, E_r, is calculated from the stiffness at the onset of the unloading S and the projected area of contact between the probe and the material A_C as Eq. 1:

$$E_r = \frac{\sqrt{\pi}}{2\beta} \cdot \frac{S}{\sqrt{A_c}} \tag{1}$$

where β is the correction factor for the shape of the indenter (whilst there is some debate, β is commonly taken as 1.034 for the Berkovich indenter geometry).

By careful calibration into a reference material of known elastic properties, the diamond area function can be determined relating the contact depth h_c, to the projected contact area. The initial unloading slope S is obtained by fitting the unloading load–displacement response.

As elastic displacements occur both in the specimen and in the indenter (as the indenter is not completely rigid), the elastic modulus of the sample is calculated from E_r using Eq. 2:

$$\frac{1}{E_r} = \frac{1 - v^2}{E} + \frac{1 - v_i^2}{E_i} \tag{2}$$

where E and E_i, v and v_i are the elastic modulus and the Poisson ratio of the tested material and indenter, respectively. For diamond indenters, E_i and v_i are 1141 GPa and 0.07, respectively. The mean pressure or hardness, H, can be calculated as in Eq. 3:

$$H = \frac{P}{A_c} \tag{3}$$

where P is the applied load. The technique has also advanced to have its own ISO standard (14577, Parts 1–4) which identifies four key parameters—force, displacement, instrument compliance, and indenter shape—that influence the quality of the results and provide the necessary methodology to accurately calibrate these. The indentation standard strictly applies to metallic and ceramic materials rather than polymers, although it does include provision for minimizing the influence of creep of metals or alloys on the accuracy of determinations of elastic modulus.

2 Nanomechanical Testing of Polymeric Materials

For polymeric materials, a similar approach to obtain E_r and H can be carried out using the Oliver-Pharr method. However, it assumes the initial unloading is purely elastic which ignores the influence of continuing time-dependent deformation which is always present to some degree in the indentation response of polymeric materials. When the creep is particularly pronounced, a nose-shape can even appear during unloading as demonstrated in Fig. 1 that result in negative S and erroneous E_r (Feng and Ngan 2002).

To counteract this, several methods were proposed in terms of the testing conditions and the analysis. Firstly, the creep behaviour at the onset of the unloading is strongly influenced by the testing conditions, such as hold period (dwell time), peak force and loading rate etc. Selection of appropriate holding period, peak force and loading rate is important (Chudoba and Richter 2001). Additionally that the severity of the effects of the residual creep on the contact stiffness, S, is also a function of the unloading rate (Chudoba and Richter 2001; Feng and Ngan 2002). The higher unloading rate is favorable to minimize creep effects. Thus, there are

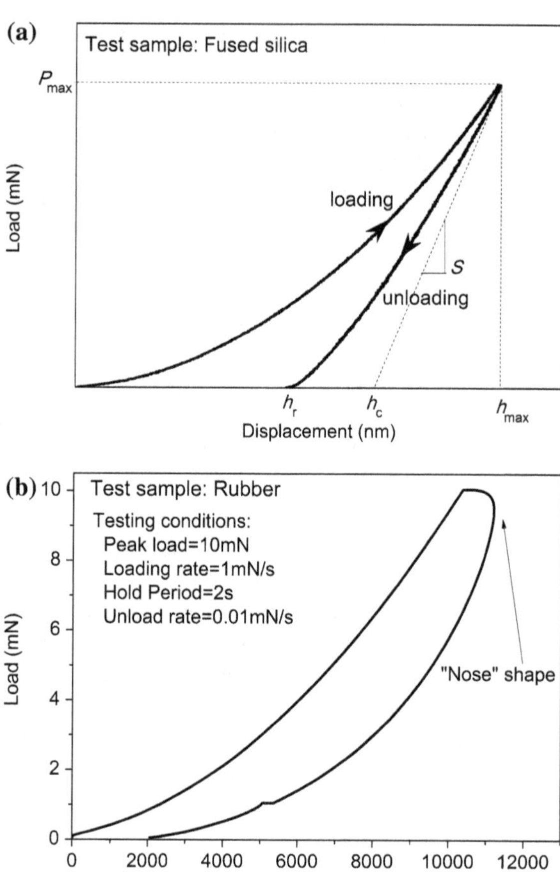

Fig. 1 Indentation load–displacement curves for (**a**) elastic–plastic fused silica, and (**b**) visco-elastic–plastic rubber material

three ways to accommodate the influence of time-dependent deformation on the contact stiffness during the testing (Dasari et al. 2009), (a) increasing the unloading rate, (b) a most common method of holding the indenter at maximum load for a long period of time resulting in a minimum residual creep rate comparing to the unloading rate, and (c) application of an oscillatory force or displacement signal to the tip-sample contact during nanoindentation and measurement of the resultant output signal and phase lag, which are used to obtain the contact stiffness and damping that are analysed to determine the viscoelastic properties of the material (loss and storage moduli, tan delta). This method is sometimes generally called dynamic indentation, or by the instrument manufacturers own terminology such as continuous stiffness measurement (Agilent), nanoDMA (Hysitron) or dynamic mechanical compliance testing (Micro Materials Ltd) etc. Despite its potential, in general good agreement between dynamic indentation and DMA has proved highly challenging for several reasons. Firstly, accurate dynamic calibration of the system is essential as extraction of the material properties requires analysis of the overall response, which includes the response of the test instrument, especially its damping behavior over the frequency range of interest (Singh et al. 2008). Secondly, reliable extraction of the loss and storage modulus relies on the application of a suitable constitutive equation. For simplicity it is common to assume that the indentation contact can be treated as a linear viscoelastic (Monclus and Jennett 2011), however, it is clear that when pointed (pyramidal, such as Berkovich) indenters are used there is often considerable residual plastic deformation of the polymer surface (Tweedie and Van Vliet 2006). Monclus and Jennett (2011) have critically examined the level of agreement between dynamic indentation and DMA and found poor agreement for a range of polymers, particularly in the loss modulus. They have suggested that more complex models are required to successfully produce loss/viscosity parameters that are equivalent.

A different approach is to account for the creep effects simply by data analysis as proposed in references (Feng and Ngan 2002; Ngan and Tang 2002, 2009). They examined their theoretical model for both linear and power law viscoelastic materials and this model has been shown good reliability for various polymeric materials according to the authors' experience. This correction can be described as in Eq. 4

$$\frac{1}{S} = \frac{1}{S_u} + \frac{\dot{d_c}}{\left|\dot{P}\right|} \tag{4}$$

where S is the contact compliance, S_u is the elastic contact stiffness at the onset of unloading, $\dot{d_c}$ is the displacement derivative at the end of the hold period, and $\left|\dot{P}\right|$ is the unloading rate at the onset of unloading so that the standard contact stiffness equation (Eq. 1) is modified to Eq. 5

$$E_r = \frac{\sqrt{\pi}}{2\beta} \cdot \frac{S_u}{\sqrt{A_c}} \tag{5}$$

The mechanical properties (hardness and elastic modulus) of various polymeric materials like PVC, PMMA, PET, polypropylene (PP), polycarbonate (PC), poly(ethylene oxide) (PEO), poly(acrylic acid) (PAC), nylon 6, nylon 66, rubber, and so forth have been investigated using the nanoindentation technique (Kranenburg et al. 2009; Gray and Beake 2007; Gray et al. 2009; Dasari et al. 2009; Beake et al. 2007). Unsurprisingly the results showed that their mechanical properties are sensitive to the testing conditions including the contact force due to their low stiffness (Kaufman and Klapperich 2009) and peak force, loading rate, unloading rate and hold time due to their rate- and time-dependent properties (Kranenburg et al. 2009). Elastic modulus determinations on polymers are often slightly higher than determined in bulk compression testing, and usually increase as the scale of the contact is reduced (Tweedie et al. 2007).

The creep behavior of the polymers at the nano-/micro-scale is also of intrinsic interest. Several approaches thus have been developed to investigate the creep data collected during nanoindentation with mostly widely used assuming constitutive

Fig. 2 Schematics of 3-element Kelvin-Voigt model

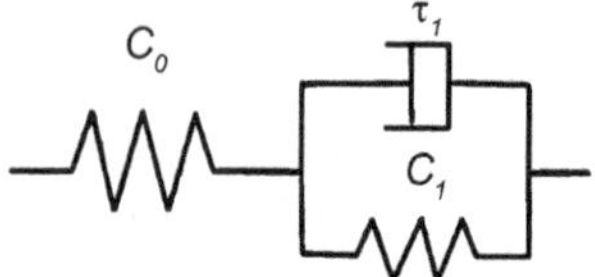

models (Mencik et al. 2009; Oyen 2005) such as linear viscoelasticity with data fitted to 3-element (Maxwell or Kelvin-Voigt) (Fig. 2) or 4-element (combined Maxwell-Voigt) models.

The instantaneous elasticity, instantaneous plasticity, viscoelasticity and visco-plasticity can be attributed to different physical elements. The creep compliance function is then deduced and the formula between the penetration depth and the applied load can be directly calculated using the Boltzmann integral operator. This methodology has been successfully to simulate the nanoindentation tests. The interested reader can find more details in the references (Mencik et al. 2009; Oyen 2005, 2006, 2007; Oyen and Cook 2009; Chen et al. 2010). Generally speaking, the accuracy of this method strongly depends on the number of the physical elements. In practice a better fit required more variables, which costs more in computation time and relies on a careful selection of the initial value to make the iteration converge.

A semi-empirical logarithmic method has also been used to analyze the dwell data collected at the hold period (Beake 2006; Berthoud et al. 1999; Chen et al. 2010). The logarithmic equation can be expressed as Eq. 6 Beake (2006); Chen et al. (2010)

$$\Delta d = A \ln\left(\frac{t_h}{\tau_L} + 1\right) \tag{6}$$

where Δd and t_h are the increase in depth and hold time during hold period, A and τ_L are termed the extent parameter and the time constant, respectively. The creep strain (ε) can be termed as $\Delta d/d(0)$ where $d(0)$ is the initial depth at the hold period. $A/d(0)$ (Beake 2006) or $(\varepsilon_e/\varepsilon(0))$ (Beake et al. 2007) is termed as the creep strain sensitivity.

The logarithmic equation has been found to closely fit short-time experimental indentation creep data of a wide range of polymeric materials under different testing conditions. The fitting is successful but only two variables—measures of extent A and rate τ_L—are used to describe the visco-deformation behavior. This method can be used not only for linear viscoelastic materials, but also non-linear viscoelastic materials, and with the quality of the fit it is possible to predict the creep response over a relatively long time. However, its limitations are that it is empirical and lacks explicit physical meaning. It is also not possible to deconvolute the particular contributions of elasticity, plasticity, viscoelasticity and visco-plasticity by this approach (Chen et al. 2010).

Similar works have been gradually carried out on the polymer nanocomposites, such as nylon 66/clay (Shen et al. 2004a, b), nylon 12/clay (Phang et al. 2005), epoxy/CNT (Li et al. 2004), nylon 6/CNT (Liu et al. 2004), photopolymer/SiO$_2$

(Xu et al. 2004), acrylonitrile/butadinene/styrene (ABS) (Beake et al. 2002) and poly(ehylene oxide)(clay/PEO) (Beake et al. 2002) in recent years. The effects of the nano- and micro-scale fillers are a subject of increasing research. In general the resistance to indentation of these nanocomposites gradually increases with increasing filler loading although there are exceptions and the response is typically non-linear (Dasari et al. 2009).

These nanomechanical tests are normally carried out under ambient conditions that may not be particularly close to the actual working conditions in the application. It is well known that the properties of polymeric materials are highly sensitive to the environment. For example, temperature can play an important role. A system may operate efficiently at one temperature and fail when the temperature is changed (Beake 2011; Chen et al. 2010; Everitt et al. 2011; Gray and Beake 2007). To address this problem, several approaches have been used to probe the surface properties of materials under conditions that mimic those which they experience in actual service. In the following sections, we will introduce these advanced nanomechanical test methods and their application on polymeric materials

3 Environmental Nanomechanical Testing of Polymeric Materials

3.1 Environmental Nanoindentation

3.1.1 Influence of Temperature

Obviously the test temperature can strongly influence the properties of polymeric materials. For example, amorphous polymers undergo a transition from a rubbery, viscous amorphous liquid, to a brittle, glassy amorphous solid. According to the viscoelasticity theory, the time-depended properties of polymeric materials depended on the free volume available for molecular (segmental) motions (Beake 2006; Beake et al. 2007). When the temperature increase around the glass transition temperature (T_g) can create sufficient free volume to allow molecules to move relative to one another (Beake et al. 2007). It follows that the creep behavior around T_g will be changed.

Beake et al. (2007); Beake (2006) used nanoindentation to systematically investigate the creep behavior of a range of polymer systems including polystyrene (PS), polypropylene (PP), polycarbonate (PC), polyethersulphone (PES), poly(methylmethacrylate) (PMMA), polytetrafluoroethylene (PTFE), poly(ethylene terephthalate) (PET), ultra-high molecular weight polyethylene (UHMWPE), low-densitypolyethylene (LDPE), acrylonitrilebutadiene-styrene (ABS) copolymer, Santoprene® containing ethylene propylene diene monomer (EPDM), Surlyn® 8140, ethylene/methacrylic acid (E/MAA) copolymer. Their creep factors, such as strain rate sensitivity ($\varepsilon_e/\varepsilon(0)$) and creep rate term ($\varepsilon_r = 1/\text{time constant}$) calculated using the logarithmic Eq. (5) were investigated in terms of their numerical distance, y, from T_g which can be defined as Eq. (7):

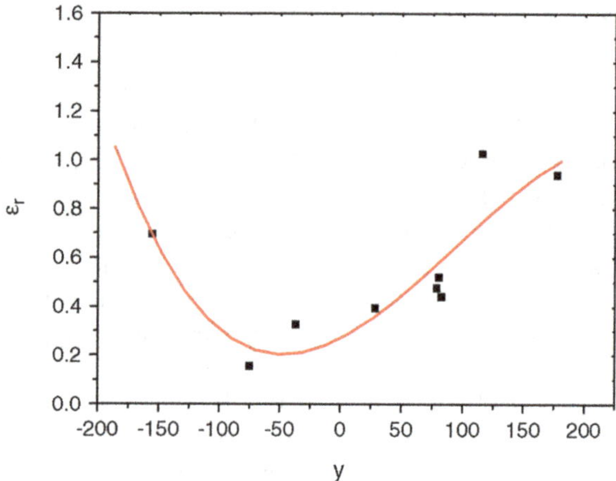

Fig. 3 Creep rate as a function of y defined by Eq. (6) (Beake et al. 2007)

$$y = T_g - T_{exper} \tag{7}$$

Therefore, negative y values correspond to rubber- or liquid-like behavior and positive y values to glassy behavior (Fig. 3).

The change of ε_r with y could be divided into three regions. Firstly, large negative values of y represent the liquid or rubbery region. This implies high v_f values, consequently large chain mobility, and thus high creep rate ε_r. Another is at large positive y values. Polymers are glassy in this region and material brittleness and crack propagation are likely to be the dominant mechanisms of creep, especially for $y > 50$ K or so. In the middle of y range, there is a minimum of ε_r which could be attributed to restricted chain mobility and reduction in brittleness.

To further investigate the effects of temperature on the mechanical properties of polymers, non-ambient temperature tests at both high temperature and low temperature were also carried out. The development of non-ambient temperature nanoindentation can be traced back to 1996 when Suzuki and Ohmura (1996) developed a prototype high-temperature ultra-micro indentation apparatus capable of testing up to 600 °C. However, the test sensitivity of in this prototype was affected by the high testing rates and temperature. Since then, many researchers have made improvements to both the instrumentation and the required experimental methodology (Duan and Hodge 2009). Currently there are three main manufacturers of commercially available systems capable of high temperature nanoindentation testing (1) Hysitron (2) Agilent (previously MTS NanoInstruments) and (3) Micro Materials. The latter's NanoTest has proved popular for this application, with a survey of published papers between 1996 and 2010 reporting around 60 % of published high temperature nanoindentation reports were using it (Everitt et al. 2011). The proportion increased as the test temperature increases above 300 °C (Everitt et al. 2011). There are distinct differences in experimental configuration in the different instruments and these have a direct influence on both the stability and reliability of the test data and the peak temperatures reachable. The Hysitron and Agilent instruments have a similar vertical loading setup. For example, the Hysitron triboindenter systems used a Peltier thermal element and a resistive heating element as the heating stage. The Peltier thermal element allows for a temperature range from -10 up to 120 °C and with the addition of the resistive element, this testing temperature could be expanded up to 200 °C. By the adoption of a liquid-cooled unit in the resistive heating element, the company has suggested that the temperature can extend to 400 °C (Hysitron 2012). However, a severe limitation of this setup is the use of sample-only heating, particularly for testing at >200 °C. Thermal equilibrium in the contact zone is achieved by holding for a long duration in contact prior to loading. The disadvantage is that as the indentation progresses, it is necessarily involves contact between the colder indenter and the hotter sample.

In the Micro Materials NanoTest nanoindentation system the sample is mounted vertically so that indentation occurs horizontally. The system can be operated to a maximum allowable temperature of up to 750 °C (MicroMaterials 2012). The displacement transducer is placed behind a water/air cooled heat shield to minimize/eliminate radiative heating, and an advantage of the horizontal loading configuration is that convection currents do not transfer significant heat to the displacement measurement electronics. Displacement calibration does not vary within the 25–750 °C temperature range. A key element in its design to reach a much higher working temperature is use of a dual heating strategy (isothermal contact method) where the sample side is heated with a resistance heater and the diamond indenter side is heated up with a small heater with a miniature thermocouple (Everitt et al. 2011; Beake and Smith 2002). This design enables independent heating and control of the sample and the indentation temperatures to minimize the heat flow between the sample and indenter (Fig. 4). Everitt et al. (2011) used finite element modeling to analyse the thermal picture under a diamond indenter with specimen with different conductivities. For the high-conductive materials, a very steep thermal gradient was formed at high temperature which must accommodate their deformation. Through experiments on fused silica up to 600 °C and annealed gold up to 300 °C, they found that the isothermal contact method maintained acceptable thermal drift and produced values of modulus of hardness that compared well with those in literature (Everitt et al. 2011).

It is clear that reliable high-temperature nanoindentation tests require a stable temperature field with minimal thermal flow between the indentation tip and sample, components and surroundings. Design strategies such as heat shielding of components and displacement measuring electronics, isothermal contact by dual heating, long hold period at peak load and relatively large heating blocks for effective heat dissipation from the rest of the sample stage assembly have been employed effectively. Moreover, typically at temperatures in excess of 500 °C, protective gas (e.g. argon purging) may also be necessary to avoid the deterioration of the diamond tip or the sample.

These non-ambient temperature nanoindentation instruments/techniques have been successfully utilized to characterize the mechanical properties of polymers, metals and hard coatings (Beake and Smith 2002; Everitt et al. 2011; Fox-Rabinovich et al. 2006; Gray and Beake 2007, Gray et al. 2009; Lu et al.

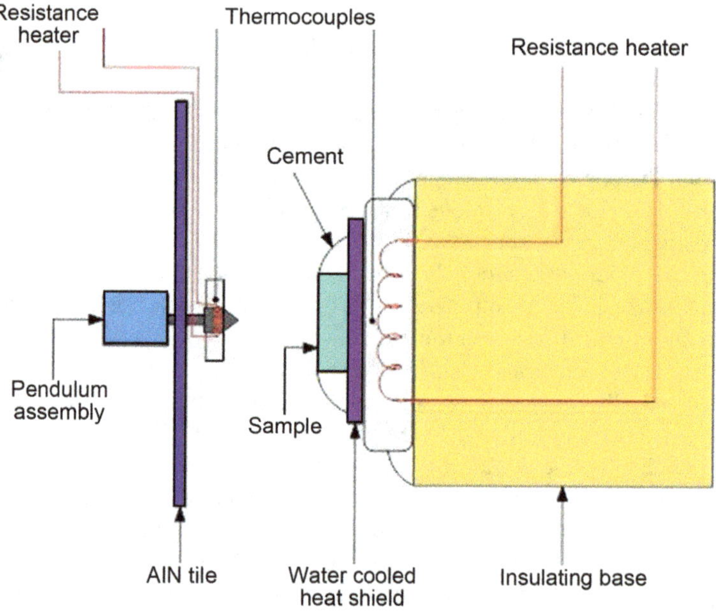

Fig. 4 Schematic of a commercially available high-temperature nanoindenter (NanoTest) employing separate indenter and sample heating and control to achieve isothermal contact

2010; Sawant and Tin 2008; Schuh et al. 2005; Xia et al. 2003; Ye et al. 2005). Encouragingly, it has been found that the properties at measured at non-ambient temperature can reflect their performance at real in-service conditions. For example, the high temperature nanomechanical and micro tribological properties of TiAlN and AlCrN coatings on cemented carbide cutting tool inserts (Fox-Rabinovich et al. 2006) have been correlated directly with their performance in extreme applications such as high speed machining.

Various polymeric materials have been studied using non-ambient temperature nanoindentation. An aerospace polymer resin, PMR-15 polyimide, was investigated by Lu et al. (2010) using a MTS nanoindenter up to 200 °C. Elastic modulus of PMR-15 showed a linear decrease with the increase of the temperature. Gray and Beake (2007), Gray et al. (2009) used the NanoTest to investigate the mechanical properties of PET films with different processing history and crystallinity over the temperature range 60–110 °C. They found that the mechanical properties of undrawn (amorphous) and uniaxially drawn (low crystallinity) PET films dropped quickly at 70–80 °C corresponding to their glass transition temperature, while the properties of biaxially oriented film showed a much more gradual decrease which can be attributed to its high crystallinity. The strain rate sensitivity parameter was also found to be able to characterise the increased time-dependent deformation around the glass transition region. Figure 5a combines measurements taken at room temperature and high temperature on a wide range of amorphous

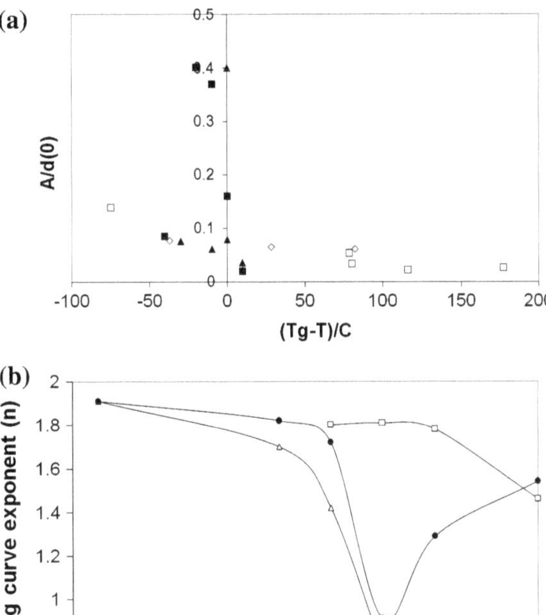

Fig. 5 **a** A/d(0) versus (Tg-T)/ °C from room temperature and high temperature measurements on semicrystalline and amorphous polymeric materials. **b** Variation in loading exponent n for PET films around their glass transition temperature range

and semi-crystalline polymers. It shows a dramatic peak at a temperature a few degrees above T_g, consistent with a maximum in the tan delta peak (which is usually offset by a similar amount from T_g determined from the inflexion of the storage modulus vs. T curve in DMA). Associated with this dramatic increase in A/d(0) is a marked decrease in the exponent n of the loading curve ($P = kd^n$) as shown in Fig. 5b.

Juliano et al. (2007) evaluated the creep compliance of three aliphatic epoxy networks with different molecular weight between crosslink over the temperature range 25–55 °C. They used a 500 μm end radius ruby indenter to generate small contact strains and maintain linear viscoelastic deformation. Tehrani et al. (2011, 2012) studied the nanoindentation creep of nanocomposites of aerospace epoxy and multi-walled carbon nanotubes over the temperature range 25–55 °C using a cBN Berkovich indenter. They determined that the strain rate sensitivity A/d(0) in the nanocomposites decreased relative to the neat epoxy, particularly at elevated temperatures. Interestingly, Li and Ngan have recently reported discrete relaxation events occurring during 600 s hold at 0.9 mN when indented with a Berkovich indenter of end radius 450 nm. Their frequency of occurrence was both crystallinity and temperature dependent. At 30 °C the discrete events (depth steps in creep curves) occurred in under 10 % of tests on LDPE but in over 55 % of tests on

HDPE. At 50 °C the percentage on HDPE was 47 % and at 70 °C the proportion reduced to 40 %. Since the probability increases with crystallinity, the authors suggested the fast relaxation events were likely to arise within the crystalline phase. The reduction in frequency with temperature was considered to be due to enhancement of the viscous flow of the amorphous phase.

Besides the high-temperature application, various engineering activities are carried out at sub-ambient temperatures. These cover winter sports, cryo-machining, marine and aerospace applications (Fink et al. 2008; Iwabuchi et al. 1996; Yoshino et al. 2001; Zhu et al. 1991). For example, the temperature of aircraft tyres can be lower than −50 °C during the aviation cycle. When landing, extremely heavy loads can be applied. Some spacecraft instruments have to be cooled to obtain an improved performance. The materials utilized in the assembly of these electronic circuits can be subjected to mechanical loading.

Adopting the dual-side thermal control approach in the high-temperature test setup, Chen et al. (2010) recently reported the development of a new cold stage accessory based on the NanoTest instrument. The novel nanoindentation capability described demonstrated its ability to investigate the local mechanical properties and the creep behavior of atactic polypropylene down to −30 °C. The sub-ambient temperature cooling system incorporates a purging chamber for eliminating condensation during cooling and two Peltier coolers as shown in Fig. 6 to achieve the isothermal contact. Extended initial contact hold period and low vibration cooling loop were adopted to get vibration free measurements.

Sub-ambient temperature nanoindentation tests have been successfully carried out on atactic polypropylene (aPP) to demonstrate the nanomechanical behaviour through the glass transition temperature, ~18 °C (Chen et al. 2010). It has been found that the hardness and elastic modulus of aPP increased as the test temperature decreased and the amorphous regions went through the glass transition as shown in Fig. 7. The derived creep extent (A) using the logarithmic method

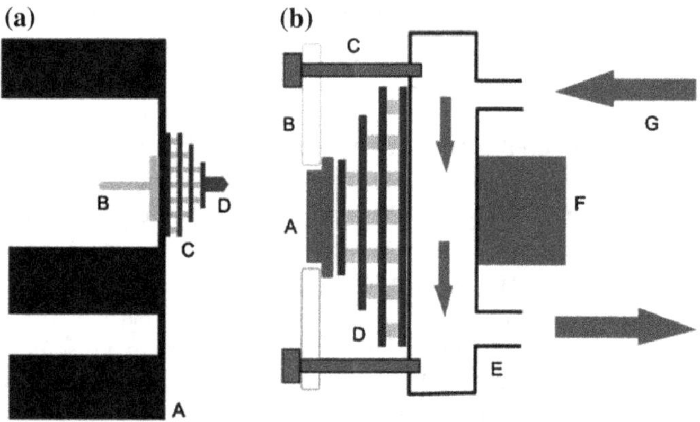

Fig. 6 Schematic diagrams for the cooling systems including (**a**) indenter cooling stack, and (**b**) sample cooling stack (Bell et al. 2011)

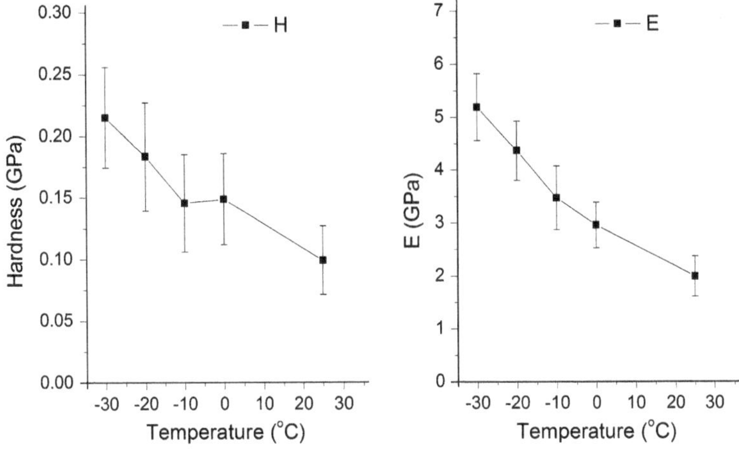

Fig. 7 Hardness and elastic modulus of aPP tested at different temperatures (Chen et al. 2010)

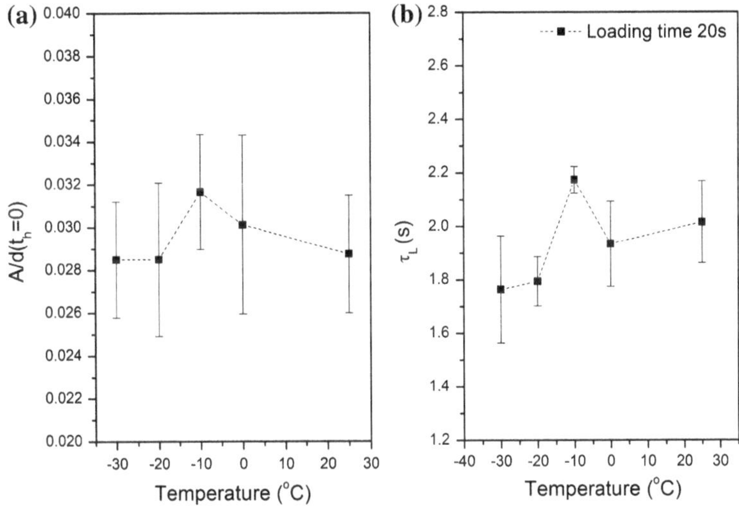

Fig. 8 **a** Creep strain rate sensitivity (A/d(0)) versus test temperature, and **b** creep time constant (τ_L) versus test temperatures fitted using logarithm equation (Chen et al. 2010)

decreased as the temperature was reduced, and for the time constants (τ_L) and strain rate sensitivity ($A/d(0)$) (Fig. 8), there were upper-limit values at -10 °C, about 8 °C above the quoted glass transition temperature.

As yet there are no reports of nanoindentation at temperatures below -50 °C. Cryogenic (down to -150 °C) temperature nanoindentation, however, could be realized using state-of-the-art Joule–Thomson cooling devices (Bell et al. 2011).

As discussed above, it can be seen that the unique ability of nanoindentation to obtain highly spatially resolved quantitative mechanical property measurements could enable microstructural changes in polymeric blends or biomaterials to be studied as a function of temperature. Maxwell and co-workers have recently used high temperature indentation to study the variation in creep compliance across the surface of a polyoxymethylene compression-moulded plate. 16×16 grids of indentations with 500 s hold for creep were performed at 23 and 50 °C. Spatially resolved normalized creep compliance maps confirmed that the edges of the moulding showed higher creep compliance. Modulus mapping with the same data showed the edges were also lower in modulus. The polymer at the edges had less time to crystallise due to more rapid cooling at the surface than the bulk, leading to lower crystallinity at the edges, as confirmed by DSC. Their work showed Arrhenius-type behavior, following the relationship between activation energy, temperature and relaxation time (τ) so that when $\ln\tau$ (determined from creep compliance data) is plotted versus $1/T$ a linear relationship is obtained. Using this approach they showed that by the appropriate Arrhenius shift the 50 °C data overlaid and extended the 23 °C data providing a route for accelerated creep testing.

3.1.2 Influence of Surrounding Media

Besides temperature, the mechanical properties of polymeric samples may vary considerably in different environmental media, such as in air with different humidity or when completely immersed in various fluids. For example, polar materials such as biopolymers [for example DNA, elastin, starch, and cutin (Round et al. 2000)] absorb water and can swell significantly at saturation. Bower (2002) explained that the molecule chain can swell or expand when there is strong attractive interaction between the solvent molecules and the polymer chain. It would be expected that their mechanical properties are interlinked with their water content. Thus it is important to test their mechanical properties and behavior in fluid media rather than to infer from measurements under normal laboratory testing conditions.

One commercial liquid cell system was designed by fitting the nanoindentation platform (Micro Materials Ltd.) with a fluid cell. The fluid cell testing photo is shown in Fig. 9. The potential benefits of the horizontal loading to the fluid testing were summarized (Bell et al. 2008) as (1) the use of an indenter adapter allowing the indenter to be fully immersed in cell, (2) all electronics are well away from the cell, so it can be heated—e.g. to body temperature and above- without risk of steaming of the capacitive displacement sensor, (3) the possibility of fluid exchange during experiment, (4) no significant buoyancy problems, (5) no large change in meniscus position during indentation. Water insoluble samples whose mechanical properties therefore do not vary on immersion in water (fused silica and polypropylene) were used to check the reliability of the setup.

Commercial Nylon-6 samples were tested by Bell and co-workers in deionized water and ambient (50 % relative humidity) conditions (Bell et al. 2008). Typical load–displacement curves are shown in Fig. 10. The hardness and elastic modulus

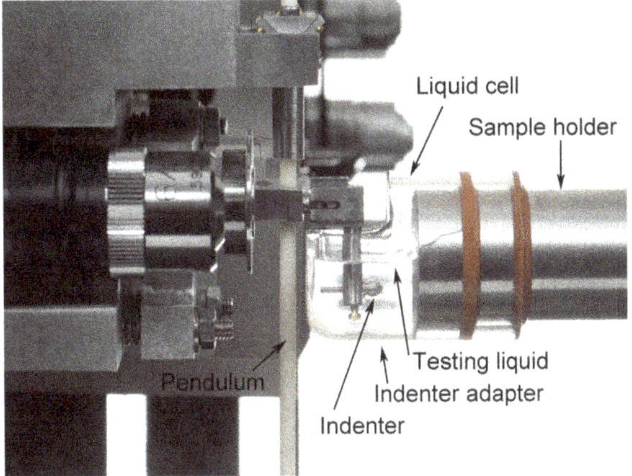

Fig. 9 Schematic of NanoTest fluid cell

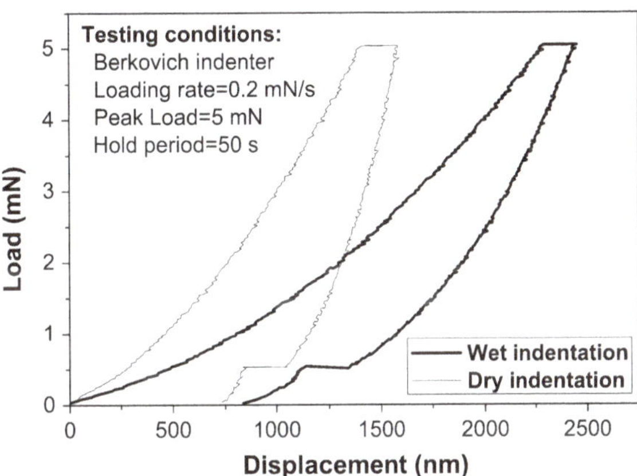

Fig. 10 Typical dry and wet indentation curves for acommercial nylon-6 polymer

measured in deionized water were significantly lower than those measured in ambient conditions. The creep rate sensitivity, $A/d(0)$, decreases significantly in water which is consistent with a decrease in the tan delta peak due to a shift in the glass transition temperature when wet.

Constantinides et al. (2008a, b) tested more compliant hydrogels, finding that the stiffness of PAAm-based electrophoresis gels decreased by a factor of about 1000 when hydrated [E(gel, water) = 270 kPa; E(gel, air) = 300 MPa]. They extended the capability of the fluid cell setup by replacing the diamond Berkovich indenter with a ruby spherical probe of 1 mm diameter chosen to approximate to linear viscoelastic deformation. Using this large radius spherical probe they conducted contact creep experiments on PAAm hydrogels and hydrated porcine skin and liver tissues. 300 s contact creep experiments on porcine skin after 1 h immersion in physiological saline were well fitted by the Kelvin-Voigt model.

Schmidt and co-workers used a NanoTest modified to act as a electrochemical call to study an electroactive polymer nanocomposite thin film containing cationic linear poly(ethyleneimine) and 68 vol % Prussian blue nanoparticles as a candidate stimulus-responsive polymer material (Schmidt et al. 2009). Electrochemical reduction of the Prussian blue particles doubled their negative charge causing an influx of water into the film to maintain electroneutrality. This resulted in swelling and a decrease in elastic modulus. The in situ nanoindentation measurements using a spherical ruby indenter of 5 μm radius showed a reversible decrease in the elastic modulus of the film from 3.4 to 1.75 GPa

Nanoindentation tests have also been performed under vacuum or in different humidities. Korte et al. (2012) have recently described the adaptation of the NanoTest to work in vacuum at temperatures up to 665 °C. Altaf and co-workers recently used the NanoTest with a humidity control unit to study the effect of moisture on the indentation response of a commercial stereolithography polymer resin, Accura 60, over the humidity range 33.5–84.5 % RH (Altaf et al. 2012). Stereo-lithography resins are highly hygroscopic and their mechanical properties are significantly affected by the level of moisture in the environment, with hardness and modulus decreasing with increased moisture in the resin. Transport of moisture from the surface to the bulk took place over a number of days so that a coupled stress-diffusion FEA was required. Gravimetric tests were performed to calculate the diffusion constants and bulk tensile, compressive and creep tensile tests to generate the mechanical material properties for the model. With appropriate modelling, the variation in hardness with (1) increasing penetration into the polymer (2) different environmental conditions could be accurately simulated.

3.2 Environmental Nanoscratch Testing

Since its inception, nanomechanical test instrumentation has developed along modular lines (Beake 2011) with nanoindentation and nanoscratch the two main modules for commercial instruments due to the simplicity to realize both these tests without particularly challenging hardware development. Polymers and polymer-based nanocomposites exhibit various deformation modes in the scratch test such as elastic contact, ironing, ductile ploughing, ductile/brittle machining, tearing, cracking, cutting, fragmentation, etc. (Dasari et al. 2009). In Briscoe and Sinha's review (2003), the most common types of material damage during scratching were illustrated. A combination of different mechanisms is usually operative in any particular contact process. Briscoe and Sinha (2003) developed scratch deformation maps for various polymers at the macro-scale that show clearly how the interplay between cone angle (which alters the contact strain) and normal load produces the different (dominant) scratch mechanisms. More recent work by Brostow and co-workers has shown that, with the exception of highly brittle polymers such as polystyrene, in general the cross-sectional area of the ridges above the scratch can be considerably lower than the cross-sectional area of the scratch groove, presumably due to appreciable densification alongside ploughing and cutting (Brostow et al. 2007). At the macro-scale there have been efforts to investigate any correlation between scratch hardness and abrasive wear with Sinha and co-workers noting that although PMMA was ~5–6 times harder than UHMWPE its wear resistance were 85 times lower under the same experimental conditions (Sinha and Lim 2006). Nevertheless, despite this, several studies have shown that mechanical properties are strongly implicated in the scratch behaviour of polymers at nano- and micro-scale. Taking advantage of the depth-sensing capability of the nano-/micromechanical test instruments viscoelastic scratch recovery has been shown to be important. A convenient measure of this is the % recovery as defined by Eq. 8:

$$\% \text{ recovery} = 100\,(h_t - h_r)\,/h_t \qquad (8)$$

where h_t is the on-load scratch depth and h_r the residual depth. Brostow and co-workers reported a wide variation in recovery after multiple pass micro-scratch experiments with a 200 μm probe (Bermudez et al. 2005a, b). For example, after 15 scans at 15 N viscoelastic recovery on nylon 6 was over 80 % but under 30 % on polystyrene. Similarly wide differences were observed by Sinha and Lim in ~200 μm deep scratches, with PMMA showing 62 % recovery and PP only 36 % (Sinha and Lim 2006). Brostow and co-workers noted that in their multiple micro-scratch tests the residual depths were greater at 1 mm/min than at 15 mm/min for all the thermoplastics they tested (Bermudez et al. 2005). They explained this as the influence of contact heating and greater chain relaxation and viscoelastic recovery at the higher speed. Recovery appears to be linked to tan delta (Brostow et al. 2006) although the relationship was highly non-linear. Beake and Leggett (2002) found that differences in H/E correlated with differences in residual depth and % recovery when scratching PET films of differing crystallinity and orientation at the nano-/micro-scale, at 1mN with a 25 μm end radius probe. When scanning at the nano-scale with a commercial AFM, Beake et al. noticed that aligned ridges formed immediately on uniaxially drawn PET of low hardness and crystallinity at contact forces over 15 nN (Beake et al. 2004) but several scans were required to create similar patterns on harder biaxially drawn PET (Beake and Leggett 2002). These authors suggested that the stick–slip process that leads to the formation of the aligned ridges on the polymer surface proceeded smoothly on the uniaxial PET as the yield threshold for localised plastic deformation is exceeded, whilst the higher crystallinity, H and H/E on biaxial PET has higher yield stress and a more gradual fatigue process over several scans is necessary before the yield stress of the damaged surface was low enough for efficient pattern formation.

In their excellent review, Dasari et al. (2009) noted that factors such as Young's modulus, yield and tensile strength and scratch hardness can all affect the scratch behaviour of polymers and nanocomposites. Surface tension has also been shown to play a role (Brostow et al. 2003). Brostow et al. demonstrated that on increasing the surface tension, friction, penetration and residual depths were also increased. The effects of the additive on the scratch resistance of nanocomposites have also been widely studied. For example, Zeng et al. used nanoindentation, nanoscratch, and nano-tensile tests to study the influence of different contents of fluoropropyl polyhedral oligomericsilsesquioxane (FP-POSS) in poly(vinylidene fluoride) (PVDF) on the mechanical properties of different systems. Compared with neat PVDF, the scratch resistance of the PVDF/FP-POSS nanocomposites was decreased due to a rougher surface derived from the bigger spherulites. A detailed review can be found in (Dasari et al. 2009).

As demonstrated above, the mechanical properties of the polymers and nano-composites depend on the specific testing environmental such as temperature, surrounding media. The mechanisms in nano-scratching polymeric materials under non-ambient conditions may vary significantly from those at ambient condition due to the change of their mechanical properties, the surface interaction etc. However, to the knowledge of the authors, the application of non-ambient nano-scratch to polymeric materials is in its infancy although studies of the nanomechanical and tribological properties of diamond-like carbon film at sub-ambient temperatures have revealed significant changes with temperature on this metastable material (Bell et al. 2011; Chen et al. 2011). At the macro-scale Burris, Perry and Sawyer used linear reciprocating pin-on-disk testing to provide evidence for thermal activation of friction (Burris et al. 2007). The friction coefficient of PTFE sliding against 304 stainless steel in nitrogen was measured over the temperature range -80 to $+140$ °C. They found that in the absence of ice the friction coefficient increased monotonically with decreasing temperature from 0.075 to 0.21, consistent with thermal activation of 5 kJ/mol.

3.3 Environmental Nano-Impact Testing

Constantinides et al. (2008a, b) have noted that whilst the mechanical response of polymeric surfaces to concentrated impact loads is relevant to a range of applications it cannot be inferred from either quasi-static or oscillatory contact loading (i.e. nanoindentation). By modifying the nano-impact module in the NanoTest they were able to assess the strain rate sensitivity of a range of amorphous and semi-crystalline polymers in the velocity range 0.7–1.5 mm/s. Polypropylene and low MW PMMA showed strain rate sensitive impact resistance whilst other polymers (PS, PC, PE, high MW PMMA) did not. They used the coefficient of restitution (e) as a convenient way to assess the energy loss during impact. In an interesting extension of the test capability they performed the nano-impact test on PS and PC over the temperature range 20–180 °C (i.e. from well below to well above their glass transition temperature ranges). They found that e decreased very slightly over the temperature range 0.2–1.0 T/T_g for both PS and PC. However, the capacity of the materials to dissipate the energy of impact greatly (e decreases) increases for temperatures exceeding the glass transition temperature (i.e. $T/T_g > 1$).

Kalcioglu and co-workers have used nano-impact (high strain rate indentation) to assess the response of fully hydrated tissues (from liver and heart) and candidate tissue surrogate materials (a commercially available tissue surrogate and styrenic block copolymer gels) (Kalcioglu et al. 2011). They were able to quantify resistance to penetration and energy dissipation constants under energy densities of interest for tissue surrogate applications. The nano-impacts were performed at impact rates of 2–20 mm/s. Although the velocity was slow, the energy strain densities were high (0.4–20 kJ/m^3) and comparable with macroscale impact tests such as pneumatic gun and falling weight impacts designed to replicate ballistic conditions (15–60 kJ/m^3) (Kalcioglu et al. 2011). They were able to determine that the energy dissipation capacities of fully hydrated soft tissues were well matched by a 50/50 triblock/diblock composition that was stable in ambient environments.

4 Concluding Remarks

Environmental, or non-ambient, nanomechanical testing has been successfully applied to the characterization of polymers and nanocomposite materials, especially by high temperature nanoindentation. Future directions in nanomechanical test technique development are likely to involve the test envelope being pushed ever outward to map onto more extreme conditions, such as cryogenic temperature, vacuum and various fluid media.

More recently developed nano-scale test techniques, such as nano-scratch or nano-impact, are less well explored than the older nanoindentation test technique. However, performing these tests under non-ambient environmental conditions is a highly promising direction for future research as it enables the tribological and dynamic properties of materials to be studied in nano/micro-scale.

References

Altaf K, Ashcroft IA, Hague R (2012) Modelling the effect of moisture on the depth sensing indentation response of a stereolithography polymer. Comput Mater Sci 52:112–117

Beake BD (2005) Evaluation of the fracture resistance of DLC coatings on tool steel under dynamic loading. Surf Coat Technol 198:90–93

Beake B (2006) Modelling indentation creep of polymers: a phenomenological approach. J Phys D Appl Phys 39:4478–4485

Beake BD (2010) Nanomechanical testing under nonambient conditions. American Scientific Publishers, Los Angeles

Beake BD, Lau SP (2005) Nanotribological and nanomechanical properties of 5–80 nm tetrahedral amorphous carbon films on silicon. Diamond Relat Mater 14:1535–1542

Beake BD, Leggett GJ (2002) Nanoindentation and nanoscratch testing of uniaxially and biaxially drawn poly(ethylene terephthalate) film. Polymer 43:319–327

Beake BD, Smith JF (2002) High-temperature nanoindentation testing of fused silica and other materials. Philos Mag A 82:2179–2186

Beake BD, Smith JF (2004) Nano-impact testing—an effective tool for assessing the resistance of advanced wear-resistant coatings to fatigue failure and delamination. Surf Coat Technol 188–189:594–598

Beake BD, Leggett GJ, Alexander MR (2002a) Characterisation of the mechanical properties of plasma-polymerised coatings by nanoindentation and nanotribology. J Mater Sci 37:4919–4927

Beake BD, Zheng S, Alexander MR (2002b) Nanoindentation testing of plasma-polymerised hexane films. J Mater Sci 37:3821–3826

Beake BD, Shipway PH, Leggett GJ (2004) Influence of mechanical properties on the nanowear of uniaxially oriented poly(ethylene terephthalate) film. Wear 256:118–125

Beake BD, Bell GA, Brostow W et al (2007) Nanoindentation creep and glass transition temperatures in polymers. Polym Int 56:773–778

Beake BD, Goodes SR, Shi B (2009) Nanomechanical and nanotribological testing of ultra-thin carbon-based and MoST films for increased MEMS durability. J Phys D Appl Phys 42:065301

Beake BD (2011) Nanomechanical testing under non-ambient conditions. In: Nalwa HS (ed) Encyclopedia of Nanoscience and Nanotechnology, 2nd edn. Vol. 18. American Scientific Publishers, Valencia, pp 115–120

Bell GA, Bielinski DM, Beake BD (2008) Influence of water on the nanoindentation creep response of Nylon 6. J Appl Polym Sci 107:577–582

Bell GA, Chen J, Dong HS et al (2011) The design of a novel cryogenic nanomechanical and tribological properties instrumentation. Int Heat Treat Surf Eng 5:21–25

Bermudez DM, Brostow W, Carrion-Vilches FJ et al (2005a) Wear of thermoplastics determined by multiple scratching. E-Polymers 001:1–9

Bermudez MD, Brostow W, Carrion-Vilches FJ et al (2005b) Scratch velocity and wear resistance. E-Polymers 003:1–10

Berthoud P, G'Sell C, Hiver JM (1999) Elastic-plastic indentation creep of glassy poly(methyl methacrylate) and polystyrene: characterization using uniaxial compression and indentation tests. J Phys D Appl Phys 32:2923–2932

Bower DI (2002) An introduction to polymer physics. Cambridge Univeristy Press, Cambridge

Briscoe BJ, Sinha SK (2003) Scratch resistance and localised damage characteristics of polymer surfaces—a review. Materialwiss Werkstofftech 34:989–1002

Brostow W, Cassidy PE, Macossay J et al (2003) Connection of surface tension with multiple tribological properties in epoxy plus fluoropolymer systems. Polym Inter 52:1498–1505

Brostow W, Clwnkaew W, Menard KP (2006) Connection between dynamic mechanical properties and sliding wear resistance of polymers. Mater Res Innovations 10:109

Brostow W, Chonkaew W, Rapoport L et al (2007) Grooves in scratch testing. J Mater Res 22:2483–2487

Burris DL, Perry SS, Sawyer WG (2007) Macroscopic evidence of thermally activated friction with polytetrafluoroethylene. Tribol Lett 27:323–328

Casellas D, Caro J, Molas S et al (2007) Fracture toughness of carbides in tool steels evaluated by nanoindentation. Acta Mater 55:4277–4286

Chen J, Lu G (2012) Finite element modeling of nanoindentation based methods for mechanical-properties of cells. J Biomec 45:2810–2816

Chen J, Bell GA, Dong HS et al (2010) A study of low temperature mechanical properties and creep behaviour of polypropylene using a new sub-ambient temperature nanoindentation test platform. J Phys D Appl Phys 43:425404

Chen J, Bell GA, Beake BD et al (2011) Low temperature nano-tribological study on a functionally graded tribological coating using nanoscratch tests. Tribol Lett 43:351–360

Chinh NQ, Gubicza J, Kovacs Z et al (2004) Depth-sensing indentation tests in studying plastic instabilities. J Mater Res 19:31–45

Chudoba T, Richter E (2001) Investigation of creep behaviour under load during indentation experiments and its influence on hardness and modulus results. Surf Coat Technol 148:191–198

Constantinides G, Kalcioglu ZI, McFarland M et al (2008a) Probing mechanical properties of fully hydrated gels and biological tissues. J Biomec 41:3285–3289

Constantinides G, Tweedie CA, Holbrook DM et al (2008b) Quantifying deformation and energy dissipation of polymeric surfaces under localized impact. Mater Sci Eng, A 489:403–412

Dasari A, Yu ZZ, Mai YW (2009) Fundamental aspects and recent progress on wear/scratch damage in polymer nanocomposites. Mater Sci Eng, R 63:31–80

Duan ZC, Hodge AM (2009) High-temperature nanoindentation: new developments and ongoing challenges. JOM 61:32–36

Everitt NM, Davies MI, Smith JF (2011) High temperature nanoindentation—the importance of isothermal contact. Philos Mag 91:1221–1244

Feng G, Ngan AHW (2002) Effects of creep and thermal drift on modulus measurement using depth-sensing indentation. J Mater Res 17:660

Fink M, Fabing T, Scheerer M et al (2008) Measurement of mechanical properties of electronic materials at temperatures down to 4.2 K. Cryogenics 48:497–510

Fischer-Cripps AC (2006) Critical review of analysis and interpretation of nanoindentation test data. Surf Coat Technol 200:4153–4165

Fox-Rabinovich GS, Beake BD, Endrino JL et al (2006) Effect of mechanical properties measured at room and elevated temperatures on the wear resistance of cutting tools with TiAlN and AlCrN coatings. Surf Coat Technol 200:5738–5742

Gray A, Beake BD (2007) Elevated temperature nanoindentation and viscoelastic behaviour of thin poly(ethylene terephthalate) films. J Nanosci Nanotechnol 7:2530–2533

Gray A, Orecchia D, Beake BD (2009) Nanoindentation of advanced polymers under non-ambient conditions: creep modelling and tan delta. J Nanosci Nanotechnol 9:4514–4519

Hysitron (2012) Temperature control stages. http://hysitron.com/products/options-upgrades/temperature-control-stages. Accessed 22 Dec 2012

Iwabuchi, A. and T. Shimizu, et al. (1996). The development of a Vickers-type hardness tester for cryogenic temperatures down to 4.2 K. Cryogenics 36: 75–81

Johnson KL (1985) Contact mechanics. Cambridge University Press, Cambridge

Juliano TF, VanLandingham MR, Tweedie CA et al (2007) Multiscale creep compliance of epoxy networks at elevated temperatures. Exp Mech 47:99–105

Kalcioglu ZI, Qu M, Strawhecker KE et al (2011) Dynamic impact indentation of hydrated biological tissues and tissue surrogate gels. Philos Mag 91:1339–1355

Kaufman JD, Klapperich CM (2009) Surface detection errors cause overestimation of the modulus in nanoindentation on soft materials. J Mech Behav Biomed Mater 2:312–317

Korte S, Stearn RJ, Wheeler JM et al (2012) High temperature microcompression and nanoindentation in vacuum. J Mater Res 27:167–176

Kranenburg JM, Tweedie CA, van Vliet KJ et al (2009) Challenges and progress in high-throughput screening of polymer mechanical properties by indentation. Adv Mater 21:3551–3561

Li XD, Gao HS, Scrivens WA et al (2004) Nanomechanical characterization of single-walled carbon nanotube reinforced epoxy composites. Nanotechnology 15:1416–1423

Liu TX, Phang IY, Shen L et al (2004) Morphology and mechanical properties of multiwalled carbon nanotubes reinforced nylon-6 composites. Macromolecules 37:7214–7222

Lu YC, Jones DC, Tandon GP et al (2010) High temperature nanoindentation of PMR-15 polyimide. Exp Mech 50:491–499

Mencik J, He LH, Swain MV (2009) Determination of viscoelastic-plastic material parameters of biomaterials by instrumented indentation. J Mech Behav Biomed Mater 2:318

MicroMaterials (2012) High and low temperature control. http://www.micromaterials.co.uk/the-nanotest/high-and-low-temperature-control. Accessed 22 Dec 2012

Monclus MA, Jennett NM (2011) In search of validated measurements of the properties of viscoelastic materials by indentation with sharp indenters. Philos Mag 91:1308–1328

Ngan AHW, Tang B (2002) Viscoelastic effects during unloading in depth-sensing indentation. J Mater Res 17:2604–2610

Ngan AHW, Tang B (2009) Response of power-law-viscoelastic and time-dependent materials to rate jumps. J Mater Res 24:853–862

Oliver WC, Pharr GM (1992) An improved technique for determining hardness and elastic modulus using load and displacement sensing indentation experiments. J Mater Res 7:1564–1583

Oyen ML (2005) Spherical indentation creep following ramp loading. J Mater Res 20:2094–2100

Oyen ML (2006) Analytical techniques for indentation of viscoelastic materials. Philos Mag 86:5625

Oyen ML (2007) Sensitivity of polymer nanoindentation creep measurements to experimental variables. Acta Mater 55:3633

Oyen ML, Cook RF (2009) A practical guide for analysis of nanoindentation data. J Mech Behav Biomed Mater 2:396–407

Phang IY, Liu TX, Mohamed A et al (2005) Morphology, thermal and mechanical properties of nylon 12/organoclay nanocomposites prepared by melt compounding. Polym Inter 54:456–464

Round AN, Yan B, Dang S et al (2000) The influence of water on the nanomechanical behavior of the plant biopolyester cutin as studied by AFM and solid-state NMR. Biophys J 79:2761–2767

Sawant A, Tin S (2008) High temperature nanoindentation of a Re-bearing single crystal Ni-base superalloy. Scripta Mater 58:275–278

Schmidt DJ, Cebeci FC, Kalcioglu ZI et al (2009) Electrochemically controlled swelling and mechanical properties of a polymer nanocomposite. ACS Nano 3:2207–2216

Schuh CA, Mason JK, Lund AC et al (2005) High temperature nanoindentation for the study of flow defects. Fundamentals of Nanoindentation and Nanotribology III, Boston

Shen L, Phang IY, Chen L et al (2004a) Nanoindentation and morphological studies on nylon 66 nanocomposites. I. Effect Clay Loading Polym 45:3341–3349

Shen L, Phang IY, Liu TX et al (2004b) Nanoindentation and morphological studies on nylon 66/organoclay nanocomposites. II. Effect Strain Rate Polym 45:8221–8229

Singh SP, Smith JF, Singh RP (2008) Characterization of the damping behavior of a nanoindentation instrument for carrying out dynamic experiments. Exp Mech 48:571–583

Sinha SK, Lim D (2006) Effects of normal load on single-pass scratching of polymer surfaces. Wear 260:751–765

Sneddon IN (1965) The relation between load and penetration in axisymmetric Boussinesq problem for punch of arbitrary profile. Int J Eng Sci 3:47–57

Suzuki T, Ohmura T (1996) Ultra-microindentation of silicon at elevated temperatures. Philos Mag A 74:1073–1084

Tehrani M, Safdari M, Al-Haik MS (2011) Nanocharacterization of creep behavior of multiwall carbon nanotubes/epoxy nanocomposite. Int J Plast 27:887–901

Tehrani M, Al-Haik M, Garmestani H et al (2012) Effect of moderate magnetic annealing on the microstructure, quasi-static, and viscoelastic mechanical behavior of a structural epoxy. J Eng, Mater Technol 134

Tweedie CA, Van Vliet KJ (2006) Contact creep compliance of viscoelastic materials via nanoindentation. J Mater Res 21:1576–1589

Tweedie CA, Constantinides G, Lehman KE et al (2007) Enhanced stiffness of amorphous polymer surfaces under confinement of localized contact loads. Adv Mater 19:2540–2546

Xia J, Li CX, Dong H (2003) Hot-stage nano-characterisations of an iron aluminide. Mater Sci Eng, A 354:112–120

Xu GC, Li AY, De Zhang L et al (2004) Nanomechanic properties of polymer-based nanocomposites with nanosilica by nanoindentation. J Reinf Plast Compos 23:1365–1372

Ye JP, Kojima N, Shimizu S et al (2005) High-temperature nanoindentation measurement for hardness and modulus evaluation of low-k films. Materials, Technology and Reliability for Advanced Interconnects, San Francisco

Yoshino Y, Iwabuchi A, Onodera R et al (2001) Vickers hardness properties of structural materials for superconducting magnet at cryogenic temperatures. Cryogenics 41:505–511

Zhu Y, Okui N, Tanaka T et al (1991) Low temperature properties of hard elastic polypropylene fibres. Polymer 32:2588–2593

Resolution Limits of Nanoindentation Testing

Ude D. Hangen, Douglas D. Stauffer and S. A. Syed Asif

Abstract As material and device length scales decrease, there must be a corresponding increase in the instrumentation resolution for accurate measurements. For these small length scale systems, including thin films, fine grained structures, and matrix composites, nanoindentation experiments provide a proven method for mechanical property measurements. Additionally, when nanoindentation is combined with scanning probe microscopy, individual tests can be placed directly in the regions of interest. However, these tests do not have infinite resolution, as they are limited by the volume probed during a test and the resulting residual damage. Here, an investigation of elastic and plastic mechanical properties is made in relation to the lateral test spacing and the mechanically probed volume. The results clearly show that closely spaced tests having residual plasticity adversely affect neighboring tests, having both poor accuracy and precision in the measurement. This is in contrast to purely elastic tests, which can be closely spaced without affecting accuracy or precision.

1 Introduction

1.1 General Principles of Indentation Testing

Nanoindentation is a generalized term that denotes a modern indentation technique in which the applied load and the measured displacement are continuously recorded throughout the experiment. The near surface mechanical properties can then be obtained from the resulting force–displacement curve. The terms *instrumented indentation testing* or *depth sensing indentation* are more accurate

U. D. Hangen (✉) · D. D. Stauffer · S. A. S. Asif
Hysitron, Inc. Technologiezentrum am Europaplatz, Dennewartstrasse, Aachen, Germany
e-mail: uhangen@hysitron.com

A. Tiwari (ed.), *Nanomechanical Analysis of High Performance Materials*,
Solid Mechanics and Its Applications 203, DOI: 10.1007/978-94-007-6919-9_5,
© Springer Science+Business Media Dordrecht 2014

classifications of the general experimental approach, as they do not impose a limitation on the measurement length scale. Indentation experiments are simple, cost effective, and reproducible in comparison to tensile or compression tests. Indentation moduli measured on a homogenous material result in a standard deviation of around 1 % at 10 nm penetration depth. These experiments can also be automated whereby hundreds of tests can be performed on a single sample. Moreover, the ability to test at very small penetration depths allows testing of thin films or regions of the sample's microstructure independently.

A typical experimental setup consists of a sample with a flat surface that establishes the boundary of a homogenous, semi-infinite, isotropic material. An indenter, typically made of diamond is then, used to penetrate the sample perpendicular to the surface, where depth of penetration at a given load is a measure of the materials resistance to deformation. The most common indenter geometry is the so called Berkovich geometry—a three-sided pyramid with an opening angle of 142.3° between one edge and the opposing face of the indenter. The Berkovich pyramidal indenter and flat sample results in a self-similar, or constant, geometry. This constant geometry results in a constant strain and similarity of the stress fields through length scales that range over 6 orders of magnitude. Depth sensing indentation experiments can thus cover a range of depth, from millimeters to nanometers, and can be used to determine material's properties averaged over a large volume or local variations in the material at the nanoscale. The forces required for this dramatic change in displacement range from kN to nN (12 orders of magnitude), requiring multiple instruments. "Nanoscale" has been defined by the ISO standard ISO14577 as 0–200 nm in penetration depth, "microscale" from 200 nm to an applied force of 2 N, and "macroscale" for forces larger than 2 N (ISO-14577-1 2002).

Indentation size can, however, affect the results of a test. Hardness frequently increases for decreasing depth for metallic materials (Tabor 1951; Gerberich et al. 2002; Nix and Gao 1998). This has been termed the indentation size effect, and has been postulated as being caused by a strain gradient or geometrically necessary dislocations. As further evidence of size scale effect, ceramics have been shown to deform ductility at the nanoscale while brittle fracture is the norm for micro to macro scaled indentations (Gerberich et al. 2009). This results in differences in the derived mechanical properties. In addition to material size scale effects, sample roughness also defines a length scale having a negative effect on the reproducibility of tests that are performed at penetration depths similar to the local roughness of the surface (Greenwood and Tripp 1967). Roughness should be reduced if at all possible, but caution is required as surface deformation may be caused by some sample preparation procedures. It is important to understand that these experimental complications can be managed for most samples. A flat sample surface is a precondition for high spatial resolution nanoindentation experiments and will not be further discussed.

It should go without saying that the stress field scales with the load, whether the stress field is elastic or plastic. A decreasing indent size results in a smaller stress field, whether elastic or plastic (Johnson 1985). The stages of elastic and plastic behavior during indentation at small scales will be discussed. Complications arise from the spherical end of the pyramidal indenter in nanoindentation, defining a lower boundary for penetration depth. A discussion on the influence of interfaces and overlapping plastic zones due to neighboring indentations will follow. The role of indenter placement within a given feature will then be highlighted in the discussion and summary.

2 Instrumentation

For nanoscale testing, both the instrument and the indenter shape play an important role. A *TI-950* TriboIndenter® by Hysitron, Inc., Minneapolis, Minnesota was used in this study. This instrument uses electrostatic actuation to generate small forces with high stability. Experiments can be performed in either force or displacement control, with a feedback loop rate of 78 kHz. A force noise floor of 30 nN (sampling@ 60 Hz) and displacement noise floor of 0.2 nm (@60 Hz) are verified. The in situ *imaging* mode can be used to raster the tip, scanning the sample surface, and generating a topographical image prior to and after testing. A TI-950 is capable of measuring modulus values from kPa to TPa with a single transducer having a force range of 30 nN to 10 mN. Loads of up to 10 N are possible with an additional transducer. The NI traceable calibrations for the force is performed by a hanging mass technique, while displacement is calibrated by interferometry. This can be compared to a cantilever based scanning probe system, where the spring constant does not provide independent force and displacement calibrations, and is not NI traceable. The moving mass of the indenter-transducer system is <500 mg. This low inertia system, combined with the low force and displacement noise floors, allows for the accurate detection of the surface during indentation experiments. This TI-950 was equipped with the dynamic modes *nanoDMA III*™ and *Modulus Mapping*™. Superimposing a dynamic force during indentation (*nanoDMA III*) allows continuous measurements to be made throughout the loading. A similar superposition of an oscillatory force over a static imaging set point during scanning probe imaging (Modulus Mapping) allows the generation of high resolution images of mechanical properties at the surface of the sample. Due to the low forces involved, this mapping measurement is performed entirely in the elastic regime. Here, electrostatic force actuation, F_{el}, is dependent on the square of applied voltage U. For the case of a dynamic test, F_{el} consists of static, U_{DC}, and sinusoidal dynamic, $U_{AC}\sin(\omega t)$, voltages;

$$F_{el} \sim (U_{DC} + U_{AC}\sin{(\omega t)})^2 = U_{DC}^2 + \frac{U_{AC}^2}{2} + 2U_{DC}U_{AC}\sin{(\omega t)} - \frac{\cos{(2\omega t)}}{2}.$$

$$(1)$$

The 2ω component vanishes for a DC component significantly larger than the AC component. The system locks in on the ω component. It is important to note that the effective dynamic force signal scales linearly with the applied static voltage. The reasoning behind this is explained in the following section.

2.1 Indenter Area Function

Machine compliance and the indenter area function are essential calibration routines for nanoindentation testing. The Oliver and Pharr (1992) analysis is valid for both purely elastic and elastic–plastic deformation. A valid area function can be generated for the entire depth range, typically on a fused quartz sample. This analysis requires the assumption of a constant reduced modulus, $E_r = 69.6$ GPa for fused silica. After calculation of the area function, any residual error in the area function fitting will result in a deviation of the reduced modulus. An observation can then be made with respect to the vast majority of publications with published depth profiles of hardness and modulus, which tolerate large deviations from a constant reduced modulus in the sub-50 nm depth range, instead emphasizing the importance of being able to test at a penetration depth of 1 μm. However, the real challenges in making accurate measurements in the sub-50 nm regime often lie within the tolerance of the measurement equipment. An indenter system that is capable of applying a 500 mN maximum load must then sacrifice in the force and displacement noise floor. In addition, sensing the surface correctly (0 nm penetration depth) is also a problem. This is because higher loads necessitate higher moving mass indenters, in which the initial surface contact results in high inertial "impact". The negative effects of slow data acquisition and slow feedback capabilities can be improved with modern controllers and computer systems, but the underlying physical principles do not change.

One other possible error in this region may be excessive oscillation amplitudes, as studied by Cordill et al. (2009) and by Oliver and Pharr (2004). This gives some insight as to the experimental challenges, as well as a possible path forward. When performing a depth profile on a Hysitron nanoindenter, the test is performed so that the superimposed dynamic load is much smaller than the static load. For a typical calibration material such as fused quartz, the 10 mN indentation head with its experimental noise floor specs (30 nN/0.2 nm) has a sufficient sensitivity to determine the surface level of the sample through changes in the static force reading. A dynamic (oscillating) approach is possible, which has been shown to help on compliant ($E_r < 10$ MPa) samples but is less effective on stiffer samples. These experiments are typically performed with a constant strain rate load function controlled using a constant ratio of the instantaneous change in force over the force, $\dot{F}/F$ [11], where the dynamic loading is initiated at a static load of 3 μN. Strain rates of 0.05/s and data acquisition of 200 Hz are typical, but data acquisition can be increased to 38 kHz for the tracking of fast changes.

The inherent coupling of the AC and DC signal, Eq. 1, is beneficial as it eliminates the need for an additional feedback loop around displacement amplitude for maintaining the displacement signal. This coupling results in a reduced time requirement for analysis of the displacement phase and amplitude signal from the lock-in amplifier. Small amplitudes are necessary to avoid

Fig. 1 Measured modulus (**a**) and hardness (**b**) of fused quartz after indenter tip area function calibration for three Berkovich indenters of varying tip radii. A dynamic indentation mode enables measurement of values for penetration depths as low as 1 nm, with a low standard deviation. © Hysitron, Inc

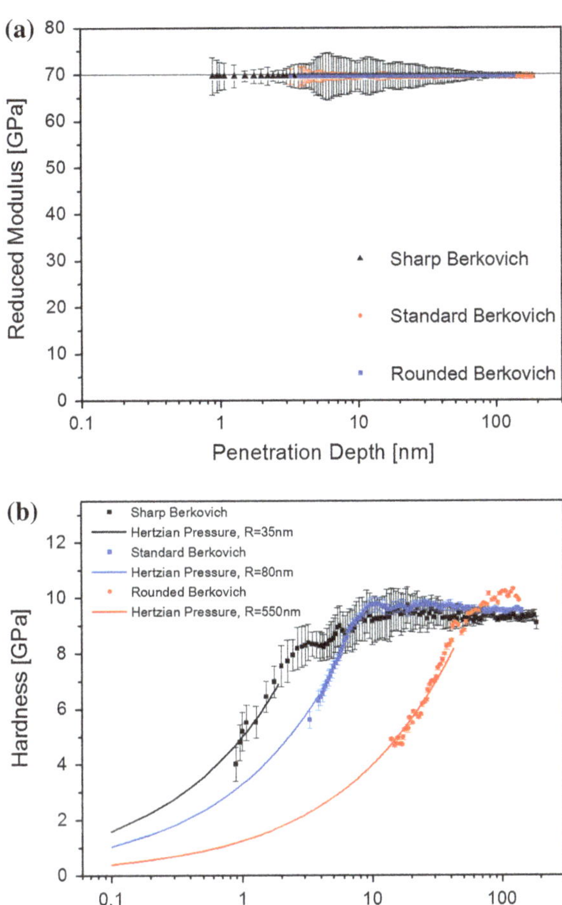

Jackhammering the sample (Cordill et al. 2009; Lucas et al. 1996) at small penetration depths. Typical displacement amplitudes used in these experiments range from 0.2 to 1.5 nm, with an oscillation frequency of 220 Hz. The maximum force or displacement is an additional input parameter, and test outputs include both static and dynamic measures of force and displacement. Determining the hardness and modulus of the sample requires: indenter displacement, h, applied static load, F, and contact stiffness, S, as relevant parameters. For a tip geometry with a well-defined tip area function, the experimental outcome on the same material as that used for the tip calibration should appear as seen in Fig. 1.

The results of Fig. 1 were generated after re-analyzing the test data on fused quartz after defining a six parameter, C_1–C_6, area function per Oliver and Pharr (1992). An assumed constant E_r of 69.6 GPa was used in calibrating the tip area, resulting in a constant value versus depth, Fig. 1a. However, hardness can only be assumed constant in the pyramidal portion of the indenter, at depths greater than approximately 1/3 that of the tip radius, R. For the initial, spherical contact, the ratio of the contact radius to the tip radius, a/R, is proportional to the strain, while the contact stress is the ratio of the load to the projected contact area, $F/\pi a^2$. It then becomes obvious that at low stress and low strain that the contact will be elastic. This reduction in contact pressure can be modeled using the Hertz theory (1896), using the relationship:

$$F = \frac{4}{3} E_r \sqrt{R} h_c^{3/2} \tag{2}$$

where, h_c is the contact depth. The pressure, P, under the indenter can then be calculated for a purely elastic deformation as:

$$P = \frac{\frac{4}{3} E_r \sqrt{R} h_c^{3/2}}{\pi R h_c} \tag{3}$$

A fit to the data using equation (3) is then used to plot the Hertzian, or average elastic contact, pressure in Fig. 1b. Increasing stress and strain eventually causes the onset of yield. This onset of plasticity can be related to the material yield strength, σ_ε, using the Tabor (1951) relationship. It is important to understand that the drop in H does not represent a reduction in the yield strength, but represents a purely elastic contact pressure. Figure 1 shows that a probe with a larger radius of curvature reaches the necessary stress for yielding at much larger displacements and therefore contact area, which serves to reduce the spatial resolution. Meaningful measurements of material hardness and modulus can be performed at more than an order of magnitude smaller penetration depth with a sharp tip.

2.2 The Volumes of Plastic and Elastic Deformation Under Contact Loading

While the goal is to obtain the highest lateral spatial resolution, an indentation occurs in three dimensions. It is therefore critical to determine the volume of material being probed. The probed volume for the elastically deformed region is not the same as that of the plastically deformed region (Johnson 1985). For an indenter with self-similar geometry, as in the case of a Berkovich indenter, the characteristic length scale is not a function of the sample or indenter size, instead is the function of the contact radius, a. The contact radius is defined as the radius of a circle of equivalent area to that of the actual contact-area, which may or may not be circular. This implies that the contact physics are similar whether the tip is a pyramidal Berkovich indenter or a sharp conical indenter with the same depth to contact radius ratio, Fig. 2. It follows that for a contact radius to contact depth ratio of 2.8, the contact diameter to contact depth ratio is 5.6. Due to the triangular

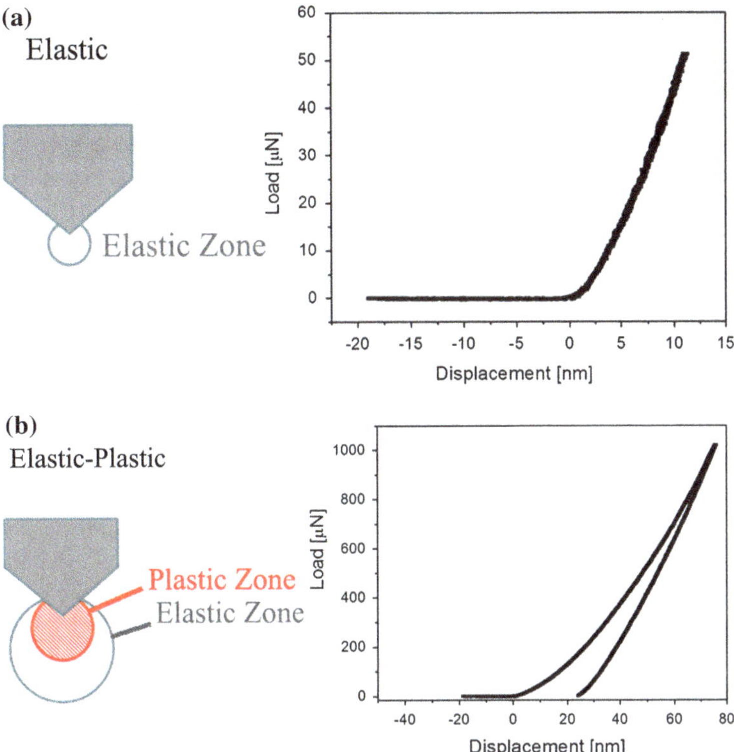

Fig. 2 Schematic of the elastic and plastic deformation zones under the indenter. During elastic indentation (**a**) in the Hertzian regime the average contact stress is less than that required to initiate plasticity. The elastic–plastic indentation (**b**) shows hysteresis in the load-depth curve. In this case a constant hardness is observed. © Hysitron, Inc

geometry, the actual widest contact point for a Berkovich indentation would result in a ratio of 7. Figure 2 illustrates the cross-sectional size of the elastic and elastic–plastic volumes under the indenter. Since these zones designate a field, their extent can only be identified by defining a minimum stress or strain. This minimum will depend on the acceptable relative change of properties in proximity to (1) a phase boundary or (2) a second indentation. Reproducibility limits resulting in 1 % standard deviations are the best case scenario of indentation testing. In this case, the limits for the elastic and plastic zones must be defined in such a way that the average mechanical properties do not change by more than 1 % when compared with undisturbed material (semi-infinite half-space situation).

When performing indentations in a close proximity with a Berkovich indenter, the concern is that the plastic zones will overlap. The material within a plastic zone could be considered to have been cold-worked by the indentation, and in the worst case scenario, the semi-infinite half-space approximation is no longer valid due to the disturbance of the surface. For stronger metallic materials, such as steel, the radius of the plastic zone measures approximately three times the contact radius, but can be as large as six for some soft metals, such as Cu (Nix and Gao 1998; Durst et al. 2005; ASTM-E384—11e1 2011; DIN ISO-6507-1 2005). Therefore, the distance between the centers of two indents should be a minimum of three times the widest contact diameter in these materials. This indent to indent spacing requirement significantly reduces the lateral resolution for determining local changes in material properties. For a Berkovich tip with a standard sharpness, plasticity sufficient for hardness measurements begins at a penetration depth of around 15 nm, Fig. 1. The resulting widest contact diameter is then approximately 105 nm. For an indent spacing of threefold the widest length, the lateral distance must be at least 315 nm. It is very difficult to predict the material behavior for overlapping plastic zones, including but not limited to work hardening and surface pile-up. It would be nearly impossible if the indents are spaced such that the residual impressions begin to overlap.

The situation improves somewhat when an indenter with steeper geometry, such as a cube corner geometry replaces the Berkovich geometry, due to the decreased contact radius to depth ratio. For a cube corner, the resulting contact diameter to contact depth ratio is approximately 1.4. As the plastic zone scales with contact radius, the plastic zone is reduced by almost a factor of 5, compared to a Berkovich indentation of the same depth. As the actual size of the plastic zone is material dependent (Durst et al. 2005), any effect of inhomogeneity in the sample is difficult to predict and decisions should be made as a result of actual testing. This can be illustrated by the presentation of two case studies: (1) the influence of one indentation on its neighbor, and (2) the influence of interfaces in the proximity of in indent.

Reducing the center to center distance, d, for an array of indents of 50 nm displacement on single crystal Al, using a Berkovich indenter shows large variations in both hardness and reduced modulus, Fig. 3a. This reduction of d in homogenous

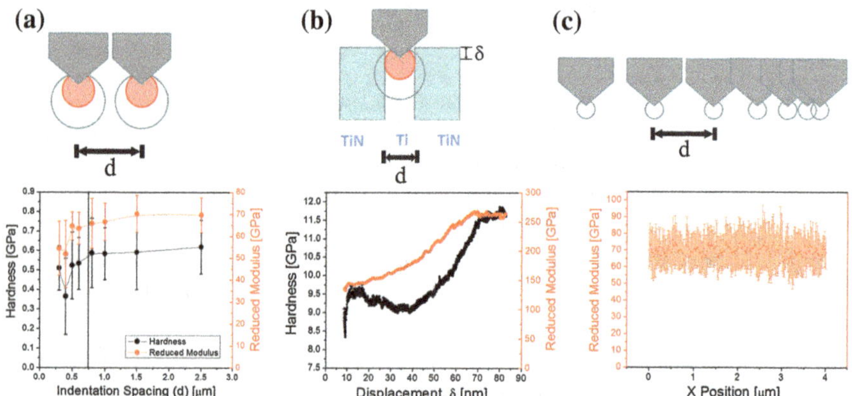

Fig. 3 **a** Hardness and modulus versus neighbor to neighbor distance observed with indentations of 50 nm penetration depth on single crystal Al. **b** Hardness and modulus of an indent performed in Ti between two rigid TiN interfaces. **c** A fully elastic contact can have arbitrary spacing between indents without adversely affecting subsequent testing; testing is the average of 25 line scans on Al single crystal with low surface roughness. © Hysitron, Inc

materials result in a decrease in hardness, not a simple increase as would be expected for work hardening. Additionally, the deviations in the modulus indicate that something is clearly wrong with the measurement. In this example, surface damage from the residual plastic zone induces error in the contact area, resulting in a flawed measurement. Deviations from "bulk" behavior begin at $d = 750$ nm, slightly less than the recommended distance of 2.8 times the contact diameter, or 840 nm for a maximum displacement of 50 nm. From this example, it can be clearly seen that the residual plastic zone in a hardness map sacrifices accuracy when trying to increase lateral resolution. While the overall size of the residual impressions can be minimized by using a sharp indenter at the lowest possible indentation depths, the overall effect is that the lateral resolution must necessarily be sacrificed. A dramatic improvement in the spatial resolution occurs when moving to the elastic regime. For a purely elastic indent, the indents can be positioned much closer together as there is no residual plastic zone, allowing for a neighboring test to be performed within the contact radius of the previous. While testing within the elastic regime forgoes a measurement of hardness, a high spatial resolution map of elastic surface properties can be made by indentation or modulus (dynamic imaging) mapping modes.

Another complication with respect to lateral resolution comes in performing indentations on small structures in proximity to a material interface. A profile of hardness and modulus as a function of depth for a 750 nm wide Ti-layer sandwiched between TiN-layers was performed with a Berkovich indenter, Fig. 3b. TiN is a very hard material, $H \sim 25$ GPa, and the interfaces can therefore be considered as rigid. The correct hardness and reduced modulus values were measured at only very shallow depths, 15–30 nm. At larger depths, the effect of the rigid interfaces causes an increase in measured reduced modulus and hardness. The placement of the test in the center of the structure was done by careful in-situ SPM imaging of the surface, utilizing a scanning piezo on which the indentation transducer is attached. The tip is then rastered along the surface at low load, acquiring height and gradient images of the surface.

This information provides us with a background for developing ideas regarding the achievable resolution of nanoindentation. Two major regimes have been identified—a regime of purely elastic deformation and one of elastic-plastic deformation, Fig. 4. The influence of a neighboring interface has been investigated both theoretically and practically in a study of the edge effect of a sample (Jakes et al. 2009; Jakes and Stone 2010). While the example of a void within the sample is a worst case scenario, it serves as a good idea of the parameters involved. The three dimensionless parameters found by Jakes, et al. are: the size of the indent in relation to the distance to the edge, $\sqrt{A}/d$, the ratio E/H, and the Poisson ratio of the material. The latter two of these are, of course, material dependent, making the ultimate lateral resolution material dependent. Using these parameters and the procedure of Jakes, et al., the minimum d can then be set for a limited modulus change, less than 2 % in this case. Since d is also related to the root of the contact area, d is a function of displacement. The smallest values of d are therefore found at the lowest penetration depths, within the fully elastic regime. This is important for understanding measurements taken on heterogeneous materials. Additionally, when testing in the elastic regime, the contacted area is not sacrificed by the

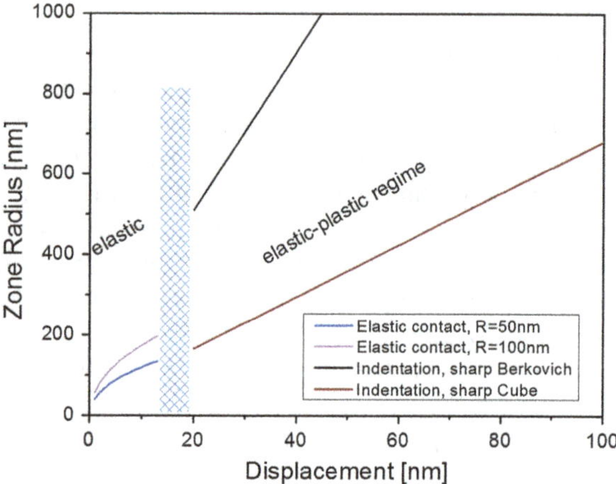

Fig. 4 Theoretical resolution for a given indentation size as a function of the distance to an interface. Material parameters are held constant for this theoretical comparison of modulus mapping to indentation resolution. A higher theoretical resolution of modulus mapping is confirmed. Following Jakes et al. (2009). © Hysitron, Inc

manufacture of a residual plastic impression. This will be confirmed by the high resolution of Modulus Mapping in the following section.

2.3 Hardness Mapping of a Multilayer Coating

The resolvability of local mechanical properties is illustrated with a Ti-Al6-V4 alloy coated with a Ti and TiN multilayer, which has subsequently been cross-sectioned and polished. The result is a sample where the Ti and TiN microstructure has been exposed, having minimal surface roughness. An in situ SPM image, Fig. 5a, shows slight topographical variation between the Ti-Al6-V4 (far left) substrate and the Ti and TiN multilayers.

Mechanical property variations across the sample surface were tested with both indentation hardness mapping and well as Modulus Mapping. The Ti layers, which are not perfectly straight due to the substrate roughness, have a thickness of approximately 750 nm. For accurate indentation measurements, small penetration depths and precise positioning in the center of the layer are of particular concern for the Ti-layer. The positioning on TiN or on the Ti-Al6-V4 substrate is not as critical as long as the distance between two indents is large enough to avoid interference. The indentations were positioned using an automated software feature, which places the indents at the desired locations after scanning with the tip. A maximum depth of 50 nm is chosen for the indentation experiments with a spacing of approximately 720 nm between the indentation experiments. The residual impressions from indentation on the surface of the sample are seen in Fig. 5b. The first two vertical lines of indents are positioned on the Ti-Al6-V4 substrate. The third vertical line is positioned on the Ti followed by four vertical lines placed on the TiN layer. The next vertical line is positioned on the Ti layer again. The maps of hardness and reduced modulus from this array are seen in Fig. 6, where the layered structure of the coating can be identified.

Fig. 5 **a** In-situ scanning probe imaging of a cross-sectioned multilayer sample, with the Ti-Al6-V4 substrate on the far left, followed by alternating layers of Ti and TiN. The static imaging load was 2 μN, using a 80 nm radius Berkovich indenter using the SPM mode on a Hysitron TriboIndenter. **b** In-situ SPM image of TiN-Ti-TiN multilayer after hardness mapping using a 12 × 12 indentation array. © Hysitron, Inc

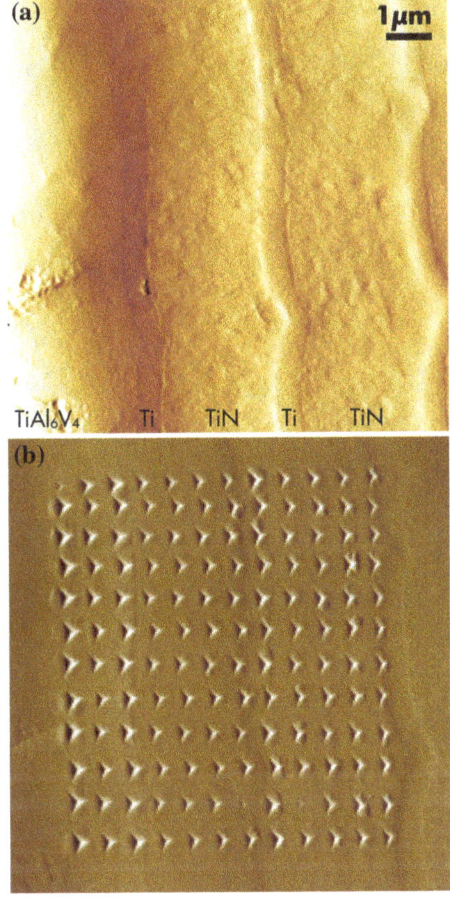

2.4 Modulus Mapping of a Multilayer Coating

As a consequence of the above findings, indent size needs to decrease in order to gain spatial resolution. Ultimately, the indentations will become purely elastic and eliminate the plastically deformed region. It was shown in Fig. 2 that a shallow indent produced no hysteresis in the load and displacement-curve, and therefore produced no plastic zone. If the indentation depth is increased, hysteresis, and residual damage will eventually occur, Fig. 2b. This lack of residual plasticity means that individual tests can be placed at any arbitrary d spacing—thereby increasing the spatial resolution. While one could closely space individual elastic indentations, Hysitron has developed a Modulus Mapping mode that combines the in situ imaging and mechanical testing functionalities. This takes advantage of very shallow depths of penetration and correspondingly small loads, resulting in a purely elastic, nondestructive algorithm. The maximum load of the test is defined

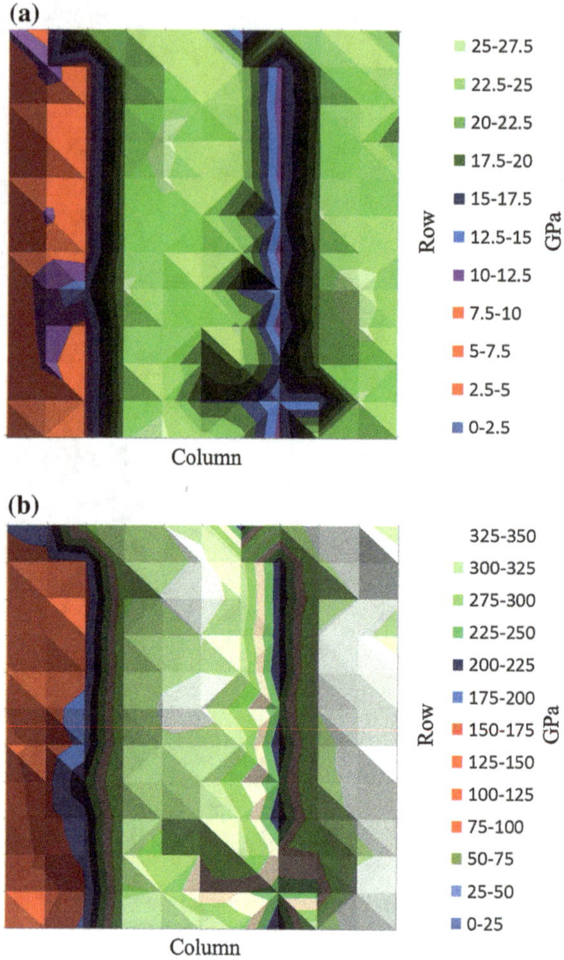

Fig. 6 **a** Hardness (**a**) and modulus (**b**) maps using indentation mapping of Ti-TiN multilayer cross-section

by the imaging setpoint, F_{DC}, and the stiffness of the contact, S, and is determined by analyzing a small superimposed mechanical oscillation using a lock-in amplifier. The reduced modulus of the sample can be found, assuming a Hertzian spherical contact. Hardness cannot be determined from this purely elastic deformation.

The results of a Modulus Map, Fig. 7, are calculated from the contact stiffness map. The map, in this case 15×15 μm, consists of 256×256 pixels which contain more than 65,000 individual mechanical measurements. The substrate can be clearly differentiated from the multilayer coating, with irregular features such as Ti droplets and thickness variation within the layers evident. This latter point is of particular interest, as technical processes often deviate from ideal, and Modulus Mapping is capable of visualizing these local variations with the highest resolution.

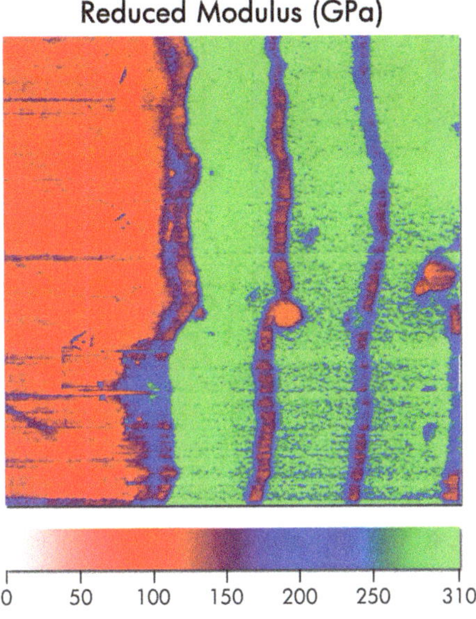

Fig. 7 A map of the modulus for the Ti-TiN cross-sectioned multilayer system in a 15 μm scan, using the combined scanning probe and dynamic oscillation technique. The scan shows the high obtainable spatial resolution using elastic contact, especially when compared to the maps of Fig. 6. © Hysitron, Inc

2.5 *Comparison of Results*

A hardness map delivers hardness and reduced modulus values; Modulus Mapping allows for determination of elastic properties only, including storage and loss modulus for viscoelastic materials. These two techniques are complementary, such that only the reduced modulus results should be compared here. From profiles of the two techniques, Fig. 8, it is obvious that the Modulus Mapping technique has a much higher data density, with spacing of 750 nm for indentation tests and only 50 nm between modulus mapping points. The magnitude of the values for large structures such as the Ti-Al6-V4 substrate with $E_r = 110$ GPa and TiN with $E_r = 310$ GPa are identical. However, it is for thinner layers and interfaces that the difference in the data is most apparent. Indentation testing struggles to adequately describe the sharpness of the transition, and fails to accuratelwy calculate the $E_r = 150$ GPa of the Ti layer. This difference is a result of the size of the deformation strain under the indenter at 50 nm displacement. The increasing interaction of the strain with the neighboring layers as a function on displacement was also shown in Fig. 3b.

3 Concluding Remarks

Indentation testing and Modulus Mapping are complementary techniques of mechanical testing. Modulus Mapping provides for measurements of elastic properties at the highest spatial resolution. In contrast, indentation testing for both elastic and

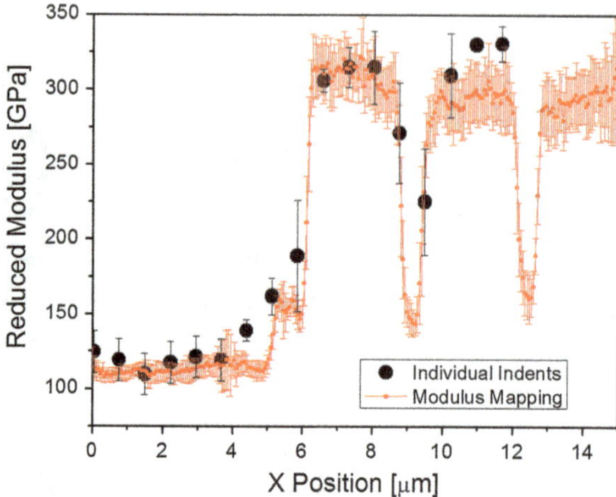

Fig. 8 Comparison of moduli measured by individual indents (hardness mapping) and by fully elastic dynamic scanning probe imaging (modulus mapping). The profile of the Ti-TiN multilayer cross-section shows the much higher spatial resolution of the elastic mapping. The Modulus Mapping data is averaged from 15 parallel line scans—with a standard deviation calculated from the 15 points at each position. The hardness values are averaged values from the 12 columns of indents, with corresponding standard deviation. © Hysitron, Inc

plastic properties is capable of testing small volumes of material when positioned with adequate spacing from previous indentations. Using a hardness mapping approach requires that the operator decide on the location of the indents, such that adequate spacing is maintained, and that simply placing a grid over the sample is unlikely to guarantee a map that accurately describes the properties of the phases. Therefore, imaging the sample in order to find the center of the locations of interest is necessary, as there might be only enough material for a single test. For homogeneous materials, indentation mapping by a grid array is an excellent means to obtain data for statistics over a large area. However, for applications where quantitative results with high spatial resolution are critical, an elastic, dynamic, imaging technique such as Modulus Mapping will satisfy the user's requirements.

4 Additional Information on Dynamic Techniques

A small dynamic load can be superimposed over a quasistatic load to the indenter, producing an oscillation of the tip with peak to peak amplitudes near 1 nm. A lock-in amplifier facilitates the detection of the material response, such as the stiffness, S, of the contact throughout the experiment. The Dynamic Mechanical Analysis (nanoDMA III®) technique is utilized for characterization of both elastic and viscoelastic materials. The data acquired is the total displacement into the sample, the dynamic displacement amplitude, the applied force, and the phase shift from the input to the response. From these measured quantities the storage and loss modulus, tan delta, and storage and loss stiffness can be calculated.

Following Asif et al. (1999, 2001) the analysis of the dynamic test can derived from the classical equation for a single degree of freedom harmonic oscillator as given in Eq. 4 where F_o is the magnitude

$$F_o \sin(\omega t) = mx'' + Cx' + kx \tag{4}$$

of the sinusoidal force, ω is the frequency of the applied force, m is the mass, C is the damping coefficient, and k is the stiffness of the system. The dynamic mechanical response of the transducer in contact with the sample is modeled using two Kelvin-Voigt mechanical equivalents, Fig. 9, from which the contact stiffness and the damping properties of the material can be accurately calculated. The solution to the differential equation, Eq. 4, is seen in Eq. 5 where a displacement amplitude response, X_o, is a function of a given F_o, ω, C, k, and m,

$$X_o = \frac{F_o}{\sqrt{\left(k - m\omega^2\right)^2 + \left((C_i + C_s)\,\omega\right)^2}} \tag{5}$$

$$\phi = \tan^{-1} \frac{(C_i + C_s)\,\omega}{k - m\omega^2} \tag{6}$$

$$k = k_s + k_i \tag{7}$$

The subscripts i and s in Eqs. 5–7, stand for indenter and sample respectively. The constants, m, C_i, and k_i, are found by dynamic calibration of the system. The variables, X_o and Φ, are measured during the experiment, from which the unknowns k_s and C_s can then be determined assuming a linear viscoelastic response. Stiffness and damping, with their respective subcomponents, can then in turn be used to calculate the storage modulus, loss modulus and tan delta using Eq. 8a–c as;

$$\text{a) } E' = \frac{k_s\sqrt{\pi}}{2\sqrt{A_c}} \quad \text{b) } E' = \frac{\omega C_s\sqrt{\pi}}{2\sqrt{A_c}} \quad \text{c) } \tan\delta = \frac{\omega C_s}{k_s} \tag{8}$$

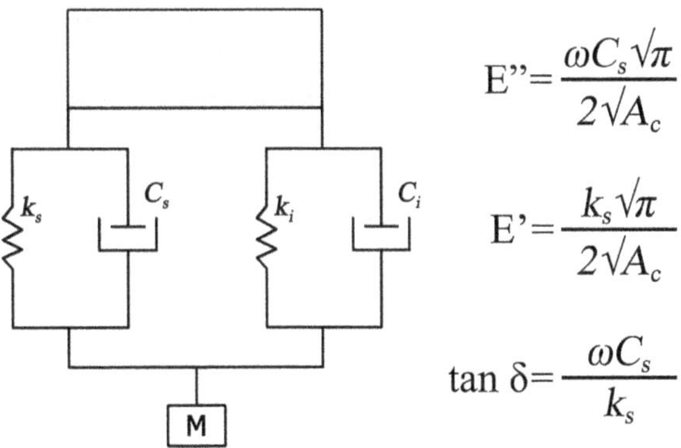

Fig. 9 Kelvin and Voigt elements that represent the instrument and the interaction with the sample. The phase difference between the force and the displacement is given in Eq. 10. © Hysitron, Inc

The relationship between the complex modulus (E*), storage modulus (E′), and loss modulus (E″):

$$E^* = E' + iE''$$

(9)

where i is defined as the square root of -1.

The nanoDMA® technique enables the usage of a number of CMX algorithms—truly continuous measurements of X (X can be hardness, storage and loss modulus etc.) as a function of indentation depth frequency and time, as seen in Fig. 3b. Moreover, the reference frequency technique is an important functionality for those tests that are performed over a longer time, such as creep or frequency sweeps (0.1–300 Hz) on viscoelastic material, which show how the mechanical properties change with applied frequency. Drift in the displacement reading could be a result of adhesion at the contact zone, or a result of material drift. The effects of both of these types of measurement drift are eliminated by recording the stiffness over time and relating it to the area of contact, and therefore penetration depth. A reference frequency experiment starts with a dynamic segment that determines the elastic constants of the sample, following Eq. 8. In the case of a fully elastic contact, k_s is equivalent to S. Once the elastic modulus of the sample is known, Eq. 8 can be rearranged and the contact area is monitored by;

$$A\,(h_c) = \frac{\pi k_s^2}{4E'^2}$$

(10)

Having calculated the indenter area function, $A(h_c)$, by calibration, displacement in the sample can then be determined. For materials with viscoelastic behavior, the storage modulus, E', is frequency dependent. Therefore, a constant and high reference frequency is chosen and applied in between changes in the frequency sweep.

The Modulus Mapping package combines the in-situ SPM imaging capability of Hysitron's nanomechanical testing instruments with the ability to perform dynamic, or nanoDMA III® tests. During the imaging process, the system continuously monitors the stiffness of the sample, and can plot this stiffness as a function of the position on the sample. At each pixel in the image, the stiffness is found, and if the geometry of the probe is known, the modulus can be calculated. This is the equivalent of performing an indentation test at each pixel in a 256 × 256 image, or 65,536 tests in a single image. While mapping, the static force, typically 1 μN, is the imaging contact setpoint. This is maintained by a feedback loop and separated from the AC modulation by a low pass filter. A lock-in amplifier is used to monitor the amplitude and phase response of the material in relation to the input signal. The dynamic load amplitude and the dynamic displacement amplitude give information about the stiffness at each point of contact in the image, done entirely in the elastic, Hertzian, contact regime. The radius of the tip is done by dynamic calibration on a known material, and is assumed to be perfectly spherical. Another approach is to use the known modulus of one of the constituent materials in the sample as the known calibration, effectively calibrating the tip and eliminating the effects of humidity on the sample. Typical frequencies of operation are near 200 Hz, but one should also be aware of avoiding the line frequency harmonics in the country where the system is being operated. This minimizes deadtime associated with the lock-in amplifier, allowing faster imaging. However, there is a tradeoff between acquiring data rapidly and having too high a strain rate. Systems that operate on the 100's of kHz or even MHz frequencies apply very high strain rates, which affect the measured value. The time required to acquire a stable image in these conditions must maintain low force drift. Possible sources of error include a water meniscus formed at the surface due to moisture, and uncertainties in the contact force. With regards to the latter, the noise floor of the Hysitron system, 30 nN, is far below the typical imaging force of 1 μN. A variety of techniques, such as dry N_2 gas or dessicant can be used to lower the humidity in the system.

Acknowledgments   More information about nanoDMA III and Modulus Mapping can be found at www.Hysitron.com.

References

Asif SAS, Wahl KJ, Colton RJ (1999) Nanoindentation and contact stiffness measurement using force modulation with a capacitive load-displacement transducer. Rev Sci Instrum 70(5):2408–2413. doi:10.1063/1.1149769

Asif SAS, Wahl KJ, Colton RJ, Warren OL (2001) Quantitative imaging of nanoscale mechanical properties using hybrid nanoindentation and force modulation. J Appl Phys 90(3):1192–1200

ASTM-E384—11e1 (2011) Standard test method for Knoop and Vickers hardness of materials. Physical testing standards and mechanical testing standards. ASTM International. West Conshohocken, USA. doi:10.1520/E0384-11E01

Cordill MJ, Lund MS, Parker J, Leighton C, Nair AK, Farkas D, Moody NR, Gerberich WW (2009) The Nano-Jackhammer effect in probing near-surface mechanical properties. Int J Plast 25(11):2045–2058. http://dx.doi.org/10.1016/j.ijplas.2008.12.015

DIN ISO-6507-1 (2005) Metallic materials: Vickers hardness test—part 1: test method

Durst K, Backes B, Göken M (2005) Indentation size effect in metallic materials: correcting for the size of the plastic zone. Scripta Mater 52(11):1093–1097. http://dx.doi.org/10.1016/j.scriptamat.2005.02.009

Gerberich WW, Tymiak NI, Grunlan JC, Horstemeyer MF, Baskes MI (2002) Interpretations of indentation size effects. J Appl Mech 69(4):433–442

Gerberich WW, Michler J, Mook WM, Ghisleni R, Östlund F, Stauffer DD, Ballarini R (2009) Scale effects for strength, ductility, and toughness in "brittle" materials. J Mater Res 24(03):898–906. doi:10.1557/jmr.2009.0143

Greenwood JA, Tripp JH (1967) The elastic contact of rough spheres. J Appl Mech 34(1):153–159

Hertz H (1896) Hertz's miscellaneous papers. Macmilan, London

ISO-14577-1 (2002) Metallic materials: instrumented indentation test for hardness and materials parameters—part 1: test method

Jakes JE, Stone DS (2010) The edge effect in nanoindentation. Phil Mag 91(7–9):1387–1399. doi:10.1080/14786435.2010.495360

Jakes JE, Frihart CR, Beecher JF, Moon RJ, Resto PJ, Melgarejo ZH, Suárez OM, Baumgart H, Elmustafa AA, Stone DS (2009) Nanoindentation near the edge. J Mater Res 24(03):1016–1031. doi:10.1557/jmr.2009.0076

Johnson KL (ed) (1985) Contact mechanics. Cambridge University Press, Cambridge

Lucas BN, Oliver WC, Pharr GM, Loubet J-L (1996) Time dependent deformation during indentation testing. MRS online proceedings library 436: null–null. doi:10.1557/PROC-436-233

Nix WD, Gao H (1998) Indentation size effects in crystalline materials: a law for strain gradient plasticity. J Mech Phys Solid 46(3):411–425. http://dx.doi.org/10.1016/S0022-5096(97)00086-0

Oliver WC, Pharr GM (1992) An improved technique for determining hardness and elastic modulus using load and displacement sensing indentation experiments. J Mater Res 7(06):1564–1583. doi:10.1557/JMR.1992.1564

Oliver WC, Pharr GM (2004) Measurement of hardness and elastic modulus by instrumented indentation: advances in understanding and refinements to methodology. J Mater Res 19(01):3–20. doi:10.1557/jmr.2004.19.1.3

Tabor D (1951) Hardness of metals. Oxford Calrendon Press, New York

Nanoindentation, Nanoscratch and Dynamic Mechanical Analysis of High Performance Silicones

Atul Tiwari and Phillip Agee

Abstract This chapter describes the basic fundamental principles that are required to understand the operation of modern nanoindentation techniques. Special attention has been paid to explain the terms that an experimentalist should know while analyzing the data from nanoindentation. Studies conducted by different researcher using nanoindentation techniques have been briefly mentioned as examples. Additionally, studies conducted on silicone quasi-ceramic coatings are also provided.

1 Introduction

Conventional polymers are evaluated using universal testing machine that operate in tensile or compressive mode. However, thin coatings or films that cannot be removed from the substrates needs to be tested using the nanoindentation technique. The use of indentation technique to determine the mechanical properties of materials was realized in late 1950 (Hutchungs 2009). During a nanoindentation test, a nanoindenter head of known geometry is pressed into the surface with a predefined load or depth of penetration and the resultant affected area is recorded. The ratio of load over area determines the value of nanoindentation hardness. The basic nanomechanical properties obtained from the nanoindentation tests are elastic modulus and indentation hardness of the material. Several nanoindentation test methods are available that can determine fracture toughness, creep, storage and loss modulus, yield stress, as well as interfacial and surface adhesion. Similarly,

A. Tiwari (✉)
Department of Mechanical Engineering, University of Hawaii, 2540 Dole Street, Holmes 302, Honolulu, HI 96822, USA
e-mail: tiwari@hawaii.edu

P. Agee
Agilent Technologies Inc., Chandler, AZ, USA
e-mail: phillip.agee@agilent.com

A. Tiwari (ed.), *Nanomechanical Analysis of High Performance Materials*, Solid Mechanics and Its Applications 203, DOI: 10.1007/978-94-007-6919-9_6,

the tribological behavior of surfaces, such as a scratch or mar resistance, friction coefficients, and wear performance can be obtained (Fischer-Cripps 2004).

In past few years, several coated and bare substrates have been tested using nanoindentation techniques. For example, the mechanism of deformation of soft coating on hard substrates (TSui et al. 1997), hard coatings on soft substrates (Chen et al. 2005; TSui et al. 1997; Charitidis et al. 2004), soft coating on soft substrates (Roche et al. 2003; Geng et al. 2008), or hard coating on hard substrates (Zhang and Huan 2005), have been broadly established. Several excellent review articles have been written on the nanoindentation (Li and Bhushan 2002; VanLandingham et al. 2000; Koleske 2006; Mammeri et al. 2005; VanLandingham 2003; FischerCripps 2006) that describes the salient features of this technique.

1.1 Nanoindenter

A standard Automated Nanoindentation Technique (ANT) equipment consists of three basic components (Fig. 1): an actuator to apply force, an indenter mounted to a rigid column through which the force is applied on the sample and a sensor that measures the displacement of the indenter (Bull 2005).

The ANT equipment can generate required forces electromagnetically using either coils and magnets or capacitors that have fixed and moving plates. In some cases, the piezoelectric actuators can be used to generate small forces. The indenters used in ANT are selected according to the information that is being collected. The indenter may have pyramidal, spherical, cube corner or conical geometry. A pyramidal shaped Berkovich indenter is most common in acquiring the nanomechanical data. The displacement in ANT can be recorded using capacitive sensors.

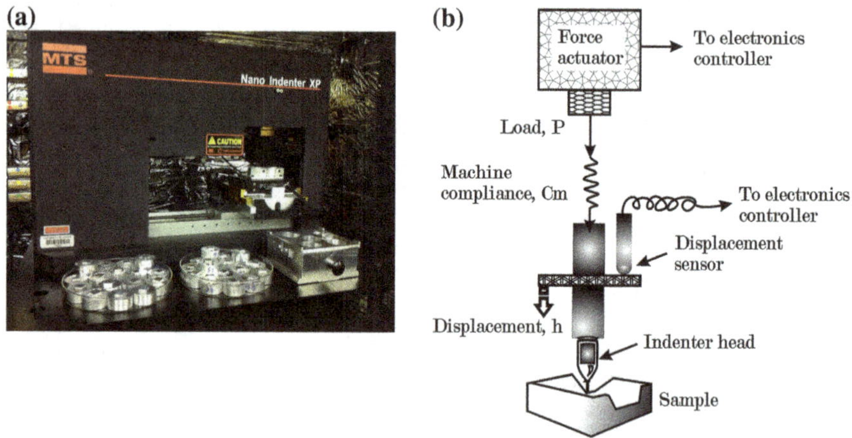

Fig. 1 **a** MTS (now Agilent) nanoindentation instrument. **b** Schematic of a typical nanoindenter setup for the nanomechanical properties measurement of coatings

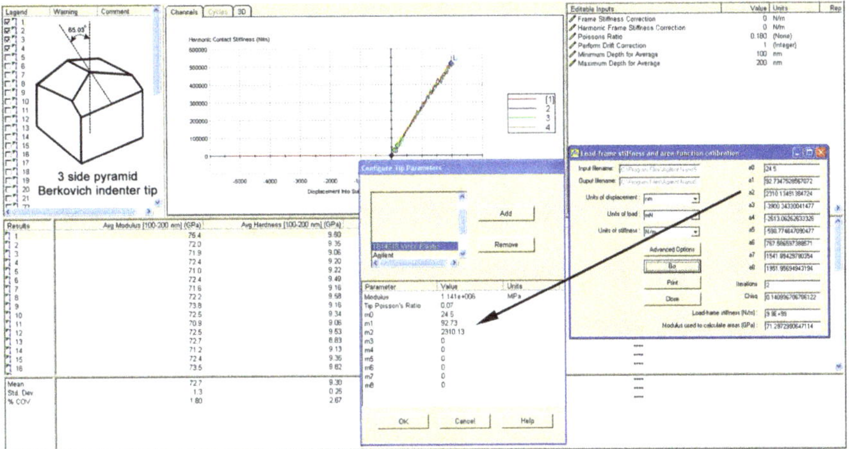

Fig. 2 A typical calibration procedure using fused silica as standard sample

1.2 Calibration in Nanoindentation

The shape and dimensions of the nanoindenter tip plays a crucial role in determining the properties of the materials. A fused silica is generally used as a standard sample, due to the known properties of this material. Data from several nanoindentation tests is recorded, and coefficients are calculated. The coefficients are then used to calculate area function of the indenter tip as shown in Fig. 2.

1.3 Stiffness Measurement

The software of modern ANT instruments may automatically detect the true surface of the sample. However, it is recommended to verify the position of the sample surface in each test. A plot of contact stiffness against displacement could be used for such an assurance as shown in Fig. 3. A sharp increase in stiffness value indicates that nanoindenter tip has reached the surface. Stiffness can also be used to evaluate the performance of the instrument. For an isotropic material, a plot of stiffness square over load (S^2/P) should give a constant value as it represents a material's property and independent of contact area. A material (such as fused silica) with a known value can be used for such an investigation.

2 Basics of Nanoindentation

The estimation of intrinsic properties of the coatings is normally influenced by the underlying substrates. However, the nanomechanical properties of the coated material are least affected by the substrate when determined from the 10 % of the

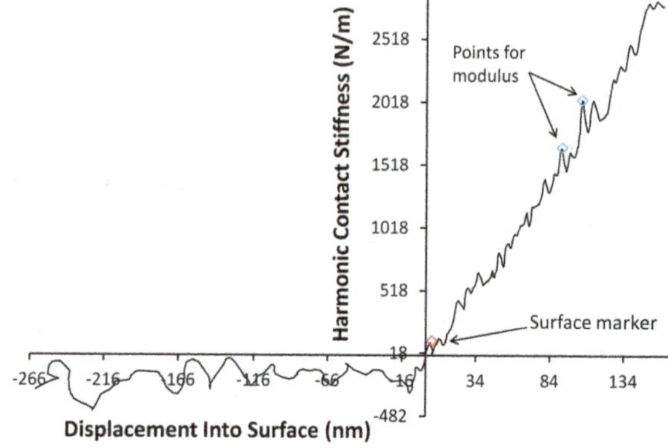

Fig. 3 Contact stiffness curve for determining the true surface contact

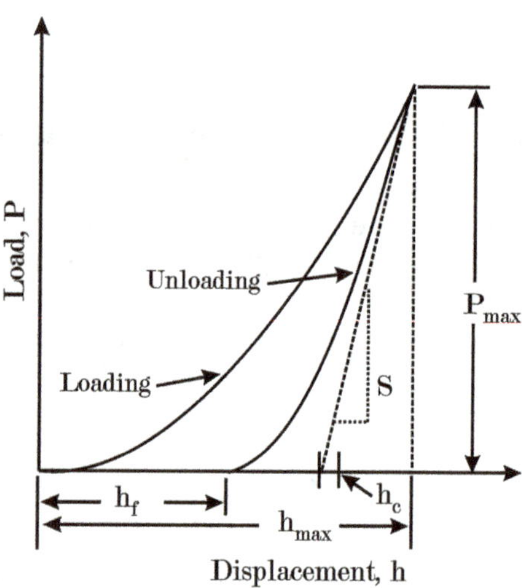

Fig. 4 Typical nanoindentation *loading* and *unloading curve*

thickness of the coating (Hay and Pharr 2000; VanLandingham 2003). In a typical nanoindentation experiment, the indenter makes contact with the material surface and then penetrates to a depth or load. A nanoindentation curve is plotted for load as a function of displacement of the indenter and shows loading and unloading pattern (Fig. 4). Any inconsistency observed in the curve indicates cracking, delamination or another failure in the coating. Figure 4 shows the unloading process and parameters associated with the contact geometry. The depth of penetration is considered to be displacement into the sample. The hardness and modulus values are determined as discussed in the following section.

The load and displacement curve can be used to determine the hardness and elastic modulus of the material (Oliver and Pharr 1992). The hardness H of the material is determined by dividing maximum load P_{max} by the projected contact area A of the indenter at maximum load as shown in Eq. (1).

$$H = \frac{P_{max}}{A} \tag{1}$$

For an ideal nanoindenter tip geometry (Fig. 5), the projected contact area, A, can be determined from the contact depth h_c at maximum load P_{max} such as in Eq. (2).

$$A = c_0 h_c^2 \tag{2}$$

The value of c_0 depends on the indenter tip. For example, the value of c_0 is 24.5 for the Berkovich pyramidal diamond tip. In the load versus displacement curve, the contact depth h_c, is different from maximum indentation depth h_{max} at the maximum load due to the elastic deformation of the area around the indenter head. The contact depth is given in Eq. (3).

$$h_c = h_{max} - \varepsilon \frac{P_{max}}{S} \tag{3}$$

where, S represents stiffness that can be calculated from the slope of the unloading curve at the maximum load. The value of ε is 0.75 for the pyramidal indenter tip.

The stiffness S_{max} is obtained from load-depth curves and assuming that the elastic modulus of the fused quartz is constant, the projected contact area A can be obtained as a function of stiffness S_{max} as given in Eq. (4).

$$A = \left(\frac{\pi}{4}\right)\left(\frac{S_{max}}{E_r}\right)^2 \tag{4}$$

where E_r is the reduced modulus, which represents the elastic deformation occurring both in the sample and indenter. The value E_r can be calculated from Eq. (5).

$$\frac{1}{E_r} = \frac{1 - \upsilon_s^2}{E_s} + \frac{1 - \upsilon_i^2}{E_i} \tag{5}$$

where E_s is the elastic modulus of the fused quartz silica, E_i is the elastic modulus of the indenter, υ_s is the Poisson's ratio of the fused quartz silica and υ_i is the Poisson's ratio of the indenter. The projected area A that was calculated using Eq. (4) can be plotted as a function of the contact depth h_c. The area function $A(h_c)$ so obtained is a fifth order polynomial that can be represented as Eq. (6).

$$A = c_0 h_c^2 + c_1 h_c + c_2 h_c^{1/2} + c_3 h_c^{1/4} + c_4 h_c^{1/6} + c_5 h_c^{1/8} \tag{6}$$

where c_0, c_1, c_2, c_3, c_4 are constant that can be determined by curve fitting of the measured area function $A(h_c)$.

The elastic modulus E_m of the material can be determined using Eqs. (7) and (5). The υ_s in Eq. (5) is taken as Poisson's ratio of the test material.

$$E_m = \frac{dp}{dh}\frac{1}{2}\frac{\sqrt{\pi}}{\sqrt{A}} = S\frac{1}{2}\frac{\sqrt{\pi}}{\sqrt{A}} \tag{7}$$

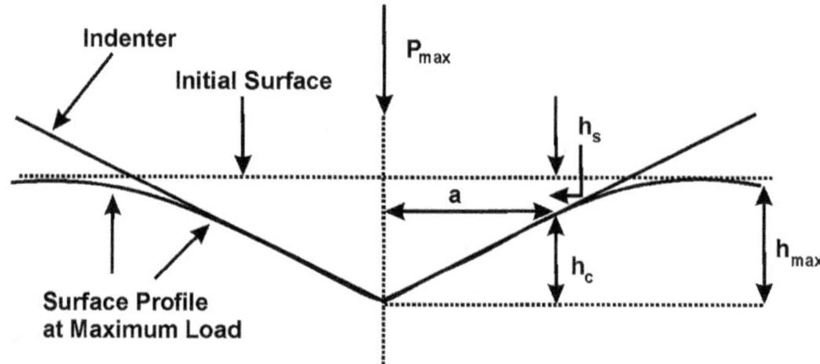

Fig. 5 Schematic of an ideal conical indenter at maximum load

3 Hardness and Modulus Analysis of Silicones

The nanomechanical analysis of materials using ANT is acquiring momentum. For example, nanoindentation studies were conducted on hybrid materials that contained higher organic branching points for crosslinking reactions. It was found that such materials possess higher hardness values than do linear polymers (Hu et al. 2005). In another study, the reduced modulus and reduced hardness found varying through the different layers of the polymeric coating (Fabes and Oliver 1990). High value of reduced modulus and reduced hardness were recorded at the surface due to a condensed morphology. Lower values were recorded in the bulk due to a porous structure while values increased again at the coating-metal interface due to a denser structure (Scherer 1987; Fabes and Oliver 1990). The inherent surface morphology of the sample plays a pivotal role in accurate determination of the nanomechanical properties of the materials. Numerical models have also been proposed for softer surfaces like PDMS so that the effect of surface roughness on the nanoindentation results could be estimated (Chen and Diebels 2012).

A plot of load on the sample as a function of displacement into surface of quasi-ceramic silicone (QCS) coating is shown in Fig. 6. The curves from traditional (SDM) and substrate independent (SIN) test methods (described in Sect. 4 ahead) are shown. The loading and unloading curves were close to each other suggesting that the material was elastic in nature. No discontinuity was observed in loading segment of the coatings suggesting that the material does not fracture, or delaminate under applied load (Chen et al. 2007). The elastic behavior observed here was due to the high volume of inorganic constituents in the backbone of the coating structure.

The hardness and modulus were determined for QCS coatings that were aged for 3 months on three different aluminium alloy substrates (Fig. 7). The hardness and modulus values were determined as a function of displacement into the coatings. The average hardness values for the QCS-coated 2024 Al, 6061 Al and 7075 Al were 0.42, 0.41 and 0.47 GPa, respectively while the average modulus values were 4.40, 4.51 and 5.45 GPa, respectively. The loading-and-unloading curves demonstrate the elastic recovery of the coatings with negligible plastic deformation. The average hardness of uncoated alloys 2024 Al, 6061 Al, and 7075 Al were 1.87, 1.47 and

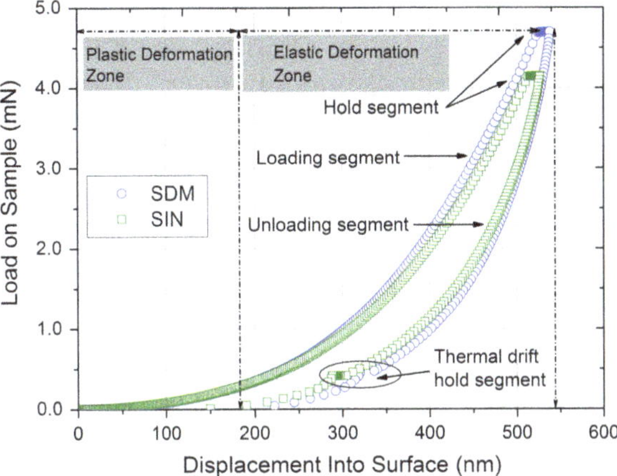

Fig. 6 Plots of load on the sample as a function of displacement into the sample surface

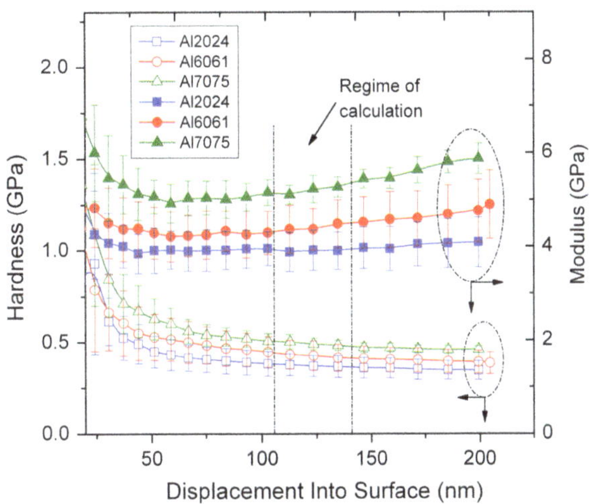

Fig. 7 Nanoindentation of hybrid silicone coating showing hardness and modulus values (Tiwari and Hihara 2010a)

2.37 GPa, respectively while the average modulus were 77, 76 and 78 GPa, respectively. The slight variation in the mechanical properties of the QCS coatings on different surfaces suggests that the coating may have been influenced to a small degree by the substrate mechanical properties or due to the solubilised alloying elements from the substrate alloy (Tiwari and Hihara 2010a, b).

In another hybrid coatings, compositions with higher organic content cause segregation in the network, resulting in high-silica regimes surrounded by hydrocarbon-rich regimes. These coating networks displayed different mechanical properties than did the pristine silica network. A silica-rich coating based on tetraethoxysilane (TEOS) display near elastic behavior while a hybrid glycidoxypropyltrimethoxysilane (GTMS) coating displayed increased penetration on loading and almost complete recovery during unloading. The TEOS coating showed the smallest amount of creep due to a densely packed, rigid silica network while the creep was higher for the GTMS coating due to the viscoelastic flow and relaxation processes associated with the long chain hydrocarbon-rich domains. Young's modulus also decreased with the increase in organic portion in the backbone of the coating structure. In fact, the modulus of GTMS was recorded 25 times less than that of TEOS (Atanacio et al. 2005).

Etienne-Calas et al. (2004) and Ferchichi et al. (2008) studied two different organic–inorganic coatings on silicone and glass substrates. They prepared the coatings through a sol–gel procedure and deposited them on substrates using the spin coating technique. The first coating composition was formulated using methyltrimethoxysilane, colloidal silica and TEOS while the second composition was based on 3-(trimethoxysilyl)-propyl-methacrylate. At lower loads, authors determined the coating hardness and modulus values from indentation curves. At higher load values, propagation of cracks were used to determine the coating toughness, residual stress and interface toughness while energy analysis was used to study chipping and delamination in the coating. It was also found that mechanical properties of the coatings were influenced by the rapid diffusion of the sodium ions from the glass substrate into the coating.

Similarly, Kim et al. (2007) prepared a coating compositions by reacting vinyl terminated polydimethylsiloxane with tetrakis(dimethylsiloxy)silane in the presence of a platinum-divinyltetramethylsiloxane complex. Authors have used a statistical, experimental design, to study the effect of different chemical constituents on resultant shear stress in the coatings. It was noticed that the modulus of the coating varied with the thickness of the coating and that the shear rate was dependent on the modulus of the coating.

The effect of inorganic fillers in the coatings was also studied using ANT. For example, Chen et al. (2007) studied AlOOH boehmite nanorods incorporated GTMS sol–gel coating. The nanorod concentration up to 40 wt % was utilized in the coating composition and was applied over a glass substrate. Authors recorded lower modulus and hardness values in the nanocoatings than in a commercially available coating composition containing boehmite nanoparticles. The hardness and modulus values for nanorods filled coating were 8.86 and 0.83 GPa while those of nanoparticles filled coating were 9.88 and 0.98 GPa. The coating composition containing nanorods with an aspect ratio of approximately 20 displayed significant improvement in the crack toughness, which was achieved by incorporating nanorods with a high aspect ratio. The orientation of nanorods in the composite coating contributed to the anisotropic toughness. The enhanced toughness was considered as an outcome of the formation of chemical bonds between boehmite nanorods and the coating. Likewise, Douce et al. (2004) performed nanoindentation experiments on silicone coatings containing various surface modified silica nanoparticles of different sizes ranging from 15 to 60 nm. The authors found that Young's modulus of the coatings increased with the increase in silica nano-fillers. However, the scratch resistance of the coating decreased with the addition of nano-fillers, probably due to a weak interaction between inorganic fillers and the coating network. In another study, nanosilica having an average diameter of 20 nm was loaded in a nanocomposite film made of methyltrimethoxysilane and applied on polymethylmethylacrylate substrate. The nanoindentation tests revealed that the scratch resistance and strength in the coating increased with the increase in nanosilica loading (Chantarachindawong et al. 2012). A polydimethylsiloxane elastomer filled with organically modified montmorillonite nanoclay was tested using ANT (Charitidis and Koumoulos 2012). Authors compared the results obtained from Oliver-Pharr and Hertizian methods. The slope of Hardness/Modulus suggested that addition of nanoclay strengthens the resultant PDMS-nanoclay nanocomposite.

The ANT has been successfully applied on the soft materials to study the tribological properties. In a study, silicone based contact lens were tested using nanoindentation technique. The experiments were conducted in both liquid as well as dehydrated conditions. It was noticed that the stiffness of the hydrogel lens changed dramatically when tested in dehydrating conditions (Zhou et al. 2013). Similarly, nanoindentation was performed on silicone contact lens using colloidal probe supported on atomic force microscope. Authors studied the lubricity of a soft hydrogel on contact lens and recorded the soft elastic modulus value of approximately 25 kPa (Dunn et al. 2013).

4 Substrate Independent Nanoindentation

The nanomechanical properties derived using Oliver-Pharr model is often influenced by the substrate effect. In order to minimize the substrate effect, the values are often extracted from the 1/10th of the film/coating surface. It is worth noting that for very thin films this 10 % rule is not effective. However, these values are not a true representation of the values from the bulk of the material. Recently, Hay and Crawford (2011) proposed a model that can accurately predict the intrinsic properties of the bulk material provided that properties of the substrate are known. Their, model is based on Song-Pharr and Gao (S–P & G) model that assumes that material under the indenter tip can be treated as a column (Fig. 8a). The film and substrate within the column are treated as two springs connected in series. This S–P & G model resulted into apparent shear modulus (μ_a) related to the shear modulus of the film (μ_f) and that of the substrate (μ_s) by the expression showing in Eq. (8)

$$\frac{1}{\mu_a} = (1 - I_0)\frac{1}{\mu_s} + I_0\frac{1}{\mu_f} \tag{8}$$

where I_0 is weighting function derived by Gao et al. (Huajian et al. 1992) that represents the ratio of the strain energy stored in the film region to the total stain energy stored in the half space. Hay and Crawford improved the S–P & G model by assuming that film can also act as a spring in parallel with the substrate (Fig. 8b). Based on this assumption, they proposed a model shown in Eq. (9).

$$\frac{1}{\mu_a} = (1 - I_0)\frac{1}{\mu_s + FI_0\mu_f} + I_0\frac{1}{\mu_f} \tag{9}$$

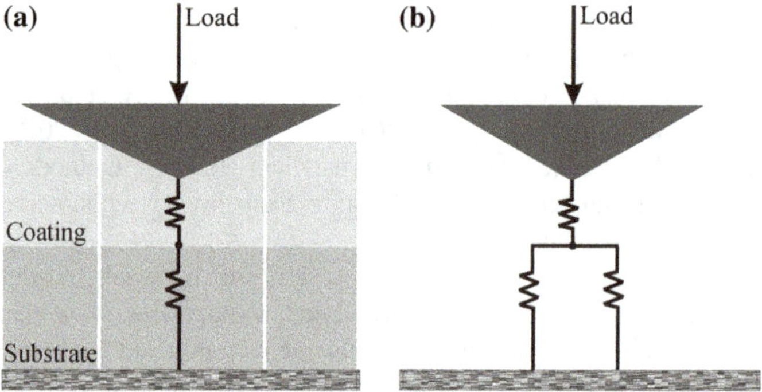

Fig. 8 Models used for the calculation of shear modulus. **a** S-P & G model. **b** H-C model. Reproduced after modification from Hay and Crawford (2011)

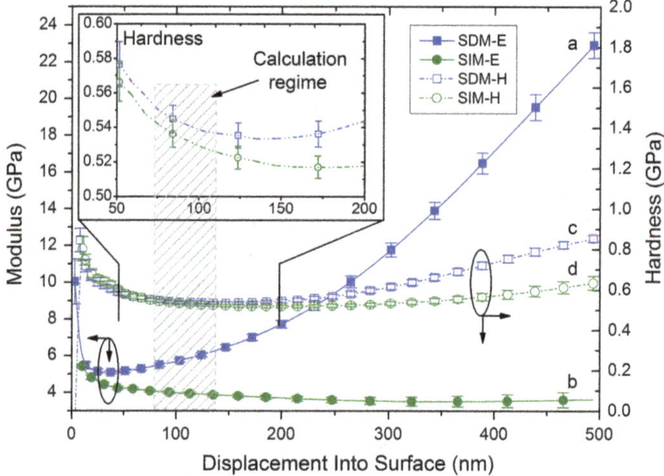

Fig. 9 Comparison between hardness and modulus values from QCS coating using traditional and surface independent nanoindentation measurement techniques

where, F is a dimensionless empirical constant. The Young's modulus of the film is calculated by solving the Eq. (9) for shear modulus and related with Poisson's ratio as in Eq. (10).

$$E_f = 2\mu_f(1 + \nu_f) \tag{10}$$

The hardness and modulus values of QCS coating were determined by traditional (surface dependent, SDM) and substrate independent methods (SIM). It can be seen from Fig. 9 that modulus value determined by SDM was affected by the substrate at higher contact depth. The hardness value determined from 1/10th of thickness of the coating and using SDM was similar to the value obtained from SIM.

5 Nano-Scratch Tests

The coating performance depends on its ability to resist mars and scratches. Researchers have proposed several techniques and test methods; however, none of them are as precise as the nanoscale scratch testing, which uses ANT. The scratch test helps determine the mechanism that causes the deformation of materials as well as the delamination of coatings. During a scratch test, the load normal to the sample is controlled and can be held constant, increased or decreased while scratch path and scratch velocity are defined in the test method.

In a typical scratch experiment, a ramp load is applied to an indenter head in the normal direction as it simultaneously moves on the sample surface in a lateral direction. The instrument controls the normal force and lateral displacements of the sample (in some cases indenter) while the lateral force and normal displacement is recorded as a function of time. Critical information such as the coefficient of friction, cross profile topography, residual deformation and pile-up of material during the scratch can be obtained as a function of scratch distance.

When a coating is subjected to a scratch test, the indenter can pass through three key regimes in the material: elastic, plastic and fracture. The fracture is followed by delamination and chipping off the coated surface. The estimation and application of the correct load required to study the above mentioned deformations is extremely valuable. The high load can fracture the coating upon contact, eliminating the appearance of the other two regimes. Generally, several different loads are applied on the sample under study and the deformations on the sample are monitored. The inception of the fracture that appears on the curve is confirmed under the high magnification/resolution microscope.

5.1 Nanoscratch Studies on the Silicones

The nanoscratch analysis using ANT is one of the superior methods available to understand the mechanical failure modes in materials. Several attempts have been made so far to investigate the surface behavior of the wide variety of substrates (Chantarachindawong et al. 2012; Dunn et al. 2013). Tanglumlert et al. (2006) prepared a hard coating suspension to improve the scratch resistance of polymethylmethacrylate surface. A coating solution was formulated by reacting silatrane with GTMS in the presence of an acid catalyst. Authors found that the scratch resistance of the coated surface increased with the increase in alkoxysilane content in the coating. Authors also discovered that the curing time and curing temperature affected the scratch resistance and adhesion properties of the coating layer. Nanoscratch tests have also been used to calculate the fracture toughness of the materials (Zhang and Zhang 2012). Atomic force microscope (AFM) assisted nanoscratch testing has been accomplished by Verma et al. (2012). The AFM images of the scratched regions demonstrated that PDMS modified polyester coating display enhanced scratch resistance compared to the pristine polyester coating.

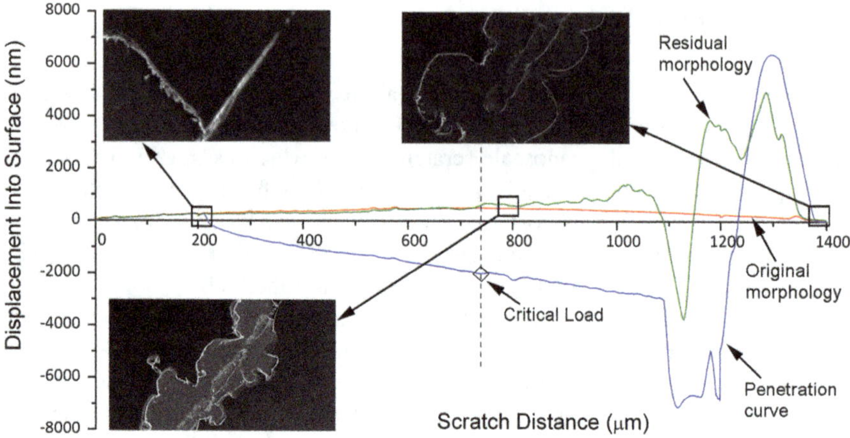

Fig. 10 Nanoscratch analysis of coated material. SEM images showing initiation, propagation and termination steps in QCS coating. Reproduced after (Tiwari and Hihara 2012)

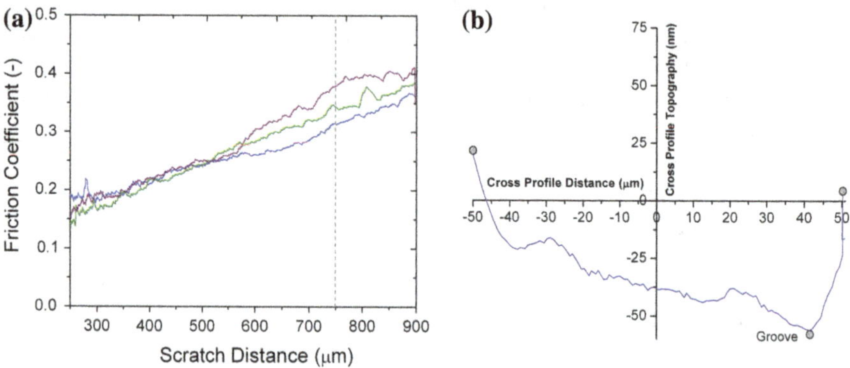

Fig. 11 Parameters derived from nanoscratch testing of silicone coating. **a** Friction coefficient as a function of scratch distance. **b** Cross profile topography

Figure 10 shows the penetration curve along with residual surface morphology as a function of the scratch distance for QCS coating. The original morphology of the coating was smooth with curvature. It appears from the penetration curve that the progressing load bearing the indenter tip pushed the material and fractured at a critical load of approximately 42 mN after travelling a scratch distance of 740 μm. The estimated thickness of the coating was approximately 5 μm, and the depth at critical load was approximately 2.0 μm. It is speculated that coating cracked and fractured simultaneously. The SEM images of the initiation, propagation and termination steps during the scratch suggested that the indenter scratched the surface after which the surface cracked. Moreover, the coating and substrate chipped-off at the termination point of the scratch test.

The coefficient of friction (COF) curves as a function of scratch distance is shown in Fig. 11a. The curves from at least three tests are shown for better clarity and COF value was determined at a scratch distance of 750 μm. The cross profile topography (CPT) of coatings as a function of cross profile distance is shown in Fig. 11b. The CPT was acquired after the end of the scratch test and when the load was 10 mN. The positive values on the X-axis show the right side of the groove while the negative values in the X-axis show the left side of the groove. Similarly, the positive values on the Y-axis show the pile-up after the scratch while the negative values in the Y-axis show penetration in the coated material. In QCS coating the depth of penetration was approximately 60 nm with random shape. The inconsistent shape of the groove and wide scratch width for QCS suggested that the material was hard and brittle in nature.

6 Dynamic Mechanical Analysis

The viscoelastic behavior of coatings can also be investigated using nanoindentation technique (Herbert et al. 2008). The stress–strain relationship of the materials displaying linear viscoelastic properties under sinusoidal loading is shown in Eq. (11):

$$\sigma = \varepsilon_o E' \sin \omega t + \varepsilon_o E'' \cos \omega t \tag{11}$$

where σ is the stress, ε_0 is the strain amplitude, ω is the angular frequency and t is the time elapsed. Rearranging Eq. (11) we get Eqs. (12) and (13):

$$E' = \frac{\sigma_o}{\varepsilon_o} \cos \phi \tag{12}$$

and

$$E'' = \frac{\sigma_o}{\varepsilon_o} \sin \phi \tag{13}$$

where σ_0 is the stress amplitude, ϕ is the phase lag between stress and strain and E' and E'' are storage and loss modulus respectively. The term E' represents the capacity of material to store energy, a component that is in phase with the applied load or displacement. The term E'' represents the capacity of material to dissipate energy, a component that is $90°$ out of phase with the applied load or displacement. The ratio E''/E' represents $\tan\phi$, also called loss factor, and is used to measure the damping characteristic of a linear viscoelastic material. The value of E' and E'' can be used to calculate the complex modulus through the Eqs. (14) and (15):

$$E = E' + iE'' \tag{14}$$

$$|E| = \sqrt{E'^2 + iE''^2} \tag{15}$$

In order to determine the value of E' and E'' from a dynamic nanoindentation experiment, the equipment supplies a controlled load to the indenter head that sets the load amplitude while the displacement amplitude and phase angle are measured (Odegard et al. 2005). At each test site, the indenter head contacts the material's surface. The indenter vibrates at a certain frequency, and the resulting response is measured. The contribution of the instrument to the total response is then subtracted to determine the response from the material.

6.1 Visco-Elastic Behavior of Silicones

The use of nanoindentation technique to determine the viscoelastic (VE) behavior of coatings is relatively new and yet to be explored in sumptuous detail (Wright and Nix 2009). Kohl et al. (2008) investigated the VE properties of three proprietary silicone-based coatings using continuous indentation technique. Authors concluded that the calculated parameters would help them in developing more durable silicone based foul release coating compositions.

The VE properties of QCS coatings were analyzed using Berkovich nanoindenter tip (Fig. 12). The storage modulus (E′) and loss modulus (E″) values were recorded as a function of frequency. The E′ value of pristine coating was seen independent of frequency while the E″ value increased slightly with the test frequency. Three concentrations (0.1, 0.3 and 0.5 wt %) of polymer coated silicon di-oxide (SiO_2) nanoparticles having an average diameter of 90 nm were added to the QCS coating composition. The three nanocoatings were again subjected to VE studies. This exercise reveals whether adding nanoparticles changes VE properties of a coating. It appears from Fig. 12 that the E′ value in nanocoatings remained

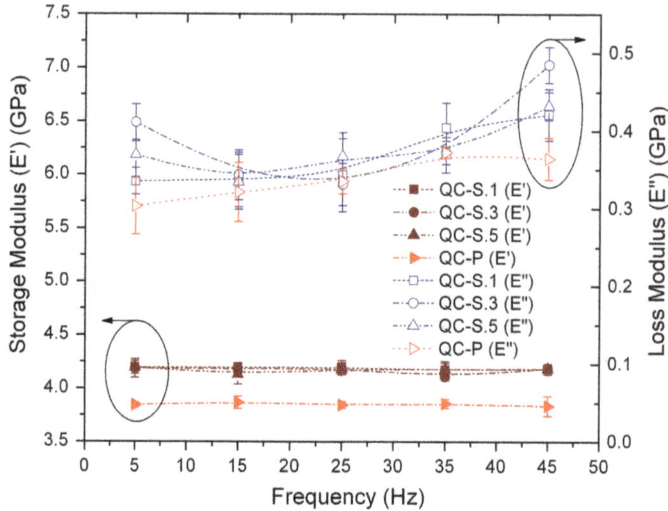

Fig. 12 Viscoelastic response obtained from QCS coating. Change in viscoelastic response was observed on incorporating silicon nanoparticles

unaffected by the test frequency. However, the E'' value for each nanocoating changed with the frequency. On comparing with pristine, E'' values were higher at lower frequencies for each nanocoating, indicating that the relaxation process associated with the regime containing nanoparticles.

It is worth mentioning that molecular motions that are associated with polymer materials are more prominent at lower frequencies. On the other hand, such transitions are not visible at higher frequencies because molecules may not have enough time to undergo rapid molecular transformations. In other words, at lower frequencies, molecules have a longer time to react and the viscous term dominates while at a higher frequency molecules may not have sufficient time to relax and the response is mainly due to elasticity with limited viscous nature.

In above mentioned QCS coatings, the quasi-linear increment in E'' for pristine as well as nanocoatings could be due to the high frequency that leads to a higher loading rate effect. VE tests suggested that the incorporation of nanoparticles in the QCS coating formulation generates free volume in the resultant nanocoating structure and the domains modified by nanoparticles were distributed in the coating network. The temperature during the analysis was kept constant; therefore, the change in glass transition temperature could not be monitored.

7 Concluding Remarks

The automated nanoindentation technique has been proven to be an invaluable tool for the estimation of mechanical properties of the materials in nanoscale regime. The use of this technique has shown tremendous growth in past 15 years. The development of new test methods that record intrinsic properties of the material while negating the substrate effect will help in accurate determination of mechanical deformations in coatings as well as substrates. This technique could help in controlling the macroscale deformations that occur inevitable with time. This in turn will help researchers in selecting suitable ingredients so that the service life span of the materiel can be enhanced.

Acknowledgments One of the authors (Atul Tiwari) is grateful to the director of Hawaii Corrosion Laboratory and the College of Engineering, University of Hawaii at Manoa for their help and cooperation.

References

Atanacio AJ, Latella BA, Barbe CJ, Swain MV (2005) Mechanical properties and adhesion characteristics of hybrid sol-gel thin films. Surf Coat Technol 192:354–364

Bull SJ (2005) Nanoindentation of coatings. J Phys D Appl Phys 38(24):R393–R413

Chantarachindawong R, Luangtip W, Chindaudom P, Osotchan T, Srikhirin T (2012) Development of the scratch resistance on acrylic sheet with basic colloidal silica (SiO 2)-methyltrimethoxysilane (MTMS) nanocomposite films by sol-gel technique. Can J Chem Eng 90(4):888–896. doi: 10.1002/cjce.21631

Charitidis CA, Koumoulos EP (2012) Nanomechanical properties and nanoscale deformation of PDMS nanocomposites. Plast Rubber Compos 41(2):88–93. doi:10.1179/17432898 10Y.0000000037

Charitidis C, Laskarakis A, Kassavetis S, Gravalidis C, Logothetidis S (2004) Optical and nanomechanical study of anti-scratch layers on polycarbonate lenses. Superlattices and Microstructures 36(1):171–179. doi:10.1016/j.spmi.2004.08.015

Chen Z, Diebels S (2012) Modelling and parameter re-identification of nanoindentation of soft polymers taking into account effects of surface roughness. Comput Math Appl 64(9):2775–2786. doi:10.1016/j.camwa.2012.04.010

Chen S, Liu L, Wang T (2005) Investigation of the mechanical properties of thin films by nanoindentation, considering the effects of thickness and different coating–substrate combinations. Surf Coat Technol 191(1):25–32. doi:10.1016/j.surfcoat.2004.03.037

Chen Q, Tan J, Shen S, Liu Y, Ng W, Zeng X (2007) Effect of boehmite nanorods on the properties of glycidoxypropyl-trimethoxysilane (GPTS) hybrid coatings. J Sol-Gel Sci Technol 44(2):125–131. doi:10.1007/s10971-007-1621-z

Douce J, Boilot JP, Biteau J, Scodellaro L, Jimenez A (2004) Effect of filler size and surface condition of nano-sized silica particles in polysiloxane coatings. Thin Solid Films 466(1–2):114–122

Dunn AC, Urueña JM, Huo Y, Perry SS, Angelini TE, Sawyer WG (2013) Lubricity of surface hydrogel layers. Tribol Lett 49:371–378. doi:10.1007/s11249-012-0076-8

Etienne-Calas S, Duri A, Etienne P (2004) Fracture study of organic-inorganic coatings using nanoindentation technique. J Non-Cryst Solids 344(1–2):60–65. doi:10.1016/j.jnoncry sol.2004.07.029

Fabes BD, Oliver WC (1990) Mechanical properties of sol-gel coatings. J Non-Cryst Solids 121:348–356

Ferchichi A, Calas-Etienne S, Smaïhi M, Etienne P (2008) Study of the mechanical properties of hybrid coating as a function of their structures using nanoindentation. J Non-Cryst Solids 354(2–9):712–716. doi:10.1016/j.jnoncrysol.2007.07.096

FischerCripps AC (2006) Critical review of analysis and interpretation of nanoindentation test data. Surf Coat Technol 200(14–15):4153–4165. doi:10.1016/j.surfcoat.2005.03.018

Fischer-Cripps AC (2004) Nanoindentation, 2nd edn. Springer, New York

Geng K, Yang F, Grulke EA (2008) Nanoindentation of submicron polymeric coating systems. Mater Sci Eng, A 479(1–2):157–163. doi:10.1016/j.msea.2007.06.042

Hay J, Crawford B (2011) Measuring substrate-independent modulus of thin films. J Mater Res 26(06):727–738. doi:10.1557/jmr.2011.8

Hay JL, Pharr GM (2000) Instrumented indentation testing. In: ASM handbook, mechanical testing and evaluation, vol 8. ASM International, USA, pp 232–243

Herbert EG, Oliver WC, Pharr GM (2008) Nanoindentation and the dynamic characterization of viscoelastic solids. J Phys D Appl Phys 41(7):074021

Hu L, Zhang X, Sun Y, Williams RJJ (2005) Hardness and elastic modulus profiles of hybrid coatings. J Sol-Gel Sci Technol 34:41–46

Huajian G, Cheng-Hsin C, Jin L (1992) Elastic contact versus indentation modeling of multi-layered materials. Int J Solids Struct 29(20):2471–2492. doi:10.1016/0020-7683(92)90004-D

Hutchungs IM (2009) The contributions of David Tabor to the science of indentation hardness. J Mater Res 24(3):581–589

Kim J, Chisholm BJ, Bahr J (2007) Adhesion study of silicone coatings: the interaction of thickness, modulus and shear rate on adhesion force. Biofouling 23(1–2):113–120

Kohl JG, Singer IL, Simonson DL (2008) Determining the viscoelastic parameters of thin elastomer based materials using continuous microindentation. Polym Testing 27(6):679–682. doi:10.1016/j.polymertesting.2008.04.010

Koleske JV (2006) Mechanical properties of solid coatings. In: Encyclopedia of analytical chemistry. Wiley, NewYork. doi:10.1002/9780470027318.a0608

Li X, Bhushan B (2002) A review of nanoindentation continuous stiffness measurement technique and its applications. Mater Charact 48(1):11–36. doi:10.1016/s1044-5803(02)00192-4

Mammeri F, Le Bourhis E, Rozes L, Sanchez C (2005) Mechanical properties of hybrid organic-inorganic materials. J Mater Chem 15(35–36):3787–3811

Odegard GM, Gates TS, Herring HM (2005) Characterization of viscoelastic properties of polymeric materials through nanoindentation. Exp Mech 45(2):130–136

Oliver WC, Pharr GM (1992) An improved technique for determining hardness and elastic modulus using load and displacement sensing indentation experiments. J Mater Res 7(6):1564–1583

Roche S, Pavan S, Loubet JL, Barbeau P, Magny B (2003) Influence of the substrate characteristics on the scratch and indentation properties of UV-cured clearcoats. Prog Org Coat 47(1):37–48

Scherer GW (1987) Drying gels: II. Film and flat plate. J Non-Cryst Solids 89(1–2):217–238

Tanglumlert W, Prasassarakich P, Supaphol P, Wongkasemjit S (2006) Hard-coating materials for poly(methyl methacrylate) from glycidoxypropyltrimethoxysilane-modified silatrane via a sol-gel process. Surf Coat Technol 200(8):2784–2790. doi:10.1016/j.surfcoat.2004.11.018

Tiwari A, Hihara LH (2010a) High performance reaction-induced quasi-ceramic silicone conversion coating for corrosion protection of aluminium alloys. Prog Org Coat 69(1):16–25. doi:10.1016/j.porgcoat.2010.04.020

Tiwari A, Hihara LH (eds) (2010b) Novel silicone ceramer coatings for aluminum protection: high performance coatings for automotive and aerospace industries. Nova Publishers, New York

Tiwari A, Hihara LH (2012) Effect of inorganic constituent on nanomechanical and tribological properties of polymer quasi-ceramic and hybrid coatings. Surf Coat Technol 206(22):4606–4618. doi:10.1016/j.surfcoat.2012.05.020

TSui TY, Ross CA, Pharr GM (1997) Nanoindentation hardness of soft films on hard substrates: effects of substrates. Mater Res Soc Symp Proc 473:57

VanLandingham MR (2003) Review of instrumented indentation. J Res Nat Inst Stand Technol 108(4):249–265

VanLandingham MR, Villarrubia JS, Meyers GF (2000) Nanoindentation of polymers: overview. Polymer Preprints 41(2):1412–1413

Verma G, Dhoke SK, Khanna AS (2012) Polyester based-siloxane modified waterborne anticorrosive hydrophobic coating on copper. Surf Coat Technol 212:101–108

Wright WJ, Nix WD (2009) Storage and loss stiffnesses and moduli as determined by dynamic nanoindentation. J Mater Res 24(3):863–871

Zhang TH, Huan Y (2005) Nanoindentation and nanoscratch behaviors of DLC coatings on different steel substrates. Compos Sci Technol 65(9):1409–1413. doi:10.1016/j.compscitech.2004.12.011

Zhang S, Zhang X (2012) Toughness evaluation of hard coatings and thin films. Thin Solid Films 520(7):2375–2389. doi:10.1016/j.tsf.2011.09.036

Zhou B, Li L, Randall N (2013) The surface tribological and mechanical behaviors of silicone-based hydrogel materials. In: Annual conference on experimental and applied mechanics, Costa Mesa, CA, pp 149–153

Part III
Contributions from Users

Nanomechanical Properties and Deformation Mechanism in Metals, Oxides and Alloys

Elias P. Koumoulos, Dimitrios A. Dragatogiannis
and Constantinos A. Charitidis

Abstract Metals, oxides and alloys are widely used in transport and industry-engineering applications, due to their functionality. In this work, the nanomechanical properties (namely hardness and elastic modulus) and nanoscale deformation of metals, oxides and alloys (elastic and plastic deformation at certain applied loads) are investigated, together with pile-up/sink-in deformation mechanism analysis, subjected to identical condition parameters, by a combined Nanoindenter—Scanning Probe Microscope system. The study of discrete events including the onset of dislocation plasticity is recorded during the nanoindentation test (extraction of high-resolution load–displacement data). A yield-type pop-in occurs upon low applied load representing the start of phase transformation, monitored through a gradual slope change in the load–displacement curve. The ratio of surface hardness to hardness in bulk is investigated, revealing a clear higher surface hardness than bulk for magnesium alloys, whereas lower surface hardness than bulk for aluminium alloys; for metals and oxides, the behavior varied. The deviation from the case of Young's modulus being equal to reduced modulus is analyzed, for all three categories of materials, along with pile-up/sink in deformation mechanism. Evidence of indentation size effect is found and quantified for all three categories of materials.

1 Introduction

Nanoindentation provides load-depth curves for a monotonically increasing load, leading to the precise determination of different properties such as yield strength, hardness, elastic modulus, wear characteristics, etc. The usefulness of

E. P. Koumoulos · D. A. Dragatogiannis · C. A. Charitidis (✉)
School of Chemical Engineering, National Technical University of Athens, 9, Heroon
Polytechneiou St., Zografos, 15780 Athens, Greece
e-mail: charitidis@chemeng.ntua.gr

A. Tiwari (ed.), *Nanomechanical Analysis of High Performance Materials*,
Solid Mechanics and Its Applications 203, DOI: 10.1007/978-94-007-6919-9_7,
© Springer Science+Business Media Dordrecht 2014

nanoindentation to obtain the fundamental mechanical properties of materials has been widely demonstrated previously (e.g.). The nanoindentation test can provide information about the mechanical behaviour of the material when it is being deformed at the sub-micron scale. With the development of nanoindenters displacement discontinuities or discrete bursts (pop-in, pop-out and elbow phenomena) have been observed. These are characteristic of energy-absorbing or energy releasing events, occurring beneath the indenter tip.

The onset of plasticity in crystals is demonstrated by discrete bursts in the load-displacement curves, which are attributed to the initial nucleation of individual lattice dislocations (Gerberich et al. 1996). For such small indentations the crystal volume probed is typically so small as to be dislocation free. Thus indentation size effects at the nanometer scale are associated with the crystal being dislocation starved and requiring the nucleation of dislocations to initiate plasticity (Nix et al. 2007; Kelchner et al. 1998; Li et al. 2002). However, the pop-in may be associated with fracture of a surface oxide layer on some materials or may be connected with phase transformations widely reported in nanoindentation experiments of Si and ceramics (Schuh 2006; Venkataraman et al. 1992). Nevertheless, pop-in event is not a prerequisite condition for a permanent plastic deformation, since as reported during nanoindentation experiments many load curves show no excursion but did show plastic deformation. Also, plastic deformation behavior (residual impression 4 nm) was observed at indentation depth that was below the critical depth conducive to the pop-in event, which indicates that some dislocations were generated even before the pop-in event (Navamathavan et al. 2008). The first pop-in separates the region of fully elastic behavior, at lower loads from the region of elasto-plastic behavior at higher loads (Rabkin et al. 2010). When the first pop-in occurs, the maximum shear stress in the specimen is in the range of G/30–G/5 (G-shear modulus), which is very close to the theoretical strength (Ogata et al. 2004). Pop-in loads vary over a wide range, and theoretical predictions based on a stress-assisted, thermally-activated, homogeneous-dislocation-nucleation model agree well with the experimentally measured statistical values (Chiu and Ngan 2002; Schuh and Lund 2004; Bei et al. 2008; Vliet et al. 2003; Schuh et al. 2005; Mason et al. 2006).

The strength of metals, oxides and alloys is strongly influenced by grain size (Kumar et al. 2003; Gleiter 2000). Materials in the nanocrystalline regime are characterized by superior yield and fracture strength, improved wear resistance, superplasticity observed at relatively low temperatures and high strain rates as compared with their microcrystalline counterparts (Masumura et al. 1998). According to the classical Hall–Petch law the yield or flow stress required for continuous plastic deformation increased with decreasing grain size. This phenomenon which observed for conventional grain size materials (1–100 μm diameter) is reversed in nanoscale since the yield stress decreased with decreasing grain size, below a critical grain size (d $\approx$ 10 nm) for a variety of metals and alloys (Aifantis and Konstantinidis 2009). In fact, microstructures resulting in high long-life fatigue resistance will generally yield lower thresholds for fatigue crack growth, especially in ultra-fine and nanocrystalline regimes (Cavaliere 2009). Although mechanical properties of metals and oxides differ markedly in the bulk, they have an intriguingly similar response to nanoindentations. Into nanoindentation regime, oxides, under certain conditions show a ductile response such as metals (Navarro et al. 2008). The onset of plasticity of engineering materials has been systematically investigated by means of two complementary techniques: macroscopic tensile or compression tests and depth-sensing nanoindentation. The scope of this research effort is to gain insight into the deformation mechanisms involved in local plasticity during nanoindentation of metals, oxides and alloys through mechanism investigation and correlation of ideal elastic modulus, hardness and Poisson ratio using nanoindentation data.

2 Experimental Details

2.1 Materials

The materials used in this work were selected on the basis of representing each main category of materials, i.e. metals (namely Cu, Co, Al, Ni, Pb and Si), aluminum (namely AA2024, AA5083, AA6082 and AA7075) and magnesium (namely AZ31, ZK10 and ZK30) alloys and oxides (namely Al_2O_3, TiO_2, SiO_2, Co_3O_4 and NiO).

2.2 Instrumentation: Approach

Nanoindentation testing was performed with Hysitron TriboLab® Nanomechanical Test Instrument, which allows the application of loads from 1 to 30,000.00 μN and records the displacement as a function of applied loads with a high load resolution (1 nN) and a high displacement resolution (0.04 nm). The TriboLab® employed in this study is equipped with a scanning probe microscope (SPM), in which a sharp probe tip moves in a raster scan pattern across a sample surface using a three-axis piezo positioner. In all nanoindentation tests a total of 10 indents are averaged to determine the mean hardness (H) and elastic modulus (E) values for statistical purposes, with a spacing of 50 μm, in a clean area environment with 45 % humidity and 23 °C ambient temperature. In order to operate under closed loop load or displacement control, feedback control option was used. All nanoindentation measurements have been performed with the standard three-sided pyramidal Berkovich probe, with an average radius of curvature of about 100 nm (Charitidis 2010).

Based on the half-space elastic deformation theory, H and E values can be extracted from the experimental data (load displacement curves) using the Oliver-Pharr (O&P) method (Cheng et al. 2002), where derived expressions for calculating the elastic modulus from indentation experiments are based on Sneddon's elastic contact theory (Eq. 1) (Sneddon 1948).

$$E_r = \frac{S\sqrt{\pi}}{2\beta\sqrt{A_c}} \tag{1}$$

where S is the unloading stiffness [initial slope of the unloading load-displacement curve at the maximum depth of penetration (or peak load)], A is the projected contact area between the tip and the substrate and β is a constant that depends on the geometry of the indenter [$\beta = 1.167$ for Berkovich tip (Oliver and Pharr 1992)]. Conventional nanoindentation hardness refers to the mean contact pressure; this hardness, which is the contact hardness H_c is actually dependent upon the geometry of the indenter (Eqs. 2–4).

$$H_c = \frac{F}{A} \tag{2}$$

where,

$$A(h_c) = 24, 5h^2 + a_1 h + a_{1/2} h^{1/2} + \ldots + a_{1/16} h^{1/16} \tag{3}$$

and

$$h_c = h_m - \varepsilon \frac{P_m}{S_m} \tag{4}$$

where h_m is the total penetration depth of the indenter at peak load, P_m is the peak load at the indenter displacement depth h_m, and ε is an indenter geometry constant, equal to 0.75 for Berkovich indenter. Prior to indentation, the area function of the indenter tip was calibrated in a fused silica, a standard material for this purpose (Bei et al. 2005).

Plasticity is quantified based in the relation $\frac{W_{tot}-W_u}{W_{tot}}$ (where W_{tot} is the work of total indentation process and W_u is the work during unloading). It has recently established an approximately linear correlation ($\frac{H}{E^*} = k\frac{W_u}{W_{tot}}$ where k is a function of θ) between $\frac{W_{tot}-W_u}{W_{tot}}$ and $\frac{H}{E^*}$, first for a given indenter geometry (i.e. $\theta=90°$) and later for conical indenters of a range of angles $60°<\theta<80°$ (Cheng et al. 2002). From this relationship, the ratio $\frac{H}{E^*}$ can be obtained readily by integrating the loading–unloading curves to obtain $\frac{W_u}{W_{tot}}$. The ratio $\frac{H}{E^*}$ is of significant interest in tribology. This ratio multiplied by a geometric factor is the "plasticity index" that describes the deformation properties of a rough surface in contact with a smooth surface (Williams 1994). When the plasticity index is much less than unity, the deformation of asperities is likely to be entirely elastic.

3 Results and Discussion

3.1 Input Functions, Load–Displacement Curves

The relation (input function) of displacement change to time for the materials examined in this work is plotted in Fig. 1 (schematic trapezoidal load-time $P = P(t)$ input function). The loading–unloading curves of the probed materials (representative) are presented in Fig. 2 (comparison of probed materials is presented, for applied load of 5,000 μN); pure Ni exhibits higher resistance to applied

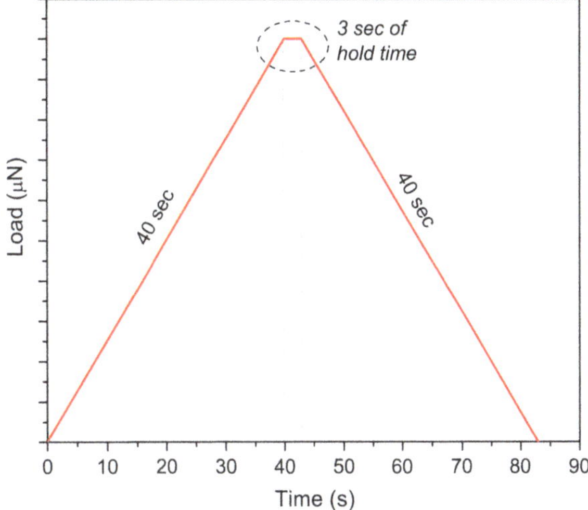

Fig. 1 Schematic trapezoidal of load-time $P = P(t)$ function for nanoindentation experiment

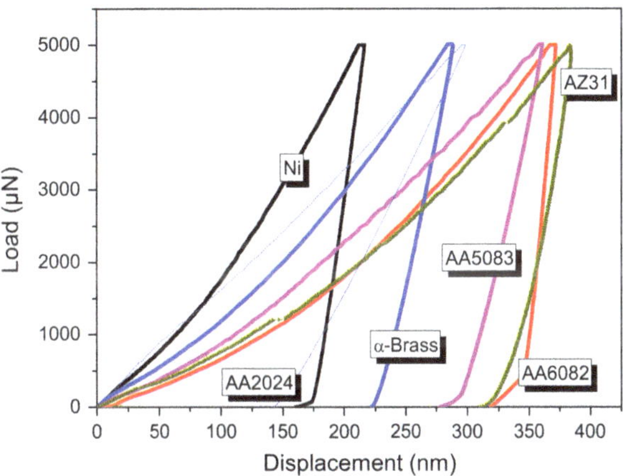

Fig. 2 Representative loading–unloading curves of metals and alloys for applied load of 5,000 μN

load (higher applied load values were needed for Ni to reach the same displacement of the rest of the materials). In the case of AA6082-T6, greater plasticity is revealed, i.e. energy stored at the material after the indentation is over (total integration of curve area), with AA2024 exhibiting higher elastic recovery (Fig. 3), in case of 5,000 μN of applied load.

In Fig. 3, the comparison of plastic deformation of metals reveals a low plasticity at low loads for the metalloid Si, implying enhancement of elasticity for weaker loads, while Cu, Al and Pb being the most plastic. This phenomenon at low loads may be attributed to the real physical effect of superelastic behavior of materials under microNewton scale forces, due to the inactivation of dislocations. It may also be an artifact due to the piling-up of the surface during indentation. Pile-up is the tendency of softer materials to overflow plastically out of the indented region

E. P. Koumoulos et al.

Fig. 3 Plasticity of (**a**) metals, (**b**) oxides and (**c**) alloys

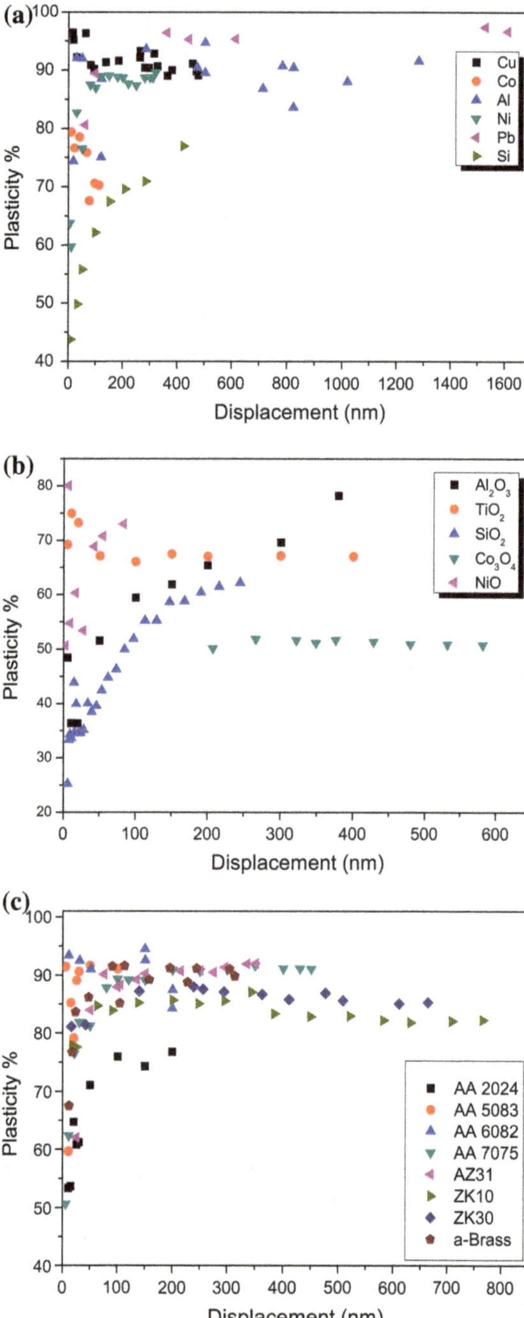

(further analysis below); this phenomenon is connected to the material, which is really soft, and also to the geometry of the tip (decreasing the corner angle of the tip increases the plastic deformation of the material and thus its pile-up) and it leads to an overestimation of H and E (underestimation of the contact area). In oxides, Co_3O_4 exhibits a more stable plastic deformation across various displacements (~50 %), while Al_2O_3 ends up (at greater displacements) with the highest plastic deformation. Brass, aluminum and magnesium alloys exhibit an almost similar behaviour in plastic deformation, with AA2024 being more elastic at greater displacements.

3.2 SPM Imaging, Nanomechanical Properties

In Fig. 4, SPM imaging (10×10 μm) of all samples is presented.

At each imposed depth E and H can be deduced from the curves. The graphs in Figs. 5, 6, 7 show the mean value of the H measurements as a function of the imposed displacement. As the indentation depth decreases below 100 nm, a rapid increase of the H value is observed. This rapid increase is probably a combination of either the real effect of a native oxide at the surface or an effect of the polishing procedure, or an artefact of the shape of the indenter tip for shallow displacements (Bei et al. 2005; Charitidis et al. 2012; Sangwal 2000). The high hydrostatic pressure exerted by the surrounding material allows plastic deformation at room temperature when conventional mechanical testing only leads to fracture. At low loads one phenomenon is very much prominent which is called Indentation Size Effect (ISE) due to imperfection in tip geometry. In several studies of materials ISE is revealed, which shows an increase in hardness with decreasing applied load. In a few cases, the hardness has been observed to decrease with decreasing indentation depth—the reverse ISE (Sangwal 2000). Apparently, the existence of ISE may hamper the accurate measurement of hardness value, and is attributed to experimental artifact, a consequence of inadequate measurement capability or presence of oxides on the surface. Other explanations include indenter-specimen

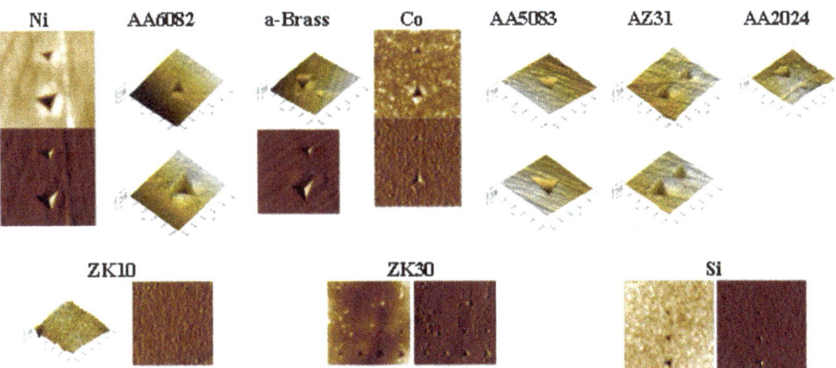

Fig. 4 Pile-up of final imprints through SPM imaging

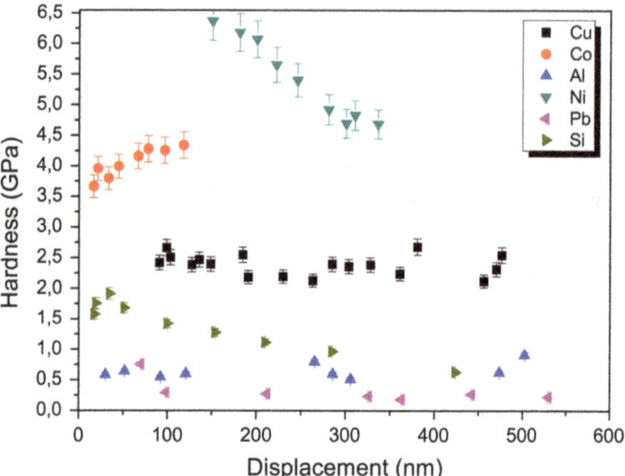

Fig. 5 Hardness of metals for various displacements

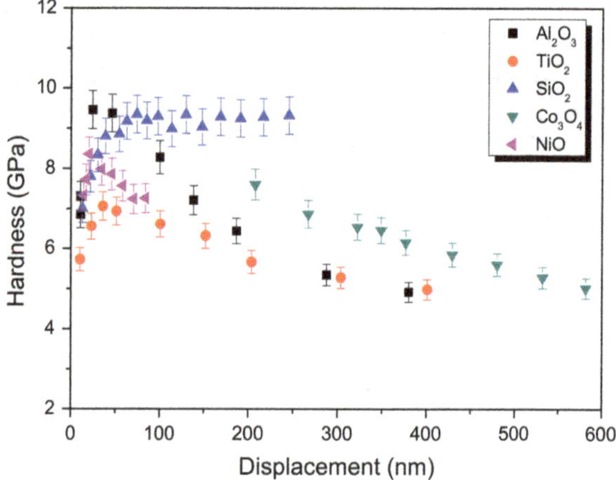

Fig. 6 Hardness of oxides for various displacements

friction, and changing dislocation density for shallow indents due to the presence, for instance, of geometrically necessary dislocations. Most of the dislocations stay generally confined around the residual imprint in a dense structure with many dislocation interactions (Charitidis et al. 2012). The reverse ISE phenomenon essentially takes place in crystals which readily undergo plastic deformation. The reverse ISE can be caused by: (1) the relative predominance of nucleation and multiplication of dislocations and (2) the relative predominance of the activity of either two sets of slip planes of a particular slip system or two slip systems below and above a particular load (Sangwal 2000).

Göken et al. proposed a method correcting pile-up effects and possible surface roughness (Göken et al. 2001). This method allows determining a correction factor for H based on the relation between the indentation modulus and Young's modulus, where Vlassak and Nix reported that E is highly dependent on the crystallographic orientation but that H is not (Vlassak and Nix 1994).

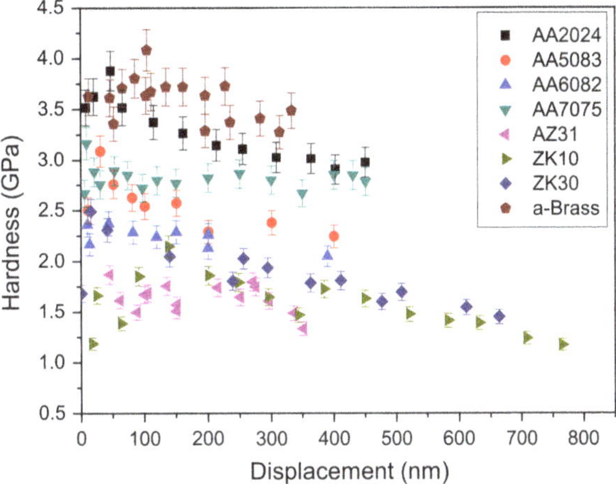

Fig. 7 Hardness of alloys for various displacements

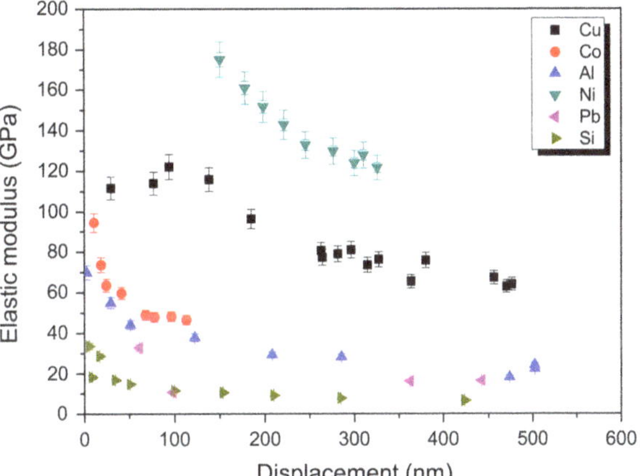

Fig. 8 Elastic modulus of metals for various displacements

Kese and Li proposed a method for accounting for the pile-up by considering the added pile-up contact area as semi-ellipses around Berkovich triangular impression (Kese and Li 2006), performed with post AFM scanning of the indented surface and measurement of the pile-up contact width for each of the three possible pile-up lobes. Lee et al. proposed a different approach by measuring the modulus of the material from early Hertzian loading analysis and using it to predict the pile-up (Lee et al. 2007). Some recent works in this field have reported study of pile-up around spherical indenters spherical-conical indenters, (Taljat and Pharr 2000; Maneiro and Rodriguez 2005) and also the effect of pile-up on thin film system measurements (Zhou et al. 2003).

In Figs. 8, 9, 10, the mean value of the elastic modulus is plotted versus the depth of the indentation. For aluminum alloys, an almost constant value for E (E_b for constant values over displacement) is obtained over the different applied test conditions, which seems reasonable. A range of values of ~50–100 GPa is

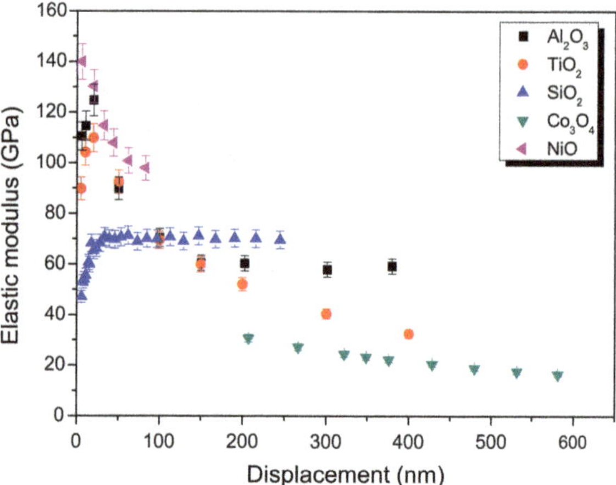

Fig. 9 Elastic modulus of oxides for various displacements

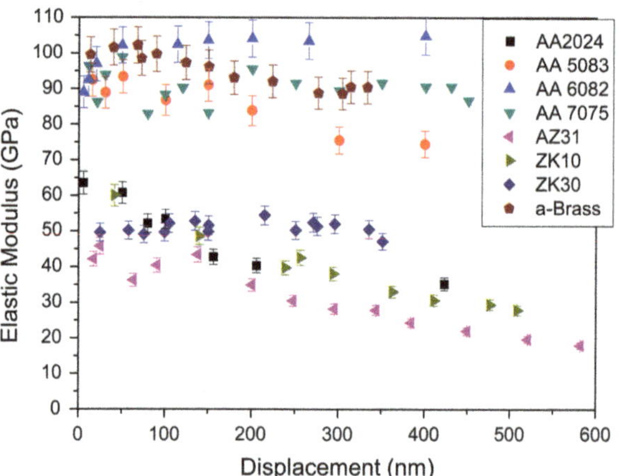

Fig. 10 Elastic modulus of alloys for various displacements

obtained which is in fairly good agreement with the reported Young's modulus in literature (Calister 1990). The same is true for brass that leads to a value of ~100 GPa, relatively close to the 110 GPa reported (Sevillano et al. 2000). The deviation from these standard values is attributed to the elastic modulus values which are calculated by using the contact area from the indentation displacement considered since the first contact point with the initial flat surface, as described by the Oliver and Pharr method (1992). However, the plastically deformed zone around the indented area can pile-up against the indenter or sink-in, depending on the material's work hardening. The consequence of this behaviour is a slight deviation from the standard values of the elastic modulus (Rodriguez and Gutierrez 2003). For brass, Vlassak and Nix determine 25 % of scatter between indentation modulus determined for different crystallographic orientations, but even higher discrepancies can be found, depending on the material (Vlassak and Nix 1994).

Fig. 11 **a** Square of the nanohardness value against the inverse of the depth. **b** Comparison of hardness at bulk and extrapolated hardness. **c** Hardness change of metals

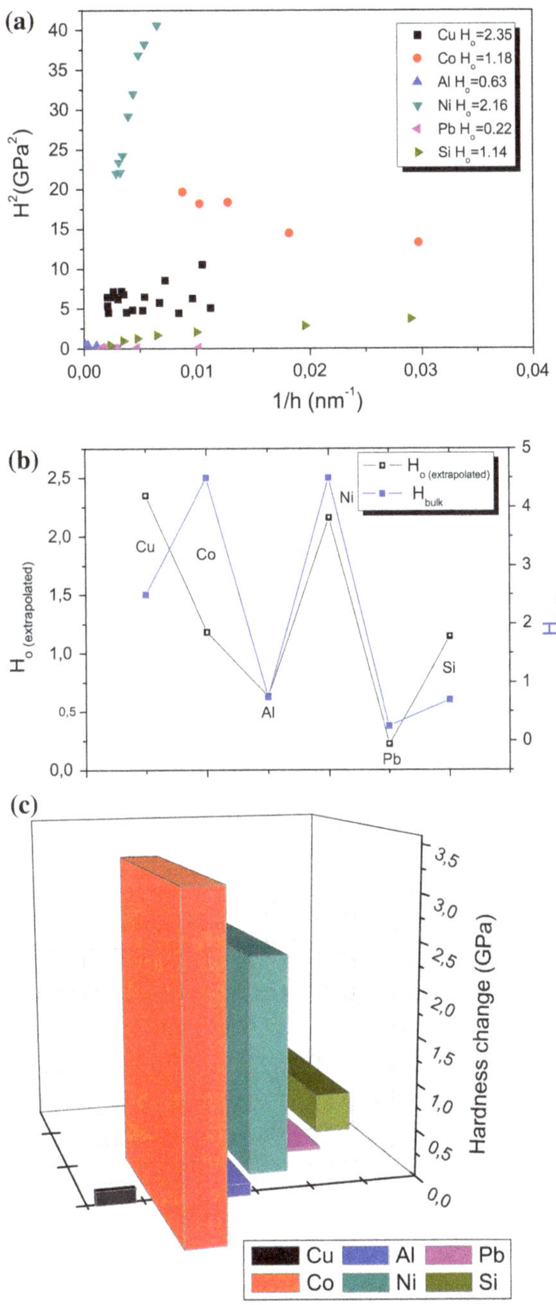

In Figs. 11, 12, 13a, the square of the nanohardness value obtained in the indentation tests is plotted as a function of the reciprocal of the indentation depth. It can be seen that a linear relation is closely followed for the most of the

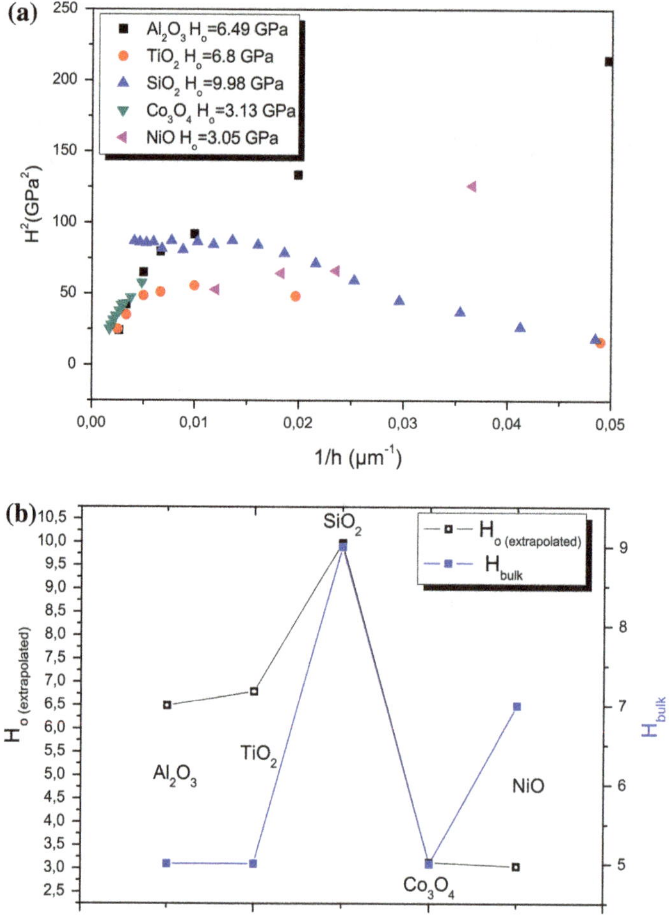

Fig. 12 **a** Square of the nanohardness value against the inverse of the depth. **b** Comparison of hardness at bulk and extrapolated hardness. **c** Hardness change of oxides

materials, in agreement with literature (Schwaiger et al. 2003). The value of the hardness at infinite depth, H_0, can be estimated by extrapolating the mentioned linear relations to $\frac{1}{h} = 0$. In the literature, however, it is reported that nanoindentation hardness data do not follow this linear trend over the whole measurement range (Lim and Chaudhri 1999; Swadener et al. 2002), an evident phenomenon observed also in this work. Instead, at small indentation depths they start to deviate from the predicted linear curve. The obtained values for hardness at infinite depth are shown in the table embedded in Fig. 13a. The comparison of H_0 and mean H at surface region (0–400 nm) (H_s) is presented in Figs. 11, 12, 13b–c, where hardness at infinite displacement clearly matches with H_s for AA6082-T6, brass and AA5083-H111. However, comparison of both hardness values of e.g. pure Ni and AA2024 exhibit great deviation (reduced plastic deformation in higher applied loads, dominated by sink-in), revealing that in order to reach constant

Fig. 13 **a** Square of the
nanohardness value against
the inverse of the depth. **b**
Comparison of hardness
at bulk and extrapolated
hardness. **c** Hardness change
of alloys

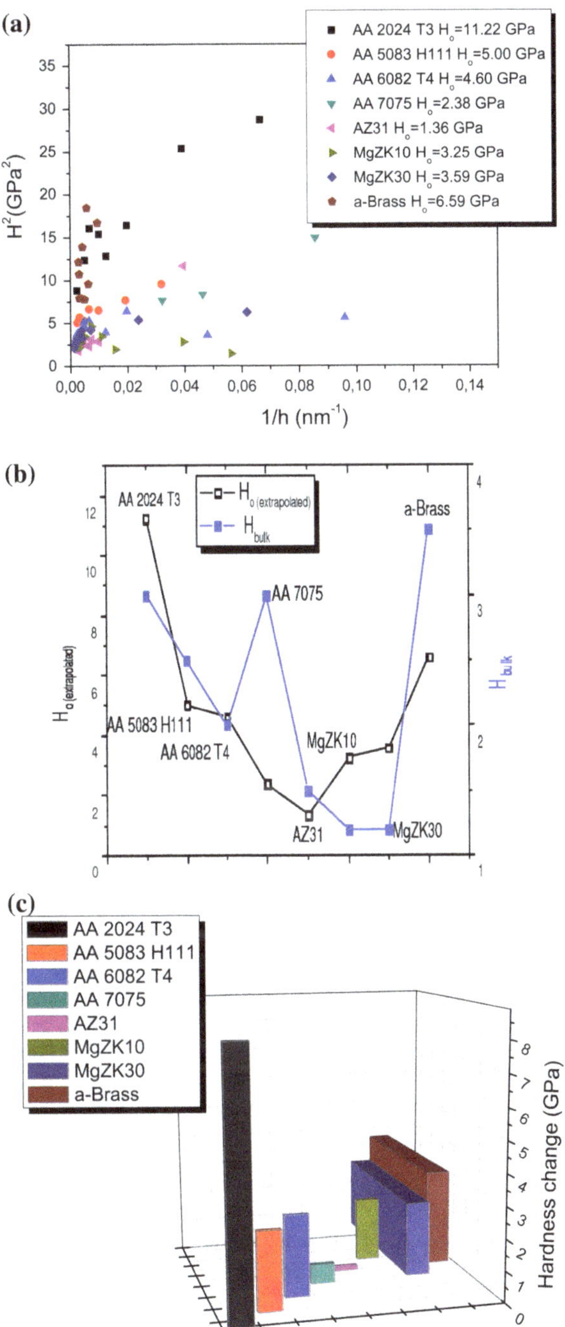

nanomechanical properties (of bulk material), indenting in greater displacement is
needed.

Although mechanical properties of metals, alloys and oxides differ markedly
in the bulk, they have an intriguingly similar response to nanoindentations. When

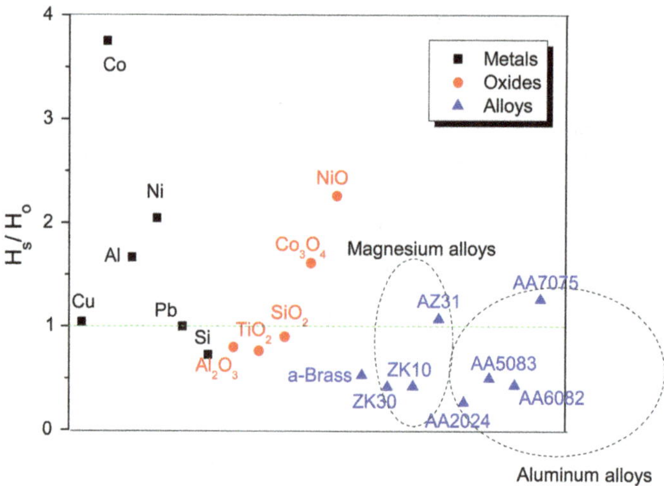

Fig. 14 H_{bulk}/H_0 ratio for all categories

scaling down into the nanoindentation regime, e.g. oxides, which in the bulk are known to be brittle, under certain conditions show a ductile response in which the nucleation of cracks is averted (Rhee et al. 2001). Some of these surfaces can even transit between as the external load increases, so does the average pressure. Once the resolved shear stress attains a certain threshold value the crystal yields and dislocation loops are generated. Then, the pressure is partially released and a low-energy-barrier dislocation loop source begins to operate. Comparing metals and oxides, we remark that in the nanoindentation region, the ratio between the respective hardnesses $\frac{H_{oxide}}{H_{metal}}$ varies (Navarro et al. 2008) (Fig. 14).

We turn now to the different behaviours of the three categories of materials using the ratio q between their hardness's in the surface region and in the bulk (i.e. extrapolated hardness at infinite displacement), i.e. $q = \frac{H_s}{H_0}$. During the nanoindentation process of metals, alloys and oxides surface the effective volume that is probed is small enough to avoid encompassing any pre-existing dislocations. Under these conditions, following the nucleation of the first yield point, the hardness in the region of constant low hardness corresponds to the operation of a low-barrier dislocation source. In the case of $q > 1$, as the tip excursion progresses into the bulk, pre-existing dislocations able to glide start being activated. The hardness is now controlled, Taylor-type, by the work hardening related to the dislocation density. As is well known, this results in a rather low value of H_0, smaller than the one corresponding to the operation of the loop source, and one ends up with a value of $q \gg 1$. On the other hand, the case of $q < 1$ is probably attributed to the fact that in the bulk, pre-existing dislocations are pinned and stresses cannot make them glide. The reason why low-barrier dislocation sources, similar to the ones proposed to control hardness in the low effective volume limit, do not seem to operate in the bulk is not yet clear. The following explanation may exist; dislocations introduced by the source can interact with pre-existing pinned dislocations -or even among themselves-resulting in very immobile locks, which further impede the motion of subsequently nucleated dislocations. If the effective volume is small, no pre-existing dislocations exist in it and the locking does not take place (Navarro et al. 2008).

3.3 Pop ins-Power Law

Figure 15 shows that the loading–unloading curves for all samples, exhibit interesting local discontinuities measured in the load-controlled test of this work; these are characteristic of energy-absorbing or energy-releasing events occurring beneath the indenter tip. Three different physical phenomena usually occur in nanoindentation testing of metals of various states of bonding and structural order; dislocation activity during a shallow indentation, shear localization into 'shear bands', and phase transformation with a significant volume increase during unloading of indentation (Schuh 2006).

Many materials undergo phase transformations when subjected to large hydrostatic stresses, and the pressure beneath a nanoindenter is generally quite high (on the order of several GPas) (Schuh 2006). Association of pop-in events with the beginning of material phase transformation is simple a revelation of sudden extrusion of highly plastic transformed material from underneath the indenter. The sudden displacement discontinuities, i.e. the pop-ins, were observed in the loading part. The first pop-in (referred as yield type pop-in in literature) implies that strain is accommodated by an abrupt existence of atomic activity beneath the indenter, that could be attributed to activation of a dislocation source (Schuh 2006).

Small displacements are equivalent to a linear-elastic response; the curvature of the free energy diagram at an equilibrium position gives rise to a particular elastic constant, represented by the initial slope in a force distance diagram. Larger displacements elicit a non-linear response. However, one property of particular interest is the theoretical shear strength of the material: the limit where a further increase in displacements elicits no further increase in the restoring force. The general relationship between the applied load (P) and the penetration depth (h) of an indenter may be described by power law as $P = ah^b$ where the constants a and b are geometric and material parameters, respectively (Charitidis 2010).

Load-unload curves often reveal discontinuities or pop-ins in the loading part. Various examples are reported in the literature for metals, although a complete explanation of this behaviour has been under investigation (Schuh et al. 2005). While yield points reported for Si and some other ceramics may be connected to phase transformations [when the mean contact pressure of hardness indentations closely matches the critical pressure a possible structural transformation is triggered (Schuh 2006; Johnson 1970)], the exhibited yield points in most of the known metals clearly reveal the beginning of dislocation plasticity (the plastic deformation of metals occurs by the motion of dislocations). Evaluations of the maximum stress under the tip reveal that stress values (almost equal to theoretical shear stress) occur in the surface of the metal (dislocation activity starts first at

Fig. 15 Hertzian elastic fit to representative indentations on metals and alloys

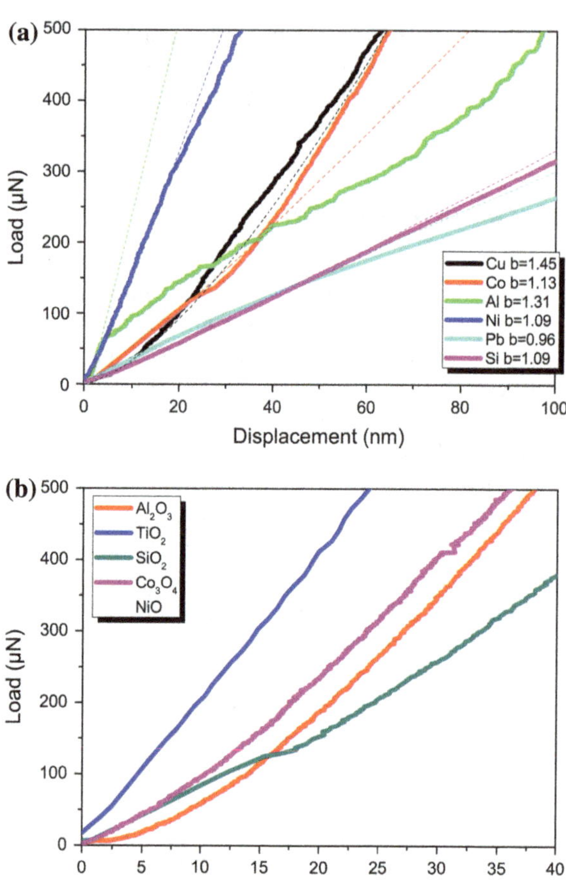

the yield type pop-in load), as generation of dislocations in the nano-scale stressed volume from the perfect crystal environment (Wo and Ngan 2005).

As shown in Fig. 15, the transition from purely elastic to elastic/plastic deformation i.e. gradual slope change (yield-type 'pop-in') of all materials occurs in the load–displacement curves, at approximately 10–30 nm. The onset-stress for plastic deformation of fcc single crystals of pure metals is very low and their flow stress exhibits an extraordinary high hardening capacity. The dislocation structure developed in a single crystal depends significantly on the applied strain and the path the straining is accomplished. It usually starts by the formation of micro and macro slip bands and proceeds by the generation of cells and cell block structures. Further deforming, keeps the process of fragmentation on. Finally, at very large strains a saturation structure with a minimum crystallite size is reached. For copper single crystals deformed at room temperature, this crystallite size is in the order of a few 100 nm.

The indenter displacement in most of the cases is accommodated plastically, and only a small portion is elastically recovered on unloading. Discontinuity in load

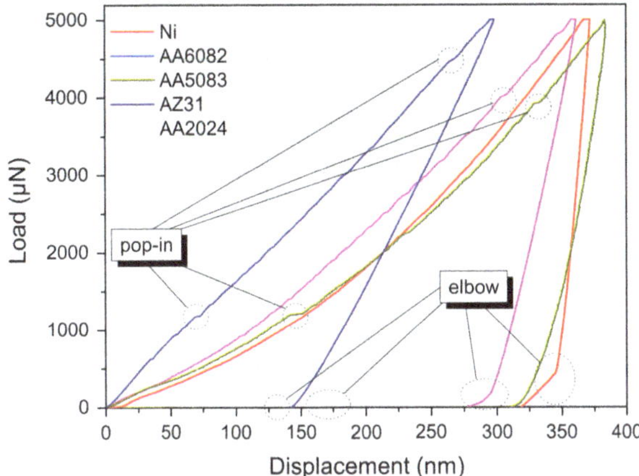

Fig. 16 Pop-ins and elbows indicated in nanoindentation loading–unloading curves

displacement, which is commonly referred as 'pop-in' effect (Fig. 16), was observed frequently in aluminum, indicating a process of producing mobile dislocations. The initial pop-in is usually associated with homogeneous dislocation nucleation, while subsequent similar events often involve avalanches of dislocation activity (Wo and Ngan 2005). In addition to defects, there may be residual stresses in the surface which influence the occurrence of the initial discontinuities in the load–displacement curve (Charitidis et al. 2012). Additionally, roughness such as surface steps could act as stress concentrators or alternatively exhibit a long-range effect of the order of several contact radius (Zimmerman et al. 2001). Gogotsi et al. (2000), Domnich and Gogotsi (2002) and Juliano et al. (2003) proposed that the "pop-out" behavior corresponds to the formation of metastable Si-XII/Si-III crystalline phases, in case of silicon nanoindentation. High stresses can cause plastic deformation not only by dislocation activity, but also by pressure-induced phase transformations to denser crystalline and amorphous forms (Ge et al. 2004). The transformation mechanisms are dependent on the indentation testing conditions e.g. peak load and loading/unloading rate or indenter angle (Jang et al. 2005). This results in a change in the unloading curve either as a "pop-out" or elbow phenomenon (Fig. 16), which indicates a lower contact depth, h_c, and therefore may influence the calculation of hardness. Thus, the experimental measurement of the hardness of indented materials is slightly higher than that of the ideal value.

3.4 Pile-up/Sink-in Deformation

The contact area is influenced by the formation of pile-ups and sink-ins during the indentation process. To accurately measure the indentation contact area, pile-ups/ sinks-ins should be appropriately accounted for. The presence of creep (time/rate dependent property of materials) during nanoindentation has an effect on pile-up, which results in incorrect measurement of the material properties. Fischer-Cripps observed this behaviour in aluminum where the measured elastic modulus was much less than expected (Fischer-Cripps 2004). Rar et al. observed that the same material when allowed to creep for a long duration produced a higher value of pile-up/sink-in indicating a switch from an initial elastic sink-into a plastic pile-up (Rar et al. 2005). Significant pile-up forms for materials start from an $\frac{h_c}{h_m}$ value of 0.7–0.88 (Khan et al. 2010).

In Fig. 17a–c, the normalized pile-up/sink-in height $\frac{h_c}{h_m}$ is plotted versus the normalized hardness $\frac{H}{E}$ for all samples. It is reported that materials with high $\frac{H}{E}$, i.e. hard materials, undergo sink-in whereas materials pile-up for low $\frac{H}{E}$, i.e. soft materials. In general it is also observed that in the case when $\frac{H}{E}$ is high (hard materials), materials undergo sink-ins regardless of work hardening and strain rate sensitivity and all materials collapse to a single curve. In addition, for materials with low $\frac{H}{E}$, soft materials, pile-up depends on the degree of work hardening (Charitidis et al. 2012). Softer materials, i.e., low $\frac{H}{E}$, possess a plastic zone, which is hemispherical in shape and meet the surface well outside the radius of the circle of contact and pile-up is expected. On the other hand, for materials with high values of $\frac{H}{E}$, i.e. harder materials, the plastic zone is contained within the boundary of the circle of contact and the elastic deformations that accommodate the volume of indentation are spread at a greater distance from the indenter. Higher stresses are expected in high $\frac{H}{E}$, hard materials, and high stress concentrations develop towards the indenter tip, whereas in case of low $\frac{H}{E}$, soft materials, the stresses are lower and are distributed more evenly across the cross-section of the material (Charitidis et al. 2012). Rate sensitive materials experience less pile-up compared to rate insensitive materials due strain hardening. Cheng and Cheng reported a 22 % pile-up for a work hardening exponent (Cheng and Cheng 1998). This is consistent with the fact that when $\frac{h_c}{h_m}$ approaches 1 for small $\frac{H}{E}$, deformation is intimately dominated by pile-up (Hill et al. 1989). On the other hand when $\frac{h_c}{h_m}$ approaches 0 for large $\frac{H}{E}$ it corresponds to purely elastic deformation and is apparently dominated by sink-in in a manner prescribed by Hertzian contact mechanics (Hertz 1986). The dependence on the elastic–plastic behaviour of the material is related to the response of the material being indented (Hertz 1986). The degree of sink-in or pile-up of the materials is reported to be expressed as a function of the work hardening exponent (Norbury and Samuel 1928). During nanoindentation, materials with a low work hardening exponent accommodate the volume of material ejecting it to the sides of the tip (pile-up). In the same way, in materials with a high value of n ($n > 0.3$) the sink-in effect is revealed. In both cases, the contact area is different from the cross-sectional area estimated by the method described by Oliver and Pharr (1992). Consequently, there is a deviation between the real and the computed area that is controlled by the elastic–plastic behavior.

Fig. 17 Normalised pile-up/sink-in height h_c/h_m of metals, oxides and alloys versus H/E ratio

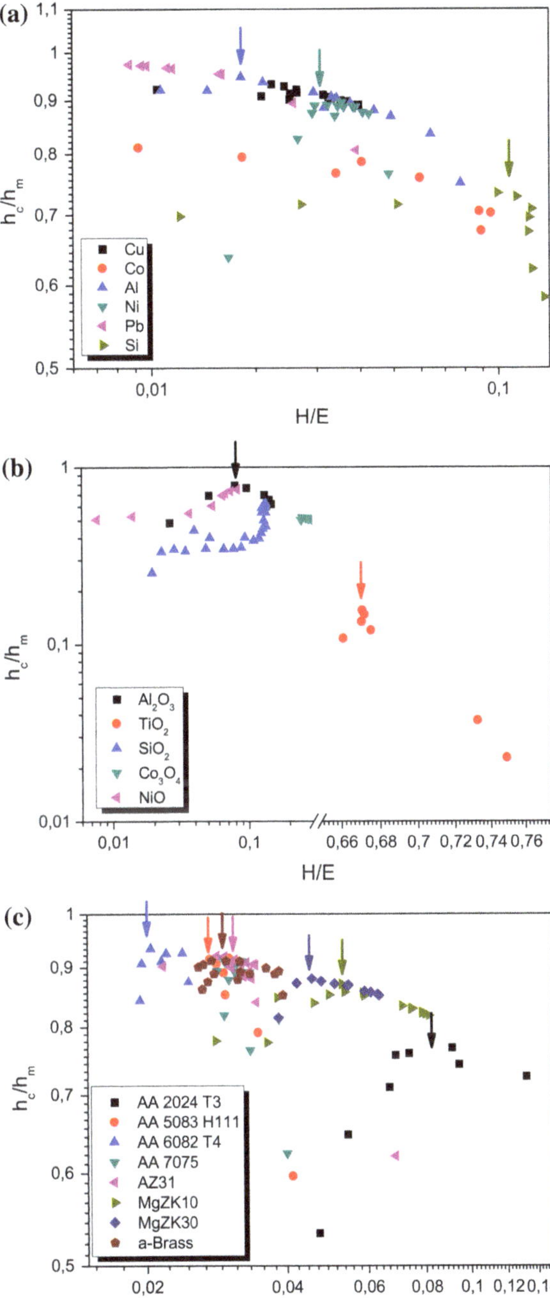

3.5 Poisson Ratio and Modulus Correlation

Elastic properties of materials are usually characterized by Young's modulus, shear modulus, bulk modulus and Poisson's ratio. For isotropic materials only two of these elastic constants are independent and other constants are calculated by using the relations given by the theory of elasticity. The best choice for these two independent moduli is considered as the shear modulus (G) and the bulk modulus (B) (Ledbetter 1977). The shear modulus relates to strain response of a body to shear or torsional stress. It involves change of shape without change of volume. On the other hand, the B describes the strain response of a body to hydrostatic stress involving change in volume without change of shape. However, the best-known elastic constant is E, which is most commonly used in engineering design (Phani and Sanyal 2008).

Materials with different Poisson's ratios behave very differently mechanically. Properties range from 'rubbery' to 'dilatational', between which are 'stiff' materials like metals and minerals, 'compliant' materials like polymers and 'spongy' materials like foams. The physical significance of v is revealed by various interrelations between theoretical elastic properties (90). At different temperatures and pressures, crystalline materials can undergo phase transitions and, attracting considerable debate, so too can glasses and liquids Bridgman 1949; Greaves et al. 2008; Poole 1997. For ceramics, glasses and semiconductors (Zhang et al. 1985; Perottoni and Jornada 2002), $\frac{B}{G} \approx \frac{5}{3}$ and $v \to \frac{1}{4}$. Likewise, metals are stiff (Cottrell 1990; Kelly et al. 1967), $\frac{B}{G}$ ranging from 1.7 to 5.6 and v from 0.25 to 0.42 (Kelly et al. 1967). In sharp contrast, polymers are compliant and yet they share similar values (Lakes and Wineman 2006; Lu et al. 1997): that is, $\frac{B}{G} \approx \frac{8}{3}$ and $v \approx 0.33$, the difference relating to the magnitude of the elastic module, decades smaller than for inorganic materials. Poisson's ratio can be followed through abrupt changes in mechanical properties. For example, when metals melt, v increases from ~0.3 to 0.5 (Santamaria et al. 2009; Jensen et al. 2010). During the collapse of microporous crystals, v rises from directionally auxetic values to isotropic values of 0.2 typical of many glasses (Valle et al. 2008; Grima et al. 2000). With densification Poisson's ratio for glasses continues to rise, for silica increasing from 0.19 to 0.33 (Zha et al. 1994). Poisson's ratio is intimately connected with the way structural elements are packed. For gold or platinum-based bulk metallic glasses, for example, which represent some of the densest metals because of the variety of atom sizes, $v \to \frac{1}{2}$. Crystalline metals are less densely packed, typified by hard metals like steel for which $v \approx \frac{1}{3}$. By contrast, the density of covalent solids is less and so is Poisson's ratio (Greaves et al. 2011).

Through contact analysis, Eq. 5 is derived:

$$\frac{1}{E^*} = \frac{1 - v_i^2}{E_i} + \frac{1 - v_s^2}{E_s} \tag{5}$$

where E^* is the reduced modulus, E_i is the Young's modulus of the indenter, v_i is the Poisson ratio of the indenter, E_s is the Young's modulus of the sample and v_s is the Poisson ratio of the sample.

Fig. 18 Deviation from $E^* = E_s$ (*red dashed line*) for (**a**) metals, (**b**) oxides and (**c**) alloys through nanoindentation

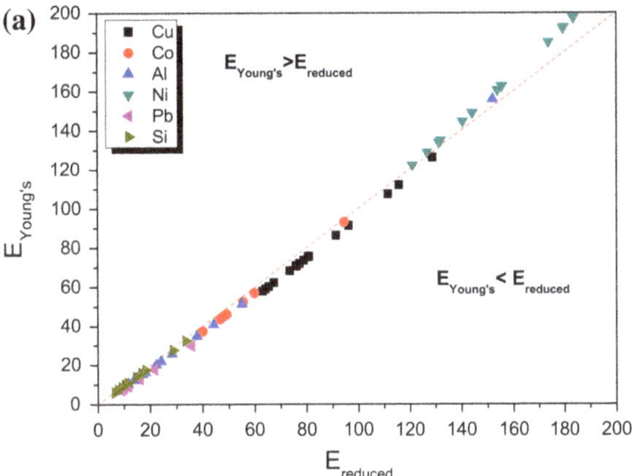

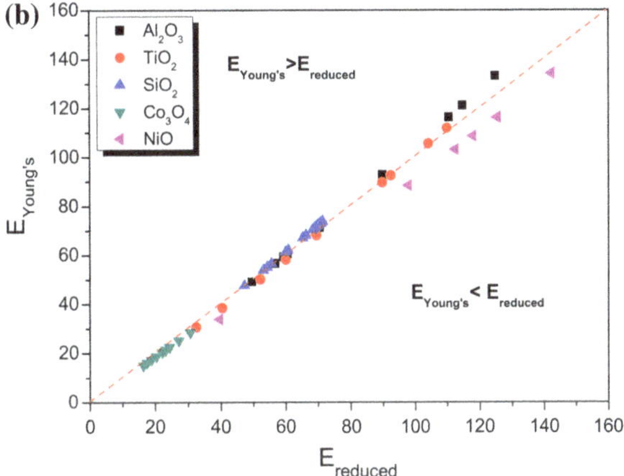

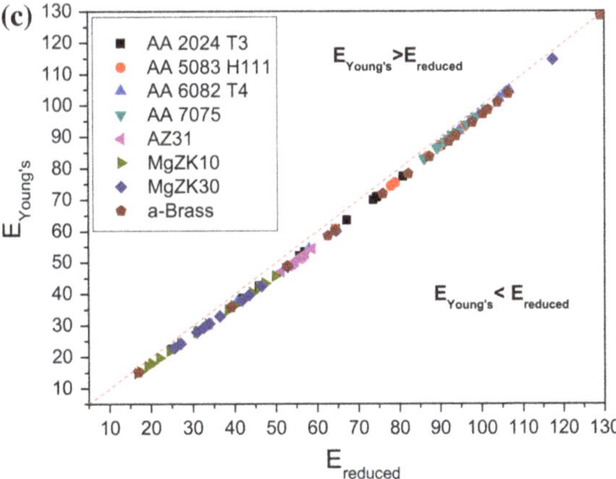

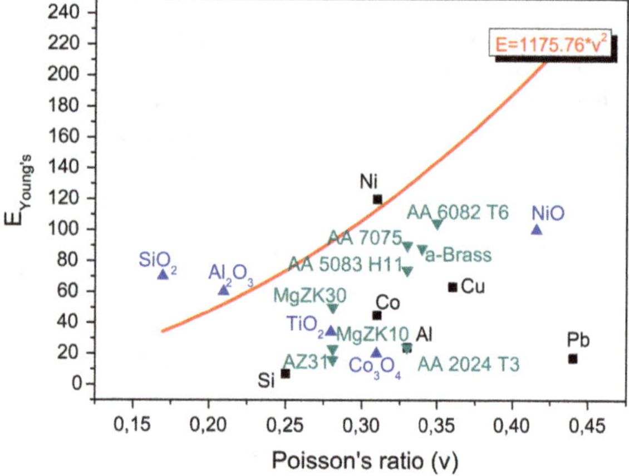

Fig. 19 Young's modulus correlation with Poisson ratio, following obtained values

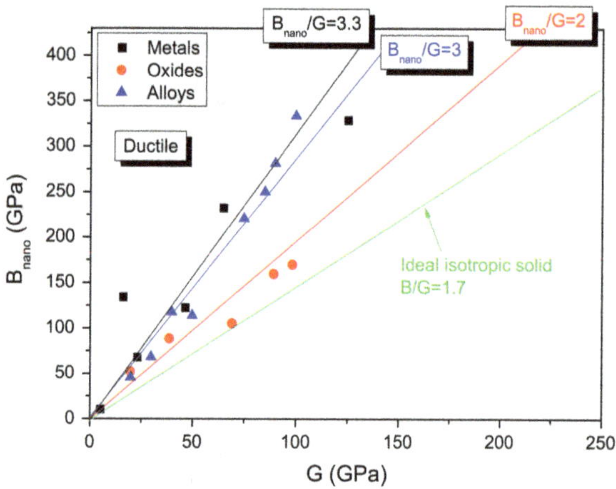

Fig. 20 Plot of bulk modulus versus shear modulus for metals, oxides and alloys through nanoindentation

Analytically, when the so-called reduced modulus is equal to Young's modulus ($E^* = E_s$), (Fig. 18) is assumed, then using Eqs. 5 and 6 are derived (for $E_i = 1140$ GPa and $v_i = 0.07$ of diamond indenter): (Fig. 19)

$$E_s = \frac{v_s^2}{1 - v_i^2} E_i = 1175.76\, v_s^2 \tag{6}$$

Figure 20 shows the variation of bulk moduli of three categories of samples with Young's moduli. Bulk modulus values (B_{nano} in Table 1) were estimated from nanoindentation Young's moduli values using Eq. (7) (Barrett et al. 1973):

$$E = 3^* B_{nano}(1 - 2v) \tag{7}$$

where E is Young's modulus, B_{nano} is bulk modulus, and v is Poisson's ratio.

Table 1 Mechanical properties and Poisson ratio of studied materials

		E_b (GPa)	H_s (GPa)	G (GPa)	E_b/G	H_s/E_b	H_o (GPa)	q (H_s/H_o)	v	B_{nano} (GPa)
Metals	Cu	65	2.5	44.7	1.45414	0.03846	2.4	1.04167	0.36	232
	Co	46.5	4.5	75	0.62	0.09677	1.2	3.75	0.31	122
	Al	23.06	1	26	0.88692	0.04337	0.6	1.66667	0.33	67
	Ni	125	4.5	76	1.64474	0.036	2.2	2.04545	0.31	328
	Pb	16.07	0.2	13.1	1.22672	0.01245	0.2	1	0.44	133
	Si	5.125	0.8	50.9	0.10069	0.1561	1.1	0.72727	0.25	10
Alloys	a-Brass	90	3.5	40	2.25	0.03889	6.6	0.5303	0.34	281
	ZK30	50	1.5	17	2.94118	0.03	3.6	0.41667	0.28	113
	ZK10	30	1.4	17	1.76471	0.04667	3.3	0.42424	0.28	68
	AZ31	20	1.5	17	1.17647	0.075	1.4	1.07143	0.28	45
	AA2024	40	3	27	1.48148	0.075	11.2	0.26786	0.33	117
	AA5083	75	2.5	28	2.67857	0.03333	5	0.5	0.33	220
	AA6082	100	2	27	3.7037	0.02	4.6	0.43478	0.35	333
	AA7075	85	3	27	3.14815	0.03529	2.38	1.2605	0.33	250
Metal oxides	Al_2O_3	89.37	5.2	18	4.965	0.05819	6.5	0.8	0.22	159
	TiO_2	38.87	5.2	90	0.43189	0.13378	6.8	0.76471	0.28	88
	SiO_2	69.41	9	31	2.23903	0.12966	10	0.9	0.17	105
	Co_3O_4	19.83	5	83.71	0.23689	0.25214	3.1	1.6129	0.31	52
	NiO	98.46	7	–	–	0.07109	3.1	2.25806	0.21	169

Ductility and brittleness relate to the extreme response of materials strained outside their elastic limits, so any relationship with v would seem at first non-intuitive. However, Poisson's ratio measures the resistance of a material to volume change (B—Bulk modulus) balanced against the resistance to shape change (G—shear modulus). Occurring within the elastic regime, any links with properties beyond the yield point must necessarily involve the time-dependent processes of densification and/or flow, already discussed for glasses above. Just as viscoelastic behaviour is expressed in terms of time-dependent bulk and shear modulus, with Poisson's ratio $v(t)$ gradually changing between elastic values, we might expect the starting value of v to provide a metric for anticipating mechanical changes, not just in glasses but also in crystalline materials, resulting in ductility (starting from a high v) or embrittlement (starting from a low v). At either extreme the microstructure will play a part, whether through cracks, dislocations, shear bands, impurities, inclusions or other means (Xi et al. 2005). Although there is no simple link between interatomic potentials and mechanical toughness in polycrystalline materials, Poisson's ratio v has proved valuable for many years as a criterion for the brittle-ductile transition exhibited by metals (Cottrell 1990; Jiang et al. 2010). The old proposal that grains in polycrystalline materials might be cemented together by a thin layer of amorphous material "analogous to the condition of a greatly under cooled liquid" has often been challenged (Rosenhain et al. 1913). However, recent atomistic simulations of crystalline grains and grain boundaries seem to confirm the dynamic consequences of this idea in many details (Zhang et al. 2009). The strengths of pure metals at low temperatures are known to be mainly governed by the strengths of grain boundaries (Cottrell 1990). These usually exceed the crystalline cleavage strength, which is governed by the dynamics of dislocations generated at the crack tip (Kelly et al. 1967). This would suggest that soft metals like gold, silver or copper might be ductile because they originate from melts that are fragile.

Conversely, hard metals such as tungsten, iridium or chromium might be brittle because, as melts, they are stronger. If this were the case then there would be consequences for the density fluctuations frozen into the grain boundaries, which will be weaker for soft metals than for brittle metals. By the same token, grain boundaries would be more ergodic in soft compared with brittle metals.

3.6 Indentation Size Effect and Onset of Plasticity

Due to the very low contact area between the indenter and the sample, very high stresses can be developed. The high hydrostatic pressure exerted by the surrounding material allows plastic deformation at room temperature when conventional mechanical testing only leads to fracture. It is revealed that some materials exhibit ISE, which shows an increase in hardness with decreasing applied load (Samuels 1989). Apparently, the existence of ISE may hamper the accurate measurement of hardness value, and is attributed to experimental artifact, a consequence of inadequate measurement capability or presence of oxides on the surface (Li et al. 1993). Other explanations include indenter-specimen friction (Li et al. 1993), and changing dislocation density for shallow indents due to the presence, for instance, of geometrically necessary dislocations (Gaillard et al. 2006).

The Berkovich indenter generates dislocations organized in a quite complex way during a nanoindentation test, even for very low deformations (Leipner et al. 2001), making difficult the formulation for the stress field generated, even during an elastic deformation, as well as its modelling. Most of the dislocations stay generally confined around the residual imprint in a dense structure with many dislocation interactions (Tromas and Gaillard 2004).

Firstly, we used the empirical equation for describing the ISE in the Meyer's law (Kolemen 2006; Sahin et al. 2007), which uses a correlation technique between the applied indentation test load and the resultant indentation size using a simple power law, $P_{max} = Ch_c^n$, where C and n are constants derived directly from curve fitting of the experimental data. In particular, the constant C is a measure of materials resistant to plastic deformation and the exponent n, sometimes referred to as the Meyer index, is usually considered as a measure of ISE. Compared to the definition of the apparent hardness, no ISE would be observed for $n = 2$ (Peng et al. 2004).

The nanoindentation data for the material examined in the present study was plotted in Fig. 21. The data showed power law relationship, implying that the traditional Meyer's law was suitable for describing the nanoindentation data. The calculated n values pointed out higher apparent nanohardness values at lower loads, in other words, the presence of an ISE. In the bulk, hardness in the plastic stage, measured by micro indentations involving large effective volumes, is commonly interpreted in

Fig. 21 Nanoindentation applied load plotted contact depth for (**a**) metals, (**b**) oxides and (**c**) alloys

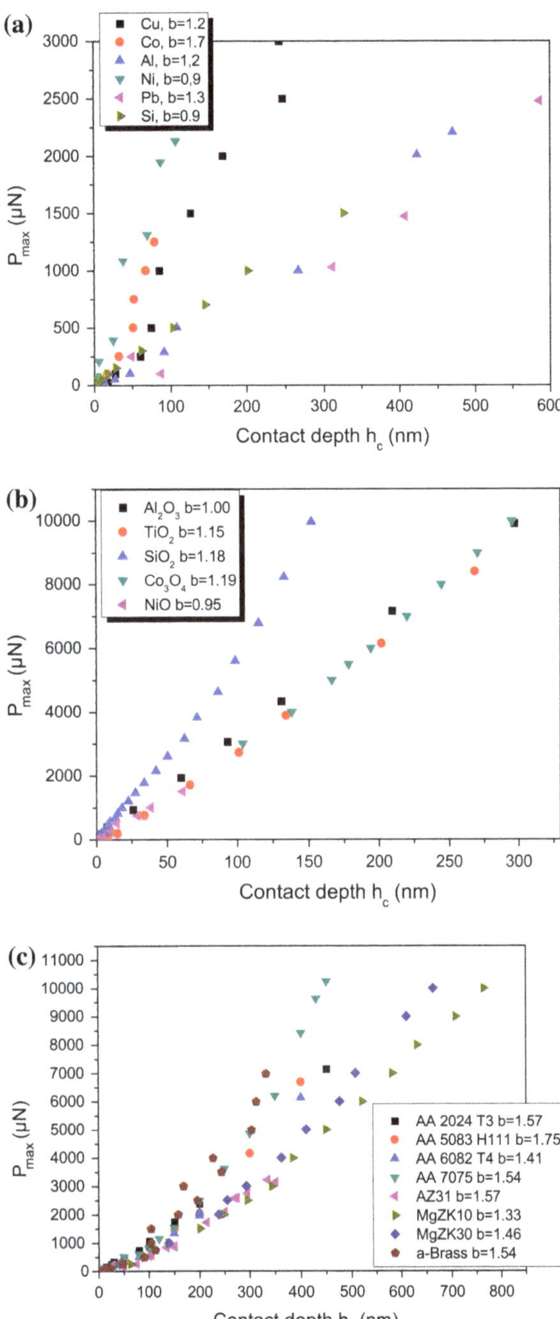

terms of a Nix and Gao mechanism (Nix and Gao 1998). This mechanism, based on the pioneering ideas of Taylor (Cottrell 1953), explains the value of hardness in terms of the work hardening of the material due to a mixture of pre-existing and geometrically necessary dislocations (GNDs). Often, this leads to the so-called indentation size effect, by which hardness increases at low penetration depths. Evidence of this ISE has been found both in oxides and metals (Peng et al. 2004). In the case of metals, even probing deep in the bulk (no ISE), the values of H are larger than the macroscopic yield strength of bulk as hardness in the plastic region is controlled by work hardening. In the case of oxides, values of yield strength of bulk are not easily found as many micro indentation studies do not intend to resolve either the elastic region or the yield point, and directly probe the plastic region. Nevertheless, some of the existing values show that H values are on the order of yield strength of bulk.

4 Concluding Remarks

Although Hertzian elastic contact theory is commonly used for the evaluation of initial load–displacement curves, it may not be an adequate model for all materials. Pure Ni exhibits higher resistance to applied load (higher applied load values were needed for Ni to reach the same displacement of the rest of the materials). All examined materials, exhibited interesting local discontinuities measured in the load-controlled test, which are characteristic of energy-absorbing or energy-releasing events occurring beneath the indenter tip. In the load–displacement curves, brass exhibited earlier transition from purely elastic to elastic/plastic deformation i.e. gradual slope change (yield-type 'pop-in'), followed by pure Ni (AZ31 and aluminum alloys exhibited late transition). Aluminum alloys (AA5083 and AA6082) exhibited significant elbow effect in unloading part. Hardness at ~400 nm displacement and extrapolated hardness show almost same behaviour for examined metals and alloys (in agreement with similar studies in literature). Comparison of both hardness values of pure Ni and AA2024 exhibit great deviation (reduced plastic deformation in higher applied loads, dominated by sink-in), revealing that in order to reach constant nano mechanical properties (of bulk material), indenting in greater displacement is needed. Additionally it has to be considered, that in case the imprint size is significantly larger than the dimension of the deformation-controlling microstructure, the hardness should be independent of the imprint size. The ratio of surface hardness to hardness in bulk was investigated, revealing a clear higher surface hardness than bulk for magnesium alloys, whereas lower surface hardness than bulk for aluminum alloys; for metals and oxides, the behaviour varied. Furthermore, the deviation from the case of Young's modulus being equal to reduced modulus was also studied, for all three categories of materials, along with pile-up/sink in deformation mechanism. Evidence of Indentation Size Effect (ISE) has been found both in oxides and metals. In the case of metals, even probing deep in the bulk (no ISE), the values of H are larger than the macroscopic yield strength of bulk as hardness in the plastic region is controlled by work hardening. In the case of oxides, values of yield strength of bulk are not easily found as many microindentation studies do not intend to resolve either the elastic region or the yield point, and directly probe the plastic region.

Acknowledgments This work was partially supported by the EU FP7 Project "Micro and Nanocrystalline Functionally Graded Materials for Transport Applications" (MATRANS) under Grant Agreement no. 228869 and partially supported by NTUA funded project for basic research PEVE-NTUA-2010/65187900.

References

Aifantis KE, Konstantinidis AA (2009) Hall-Petch revisited at the nanoscale. Mater Sci Eng, B 163:139–144

Barrett CR, Nix WD, Tetelman AS (1973) The principles of engineering materials. Printice-Hall Incorporation, New Jersey

Bei H, George EP, Hay JL, Pharr GM (2005) Influence of indenter tip geometry on elastic deformation during nanoindentation. Phys Rev Lett 95(045501):1–4

Bei H, Gao YF, Shim S, George EP, Pharr GM (2008) Strength differences arising from homogeneous versus heterogeneous dislocation nucleation. Phys Rev B 77(6060103):1–4

Bridgman PW (1949) The physics of high pressure. Bell, London

Callister WD (1990) Materials science and engineering. Wiley, New York

Cavaliere P (2009) Fatigue properties and crack behavior of ultra-fine and nanocrystalline pure metals. Int J Fatigue 31(10):1476–1489

Charitidis CA (2010) Nanomechanical and nanotribological properties of carbon-based thin films: A review. Int J Refract Metal Hard Mater 28(1):51–70

Charitidis CA, Dragatogiannis DA, Koumoulos EP, Kartsonakis IA (2012) Residual stress and deformation mechanism of friction stir welded aluminum alloys by nanoindentation. Mater Sci Eng, A 540:226–234

Cheng YT, Cheng CM (1998) Relationships between hardness, elastic modulus, and the work of indentation. Appl Phys Lett 73(5):614–616

Cheng YT, Li Z, Cheng CM (2002) Scaling relationships for indentation measurements. Philos Mag A 82(10):1822–1829

Chiu YL, Ngan AHW (2002) Time-dependent characteristics of incipient plasticity in nanoindentation of a Ni3Al single crystal. Acta Mater 50(6):1599–1611

Cottrell AH (1953) Dislocations and plastic flow in crystals. Clarendon, Oxford

Cottrell AH (1990) Advances in physical metallurgy, In: Charles JA, Smith GC (eds) Institute of metals, London

Domnich V, Gogotsi Y (2002) Phase transformations in silicon under contact loading. Rev Adv Mater Sci 3:1–36

Fischer-Cripps AC (2004) A simple phenomenological approach to nanoindentation creep. Mater Sci Eng A 385(1–2):74–82

Gaillard Y, Tromas C, Woirgard J (2006) Quantitative analysis of dislocation pile-ups nucleated during nanoindentation in MgO. Acta Mater 54:1409–1417

Ge D, Domnich V, Juliano T, Stach EA, Gogotsi Y (2004) Structural damage in boron carbide under contact loading. Acta Mater 52:3921–3927

Gerberich WW, Nelson JC, Lilleodden ET, Anderson P, Wyrobek JT (1996) Indentation induced dislocation nucleation: the initial yield point. Acta Mater 44(9):3585–3598

Gleiter H (2000) Nanostructured materials: basic concepts and microstructure. Acta Mater 48(1):1–29

Gogotsi YG, Domnich V, Dub SN, Kailer A, Nickel KG (2000) Cyclic nanoindentation and raman microspectroscopy study of phase transformations in semiconductor. J Mater Res 15:871–879

Göken M, Kempf M, Nix WD (2001) Hardness and modulus of the lamellar microstructure in PST-TiAl studied by nanoindentations and AFM. Acta Mater 49(5):901–903

Greaves GN, Meneau F, Kargl F, Ward D, Holliman P, Albergamo F (2007) Zeolite collapse and polymorphism. J Phys: Condens Matt 19(41):415102 1–17

Greaves GN, Wilding MC, Fearn S, Langstaff D, Kargl F, Cox S, Van QV, Majérus O, Benmore CJ, Weber R, Martin CM, Hennet L (2008) Detection of first-order liquid/liquid phase transitions in yttrium oxide-aluminum oxide melts. Science 322:566–570

Greaves GN, Greer AL, Lakes RS, Rouxel T (2011) Poisson's ratio and modern materials. Nat Mater 10:823–837

Grima JN, Jackson R, Alderson A, Evans KE (2000) Do Zeolites have negative poisson's ratios. Adv Mater B 12(24):1912–1917

Hertz H (1986) Miscellaneous papers. Macmillan, London

Hill R, Storåkers B, Zdunek AB (1989) A theoretical study of the brinell hardness test. Proc Royal Soc London A 423(1865):301–330

Jang JI, Lance MJ, Wen SQ, Tsui TY, Pharr GM (2005) Indentation-induced phase transformations in silicon: influences of load, rate and indenter angle on the transformation behaviour. Acta Mater 53(6):1759–1770

Jensen BJ, Cherne FJ, Cooley JC, Zhernokletov MV, Kovalev AE (2010) Shock melting of cerium. Phys Rev B 81(21):214109 1–8

Jiang MQ, Dai LH (2010) Short-range-order effects on intrinsic plasticity of metallic glasses. Philos Mag Lett 90(4):269–277

Johnson KL (1970) The correlation of indentation experiments. J Mech Phys Solids 18:115–126

Juliano T, Gogotsi Y, Domnich V (2003) Effect of indentation unloading conditions on phase transformation induced events in silicon. J Mater Res 18(05):1192–1201

Kelchner CL, Plimpton SJ, Hamilton JC (1998) Dislocation nucleation and defect structure during surface indentation. Phys Rev B 58(17):11085–11088

Kelly A, Tyson WR, Cottrell AH (1967) Ductile and brittle crystals. Phil Mag 15:567–586

Kese K, Li ZC (2006) Semi-ellipse method for accounting for the pile-up contact area during nanoindentation with the Berkovich indenter. Scripta Mater 55:699–702

Khan MK, Hainsworth SV, Fitzpatrick ME, Edwards L (2010) A combined experimental and finite element approach for determining mechanical properties of aluminium alloys by nanoindentation. Comput Mater Sci 49:4751–4760

Kolemen U (2006) Analysis of ISE in micro hardness measurements of bulk MgB2 superconductors using different models. J Alloy Compd 425:429–435

Kumar KS, Swygenhoven HV, Suresh S (2003) Mechanical behaviour of nanocrystalline metals and alloys. Acta Mater 51(19):5743–5774

Lakes RS, Wineman A (2006) On Poisson's ratio in linearly viscoelastic solids. J Elast 85(1):45–63

Ledbetter HM (1977) Ratio of the shear and Young's moduli for polycrystalline metallic elements. Mater Sci Eng 27(2):133–135

Lee YH, Baek U, Kim YI, Nahm SH (2007) On the measurement of pile-up corrected hardness based on the early Hertzian loading analysis. Mater Lett 61(19–20):4039–4042

Leipner HS, Lorenz D, Zecker A, Lei H, Grau P (2001) Nanoindentation pop-in effect in semiconductors. Phys B 308–310:446–449

Li H, Ghosh A, Han YH, Bradt RC (1993) The Frictional Component of the Indentation Size Effect in Low Hardness Testing. Journal of Materials Research 8(5):1028–1032

Li J, Vliet KJV, Zhu T, Yip S, Suresh S (2002) Atomistic mechanisms governing elastic limit and incipient plasticity in crystals. Nature 418:307–310

Lim YY, Chaudhri MM (1999) The effect of the indenter load on the nano hardness of ductile metals: an experimental study on polycrystalline work-hardened and annealed oxygen-free copper. Philos Mag A 79:2979–3000

Loerting T, Giovambattista N (2006) Amorphous ices: experiments and numerical simulations. J Phys Condens Matt 18: R919–R977

Lu H, Zhang X, Krauss WG (1997) Uniaxial, shear, and Poisson relaxation and their conversion to bulk relaxation: studies on poly (methyl methacrylate). Polym Eng Sci 37:1053–1064

Maneiro MAG, Rodriguez J (2005) Pile up effect on nanoindentation tests with spherical-conical tips. Scripta Mater 52:593–598

Mason JK, Lund AC, Schuh CA (2006) Determining the activation energy and volume for the onset of plasticity during nanoindentation. Phys Rev B 73(054102):1–15

Masumura RA, Hazzledine PM, Pande CS (1998) Yield stress of fine grained materials. Acta Mater 46(13):4527–4534

Navamathavan R, Park SJ, Hahn JH, Choi CK (2008) Nanoindentation 'pop-in' phenomenon in epitaxial ZnO thin films on sapphire substrates. Mater Charact 59:359–364

Navarro V, de la Fuente OR, Mascaraque A, Rojo JM (2008) Plastic properties of gold surfaces nanopatterned by ion beam sputtering. Phys Rev Lett B 78(224023):1–14

Nix WD, Gao H (1998) Indentation size effects in crystalline materials: a law for strain gradient plasticity. J Mech Phys Solids 46(3):411–425

Nix WD, Greer JR, Feng G, Lilleodden ET (2007) Deformation at the nanometer and micrometer length scales: effects of strain gradients and dislocation starvation. Thin Solid Films 515(6):3152–3157

Norbury AL, Samuel T (1928) The recovery and sinking-in or piling-up of material in the brinell test, and the effects of these factors on the correlation of the brinell with certain other hardness tests. J Iron Steel Ind 117:673–687

Ogata S, Li J, Hirosaki N, Shibutani Y, Yip S (2004) Ideal shear strain of metals and ceramics. Phys Rev B 70:104104

Oliver WC, Pharr GM (1992) An improved technique for determining hardness and elastic modulus using load and displacement sensing indentation experiments. J Mater Res 7:1564–1583

Peng Z, Gong J, Miao H (2004) On the description of indentation size effect in hardness testing for ceramics: Analysis of the nanoindentation data. J Eur Ceram Soc 24:2193–2201

Perottoni CA, Jornada JAHDa (2002) First-principles calculation of the structure and elastic properties of a 3D-polymerized fullerite. Phys Rev B 65(224208):1–6

Phani KK, Sanyal D (2008) The relations between the shear modulus, the bulk modulus and Young's modulus for porous isotropic ceramic materials. Mat Sci Eng, A 490(1–2):305–312

Poole PH, Grande T, Angell CA, McMillan PE (1997) Polymorphism in liquids and glasses. Science 275:322–323

Rabkin E, Deuschle JK, Baretzky B (2010) On the nature of displacement bursts during nanoindentation of ultrathin Ni films on sapphire. Acta Mater 58:1589–1598

Rar A, Sohn S, Oliver WC, Goldsby DL, Tullis TE, Pharr GM (2005) On the measurement of creep by nanoindentation with continuous stiffness techniques. In: Abstracts of symposium on fundamentals of nanoindentation and nanotribology III, Boston, Massachusetts, U.S.A November 29–December 3

Rhee YW, Kim HW, Deng Y, Lawn BR (2001) Brittle fracture versus quasiplasticity in ceramics: a simple predictive index. J Am Ceramic Soc 84:561–565

Rodriguez R, Gutierrez I (2003) Correlation between nanoindentation and tensile properties: influence of the indentation size effect. Mater Sci Eng, A 361(1–2):377–384

Rosenhain W, Ewen D (1913) The intercrystalline cohesion of metals. J Inst Metals 10:119–148

Sahin O, Uzun O, Kolemen U, Ucar N (2007) Mechanical characterization for β-Sn single crystals using nanoindentation tests. Mater Charact 59(4):427–434

Samuels LE (1989) ASTM STP 889, American society for testing and materials, Philadelphia, 1986, p 5

Sangwal K (2000) On the reverse indentation size effect and micro hardness measurement of solids. Mater Chem Phys 63:145–152

Santamaria-Perez D, Ross M, Errandonea D, Mukherjee GD, Mezouar M, Boehler R (2009) X-ray diffraction measurements of Mo melting to 119 GPa and the high pressure phase diagram. J Chem Phys 130(124509):1–8

Schuh CA (2006) Nanoindentation studies of materials. Mater Today 9(5):32–39

Schuh CA, Lund AC (2004) Application of nucleation theory to the rate dependence of incipient plasticity during nanoindentation. J Mater Res 19(07):2152–2158

Schuh CA, Mason JK, Lund AC (2005) Quantitative insight into dislocation nucleation from high-temperature nanoindentation experiments. Nat Mater 4(8):617–621

Schwaiger R, Moser B, Dao M, Chollacoop N, Suresh S (2003) Some critical experiments on the strain-rate sensitivity of nanocrystalline nickel. Acta Mater 51(17):5159–5172

Sevillano JG, Buessler P, Vrieze J, Kaluza W, Bouaziz O, Iung T, Bonifaz E, Meizoso AM, Martinez Esnaola JM, Ocaña I (2000) ECSC Steel RTD Final report, CECA7210-PR-044

Sneddon IN (1948) Boussinesq's problem for a rigid cone. Math Proc Cambridge 44:492–507

Swadener JG, George EP, Pharr GM (2002) The correlation of the indentation size effect measured with indenters of various shapes. J Mech Phys Solids 50(4):681–694

Taljat B, Pharr GM (2000) Measurement of residual stresses by load and depth sensing spherical indentation. Mater Res Symp Proc 594:519–524

Tromas C, Gaillard Y (2004) Encyclopedia of materials science and technology. Elsevier Science, Amsterdam

Valle CS, Lethbridge ZAD, Sinogeikin SV, Williams JJ, Walton R I, Evans KE, Bass JD (2008) Negative Poisson's ratios in siliceous zeolite MFI-silicalite. J Chem Phys 128(18): 184503 1–5

Venkataraman S, Kohlstedt DL, Gerberich WW (1992) Microscratch analysis of the work of adhesion for Pt thin films on NiO. Mater Res 7:1126–1132

Vlassak JJ, Nix WD (1994) Measuring the elastic properties of anisotropic materials by means of indentation experiments. J Mech Phys Solids 42(8):1223–1245

Vliet KJV, Li J, Zhu T, Yip S, Suresh S (2003) Quantifying the early stages of plasticity through nanoscale experiments and simulations. Phys Rev B 67(104105):1–15

Williams JA (1994) Engineering tribology. Oxford University Press, Oxford

Wo PC, Zuo L, Ngan AHW (2005) Time-dependent incipient plasticity inNi3Al as observed in nanoindentation. J Mater Res 20:489–495

Xi XK, Zhao DQ, Pan MX, Wang WH, Wu Y, Lewandowski JJ (2005) Fracture of brittle metallic glasses: brittleness or plasticity. Phys Rev Lett 94(12):125510 1–4

Zha CS, Hemley RJ, Mao HK, Duffy TS, Meade C (1994) Acoustic velocities and refractive index of SiO2 glass to 57.5 GPa by Brillouin scattering. Phys Rev B 50:13105–13112

Zhang SB, Cohen ML, Louie SG (1985) Interface potential changes and Schottky barriers. Phys Rev B 32(3955):3955–3957

Zhang H, Srolovitz DJ, Douglas JF, Warren JA (2009) Grain boundaries exhibit the dynamics of glass-forming liquids. In: Proceedings of the national academy science, 106:7735–7740, USA

Zhou XY, Jiang ZD, Wang HR, Yu RX (2003) Investigation on methods for dealing with pile-up errors in evaluating the mechanical properties of thin metal films at sub-micron scale on hard substrates by nanoindentation technique. Mater Sci Eng, A 488(1–2):318–322

Zimmerman JA, Kelchner CL, Klein PA, Hamilton JC, Foiles SM (2001) Surface step effects on nanoindentation. Phys Rev Lett 87(16): 165507 1–4

Nanomechanical Characterization of Soft Materials

A. H. W. Ngan

Abstract This chapter reviews the creep or viscoelastic deformation behavior of soft materials under nanoindentation-type testing. Analysis protocols of nanoindentation based on the Hertzian elastic contact theory, linear viscoelasticity analyses, and a more recent rate-jump method, are described and assessed. In addition to continuous viscoelasticity, a special type of discrete creep deformation, often observed in a wide range of materials during nanomechanical testing, is also highlighted.

1 Introduction

The advent of nanomechanical techniques including atomic force microscopy (AFM) and nanoindentation has enabled mechanical behavior of materials of micron- and smaller sizes to be characterized. However, whereas the load–displacement responses are routinely measured by these experimental platforms, these need to be deconvolved in order to obtain intrinsic material parameters, such as elastic modulus and yield properties. Despite the continuous development of the hardware over the past three decades, this step remains to be very challenging, especially for soft materials which exhibit not only purely elasto-plastic but also time-dependent deformation. Yet, examples of such materials are ample in many fronts of today's technology, including polymers, gels, low-melting metals tested at room temperature or higher melting metals tested at elevated temperatures, and also biological tissues. This chapter aims at to highlight some common experimental features and analysis methods concerning nanomechanical testing of soft materials that the author has experience with.

A. H. W. Ngan (✉)
Department of Mechanical Engineering, University of Hong Kong,
Pokfulam Road, Hong Kong, People's Republic of China
e-mail: hwngan@hku.hk

A. Tiwari (ed.), *Nanomechanical Analysis of High Performance Materials*,
Solid Mechanics and Its Applications 203, DOI: 10.1007/978-94-007-6919-9_8,

2 Common Data Analysis Protocols for Nanoindentation

Nanoindentation is the most mature nanomechanical characterization technique developed so far. Although nanoindentation was regarded as a special function of an AFM in the early days of development, as at today, there are a few commercial suppliers selling stand-alone nanoindenter machines, while AFMs took a separate line of development. In such commercial nanoindenters, a diamond probe, usually a Berkovich tip, is sent down to the sample along a vertical travel axis, and nanoscopic force and displacement data are gathered, usually by a capacitor gage, during the indentation process on the sample. As mentioned above, the most challenging step, from a user's point of view, is to deconvolve the force–displacement data to obtain material properties. Here, some common data-analysis protocols are reviewed.

Oliver and Pharr (1992) proposed a procedure to analyze nanoindentation data which has become a standard protocol available in the software of all commercial nanoindenters nowadays. Their procedure is based on the Hertzian theory for elastic contact (Hertz 1882; Johnson 1999), and is applicable to an unloading event which is purely elastic. The Hertzian theory is also widely used for analyzing nanoindentation data from experiments carried out in an AFM, especially on biological samples (Lekka et al. 1999; Rosenbluth et al. 2006; Cross et al. 2007; Li et al. 2008). For these reasons, this theory is briefly outlined here. Figure 1 shows the plan view of the contact area between two axi-symmetric elastic bodies, which represent the tip and the sample in the context of nanoindentation. A contact pressure distribution $p(s)$ is generated over this contact area, where s is the radial distance from the center. An infinitesimal area dA in this contact area is subjected to a point force $F = p(s)\,dA$ normal to the area, and from the theory of elastic stress potential of half-spaces (Johnson 1999), F produces a displacement

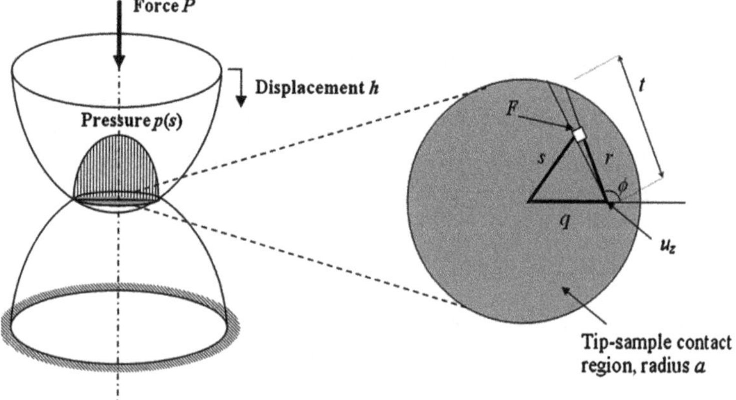

Fig. 1 Contact between two axi-symmetric elastic bodies

field $u_z(r)$ on the surface of either contacting body along the direction of contact, according to

$$u_z(r) = \frac{(1 - v^2)}{\pi E} \frac{F}{r};$$ (1)

where E and v are the Young modulus and Poisson ratio of the body, and r the radial distance from the position of F. Thus, by the principle of superposition in elasticity, the surface displacement of the body is the sum of contributions from all the point forces $F = p(s)\,dA = p(s)\,rd\phi dr$ arising from $p(s)$, and so is given by

$$u_z(q) = \frac{(1 - v^2)}{\pi E} \int\limits_{\phi=-\pi}^{\pi} \int\limits_{r=0}^{t} p(s)drd\phi, \quad q <; a.$$ (2)

In Eq. (2), q is the radial distance, from the center of the contact region, of the "field point" where u_z occurs, and a is the radius of the contact region. When applying Eqs. (1)–(2), the infinitesimal area dA is taken as $dA = p(s)rd\phi dr$, where ϕ is the angular position of the "source point" (the point where F acts) from the field point (see Fig. 1), to take advantage that the r in dA and in Eq. (1) can get cancelled. The inner integral in Eq. (2) is from $r = 0$ to t, where $t = -q\cos\phi + \sqrt{a^2 - q^2\sin^2\phi}$, and s in $p(s)$ is given by $s^2 = r^2 + q^2 + 2rq\cos\phi$.

Suppose that the two elastic bodies, i.e. the specimen surface and the tip, are both spherical with radii of curvature R_1 and R_2 respectively ($R_1 \to \infty$ if the specimen surface is flat), so that for small values of q (see Fig. 1), the two surfaces are well approximated by $z_1 \approx q^2/(2R_1)$ and $z_2 \approx q^2/(2R_2)$ respectively. The deformation requires that $z_1 + u_{z1} + z_2 + u_{z2} = h$, where h is the relative displacement of the tip into the sample surface, and u_{z1} and u_{z2} are deformations of the specimen surface and the tip according to Eq. (2). Therefore, Eq. (2) becomes

$$h - \frac{q^2}{2R} = \frac{1}{\pi E_r} \int\limits_{\phi=-\pi}^{\pi} \int\limits_{r=0}^{t} p(s)drd\phi$$ (3)

where R and E_r, given by $1/R = (1/R_1) + (1/R_2)$ and $1/E_r = [(1 - v_1^2)/E_1] + [(1 - v_2^2)/E_2]$, are respectively a reduced radius and elastic modulus of the tip-sample contact. Hertz showed that the solution to Eq. (3) is $p(s) = p_0\sqrt{1 - (s/a)^2}$, where $a = (\pi R p_0)/(2E_r)$ and $h = (\pi a p_0)/(2E_r)$ (Johnson 1999). The total indentation load is given by $P = \int\limits_{0}^{a} 2\pi s p(s)ds = (2\pi p_0 a^2/3)$. Key results of the Hertzian theory are therefore as follows:

$$a = \left(\frac{3PR}{4E_r}\right)^{1/3}; \quad h = \frac{a^2}{R} = \left(\frac{9P^2}{16RE_r^2}\right)^{1/3}; \quad p_o = \frac{3P}{2\pi a^2} = \left(\frac{6PE_r^2}{\pi^3 R^2}\right)^{1/3}.$$ (4)

                                                                A. H. W. Ngan

The second Equation in 4 gives

$$P = \left(\frac{4}{3}\sqrt{R}E_r\right) h^{3/2}$$ (5)

and hence, the reduced modulus E_r can be obtained from a fit of the load–displacement curve to the form $P \sim h^{3/2}$, provided that R is known. This method of obtaining E_r is often used in nanoindentation experiments carried out in AFMs, and is in fact incorporated into the analysis software of some commercial AFMs. However, the underlying assumption is that the loading process is purely elastic.

In the Oliver-Pharr protocol (Oliver and Pharr 1992) adopted by commercial nanoindenters, whereas the loading process of a nanoindentation experiment can involve plasticity, the onset of a subsequent unloading process is assumed to be purely elastic (Fig. 2a). Under this assumption, the tip-sample contact stiffness S at the onset of unloading process, defined as $S = dP/dh$, is obtainable from the first two Equations in (4) as

$$E_r = \frac{\sqrt{\pi}}{2} \frac{S}{\sqrt{A_c}}$$ (6)

where $A_c = \pi a^2$ is the projected area of the tip-sample contact circle (Fig. 2b). Thus, the elastic modulus E_r can be estimated if S and A_c are measured, and at the same time, the hardness can also be evaluated as $H = P_{max}/A_c$. A_c is given from the contact depth h_c at full load, through a pre-calibrated tip-shape function $A_c = f(h_c)$, but because of the elastic "sink-in" deformation of the specimen's surface (Fig. 2b), h_c is not explicitly specified in the P–h curve. For a spherical indenter, geometry gives $h_c \approx a^2/2R_2$, and if the sample is initially flat, $R = R_2$, so that the indenter displacement at full load is $h_{max} \approx a^2/R_2$ from the second Equation in (4). Therefore, $h_c \approx h_{max}/2$. However, this relation is valid only when the sample deforms purely elastically. In plastic indentation situations,

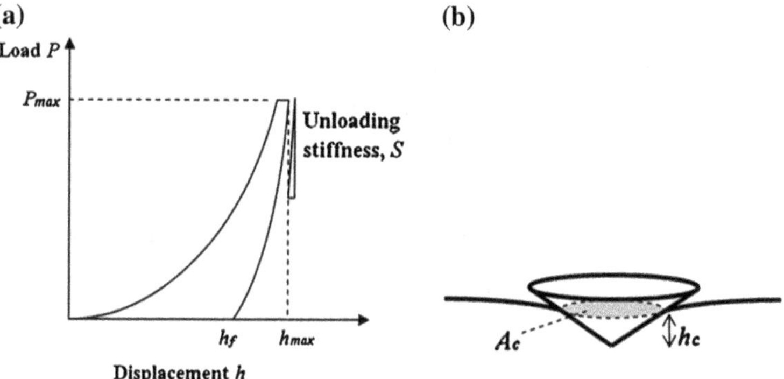

Fig. 2 **a** Schematic load–displacement graph and **b** sink-in morphology during nanoindentation

Table 1 Values of ε in Eq. (9)

Indenter shape	ε
Spherical and paraboloid	$\frac{3}{4} = 0.75$
Flat ended	1
Conical	$\frac{2}{\pi}(\pi - 2) = 0.73$

Oliver and Pharr (1992) proposed that the corresponding plastic depth, h_f, should be deducted from the h data, i.e. instead of $h_c = h_{\max}/2$, we have

$$(h_c - h_f) = \frac{(h_{\max} - h_f)}{2} \tag{7}$$

Also, from the second Equation in (4), $P \propto (h - h_f)^{3/2}$, and from this, the contact stiffness is given by

$$\frac{dP}{dh} = S = \frac{3}{2}\frac{P_{\max}}{(h_{\max} - h_f)}. \tag{8}$$

Combining Eqs. (7) and (8), we have $h_c = h_{\max} - (3/4)P_{\max}/S$, for the case of a spherical tip indenting on a flat sample. Oliver and Pharr (1992) proposed the following more general formula:

$$h_c = h_{\max} - \varepsilon\frac{P_{\max}}{S}, \tag{9}$$

where ε is a constant for a given indenter. Table 1 new gives the values of ε for different indenter geometries.

3 Viscoelastic Behavior During Nanoindentation

The analysis methods based on the Hertzian contact theory outlined above assume that the deformation is purely elastic in the relevant part of the load schedule. For soft materials, this condition is often not met, even during the unloading process. Figure 3a shows a rather extreme case of nanoindentation carried out in amorphous selenium at 24 °C (Tang and Ngan 2005), where the P–h curve bends forward during the initial part of the unload. This signifies significant creep deformation: the tip continues to sink into the specimen as it creeps under the tip load, even though the load on the tip is reducing. If the Oliver-Pharr method is applied to such a scenario, the resultant E_r would be negative as shown in Fig. 3b, since the apparent contact stiffness S at the onset of unload is negative. The creep factor in Fig. 3b is defined as

$$C = \frac{\dot{h}_h S_{corr}}{|\dot{P}_u|}, \tag{10}$$

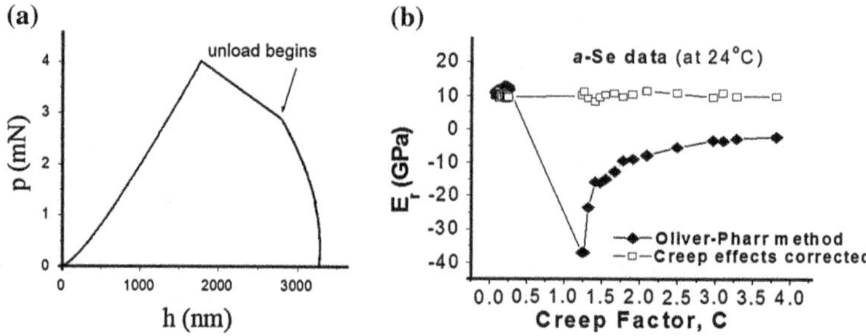

Fig. 3 **a** A typical load–displacement curve in amorphous selenium at 311 K. **b** Elastic modulus of amorphous selenium measured by the Oliver-Pharr method and after creep correction. Data from (Tang and Ngan 2005)

where $\dot{h}_h$ is the creep rate at the end of the hold period just before the unload, S_{corr} is a corrected elastic contact stiffness (see Eq. (29)), and $\dot{P}_u$ is the unloading rate. As will be seen later, C measures the relative importance of elastic and creep deformation at the onset of the unloading process (Feng and Ngan 2002). The fact that creep is significant is also represented by a high value of $\dot{h}_h$ during the load-hold before unload, and in Fig. 3a, this corresponds to a significant drop in the actual load applied onto the sample during the nominal load-hold stage prior to unload, due to the increasing spring force in the nanoindenter transducer as the tip sinks into the sample. Significant creep during unload, accompanied by a visible forward bending "nose" in the P–h curve, is a rather general behavior for soft materials, including polymers and biological (e.g. bone and cartilage) samples tested at room temperature, and even metals tested at high homologous temperatures relative to their melting points. Even though creep deformation may not be as severe as giving rise to an obvious "nose" in the P–h curve, it may still lead to significant overestimation of the contact stiffness S and hence the E_r estimated from the Oliver-Pharr analysis (Feng and Ngan 2002).

As mentioned above, certain commercial AFMs are equipped in their analysis software with the Hertzian fit protocol involving Eq. (5), where the P–h curve measured during a load ramp is fitted with a $P \sim h^{3/2}$ law to obtain the E_r value. When creep or any time-dependent deformation occurs, the measured E_r would be dependent on the load ramp rate, as shown in Fig. 4a for the case of an oral cancer cell line indented by an AFM tip (Zhou et al. 2012). Similar rate-dependent results are often seen in the literature (e.g. Li et al. 2008), and their occurrence means that the measured properties are not intrinsic to the sample.

4 Linear Viscoelasticity Analyses of Nanoindentation

The viscoelastic behaviors commonly seen in nanoindentation of soft samples have been the subject of investigation by linear viscoelasticity analyses (Feng and Ngan 2002; Sakai 2002; Cheng and Cheng 2005; Oyen 2006). In one

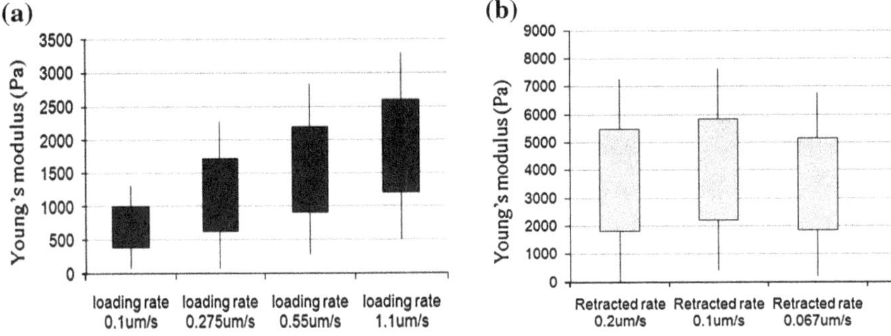

Fig. 4 Elastic modulus of UM1 oral cancer cells measured with **a** the Hertzian fit protocol in Eq. (5), **b** the rate-jump protocol in Eq. (5). Data from (Zhou et al. 2012)

approach, nanoindentation is performed in a dynamic mode, involving a small oscillatory load $\Delta P = \Delta P_0 \sin(\omega t)$ superimposed on the basic load. Because of viscosity, the displacement oscillation will in general exhibit a phase lag ϕ, i.e. $\Delta h = \Delta h_0 \sin(\omega t - \phi)$. In analogy with Eq. (6), a storage modulus E_r' and a loss modulus E_r'' can be defined as (Herbert et al. 2008):

$$E_r' = \frac{\sqrt{\pi}}{2\sqrt{A_c}} \frac{\Delta P_0}{\Delta h_0} \cos\phi; \quad E_r'' = \frac{\sqrt{\pi}}{2\sqrt{A_c}} \frac{\Delta P_0}{\Delta h_0} \sin\phi \tag{11}$$

While E_r' and E_r'' are easily measured this way, they are usually strong functions of the oscillation frequency ω.

In a second approach, a certain constitutive model is assumed as the intrinsic deformation law for the sample, and this is developed into measurables such as a P–h relation, which can then be fitted with the experimental data to obtain the coefficients in the model, which are supposed to be intrinsic material properties. Most viscoelasticity analyses carried out for nanoindentation made use of heredi-tary integrals (Sakai 2002; Cheng and Cheng 2005; Oyen 2006). Alternatively, another very useful technique is Radok's correspondence principle between lin-ear viscoelasticity and elasticity (Radok 1957). As an illustration, consider the Maxwell model of viscoelasticity, with the following constitutive relation

$$\dot{e}_{ij} = \frac{1}{2G}\dot{S}_{ij} + \frac{1}{2\eta_s}S_{ij}, \quad \sigma_{ii} = 3B\varepsilon_{ii} \text{ (viscoelasticity)}. \tag{12}$$

Here, $S_{ij} = \sigma_{ij} - \delta_{ij}\sigma_{kk}/3$ is the deviatoric stress, and $e_{ij} = \varepsilon_{ij} - \delta_{ij}\varepsilon_{kk}/3$ the deviatoric strain, G and B are the shear and bulk modulus respectively, and η_s is the shear viscosity. Laplace-transforming Eq. (12) leads to

$$e_{ij}^* = \left(\frac{1}{2G} + \frac{1}{2\eta_s} \cdot \frac{1}{s}\right) S_{ij}^*, \quad \sigma_{ii}^* = 3B\varepsilon_{ii}^* \text{ (viscoelasticity)} \tag{13}$$

where $(\ )^*$ denotes the Laplace transform of a time-dependent quantity $(\ )$, and s is the transform variable. Equation (13) is of a form analogous to Hooke's Law for a purely elastic material:

$$e_{ij} = \frac{1}{2G} S_{ij}, \quad \sigma_{ii} = 3B\varepsilon_{ii} \quad \text{(elasticity)}\,;$$

$$e_{ij}^* = \frac{1}{2G} S_{ij}^*, \quad \sigma_{ii}^* = 3B\varepsilon_{ii}^* \quad \text{(elasticity)}\,. \tag{14}$$

Comparing Eqs. (13) and (14) suggests that the Laplace transform of the viscoelastic problem can be solved by replacing the elastic constants in the purely elastic problem by the following:

$$\frac{1}{G} \rightarrow \frac{1}{G} + \frac{1}{\eta_s} \cdot \frac{1}{s}\,; \quad B \rightarrow B. \tag{15}$$

The problem of indenting on a purely elastic half-space by a conical tip with semi-apex angle α has been solved by Sneddon (1965) (also c.f. Eq. (27)) as

$$\frac{1}{E_r} P(t) = \frac{2}{\pi} \tan \alpha \, h^2(t). \tag{16}$$

Since $E_r = 4G(3B + G)/(3B + 4G)$, to obtain the corresponding P–h relation in the viscoelastic case, application of the transformations in Eq. (15) leads to the following transformation of E_r:

$$\frac{1}{E_r} \rightarrow \frac{1}{E_r} + \frac{1}{4\eta_s} \cdot \frac{1}{s} + \frac{E^2}{36B^2} \cdot \frac{1}{(3\eta_s s + E)} \tag{17}$$

where E is the Young's modulus. Therefore, the Laplace transform of the viscoelastic version of Eq. (16) is

$$\left\{ \frac{1}{E_r} + \frac{1}{4\eta_s} \cdot \frac{1}{s} + \frac{E^2}{36B^2(3\eta_s s + E)} \right\} P^* = \frac{2}{\pi} \tan \alpha \, (h^2)^*. \tag{18}$$

Inverse transforming Eq. (18) followed by differentiating with respect to t leads to

$$\frac{\dot{P}(t)}{E_r} + \frac{P(t)}{4\eta_s} + \frac{E^2}{108B^2\eta_s} \int_0^t \exp\left[-\frac{E}{3\eta_s}(t - t') \right] \dot{P}(t')dt' = \frac{2}{\pi} \tan \alpha \frac{d(h^2)}{dt} \tag{19}$$

if $P(0) = 0$.

Numerical values indicate that the third term in Eq. (19) is usually small compared with the other two terms. When this third term is ignored,

$$\frac{\dot{P}(t)}{E_r} + \frac{P(t)}{4\eta_s} \approx \frac{2}{\pi}\tan\alpha\frac{d(h^2)}{dt}. \tag{20}$$

For a load-hold process up to P_{max} following a simple load ramp up to time t_1, as shown in Fig. 5a, integrating Eq. (20) yields

$$h^2 \approx \frac{\pi\, P\mathrm{max}}{2\tan\alpha}\left[\left(\frac{1}{E_r} - \frac{t_1}{8\eta_s}\right) + \frac{t}{4\eta_s}\right] = a + bt \tag{21}$$

where a and b are constants. Figure 5b, c show load-hold nanoindentation experiments carried out in an amorphous Ge–Si thin film (Xu 2008). The h^2 versus t plot is linearly in accordance with Eq. (21). The shear viscosity η_s can be obtained from the slope of the h^2-t plot, and data measured at different test temperatures obey an Arrhenius law as shown in Fig. 5c, with activation energy of ~0.14 eV.

Constitutive models other than Maxwell can be similarly employed, and if more springs and dashpots are involved, more material constants will have to be obtained, and finding them can turn out to be a heavy curve-fitting exercise. Unfortunately, in most cases, there is simply no reliable guidance to help decide on which law is most suitable for a given material, yet the accuracy of such linear viscoelasticity analyses depends on the validity of the assumed constitutive law.

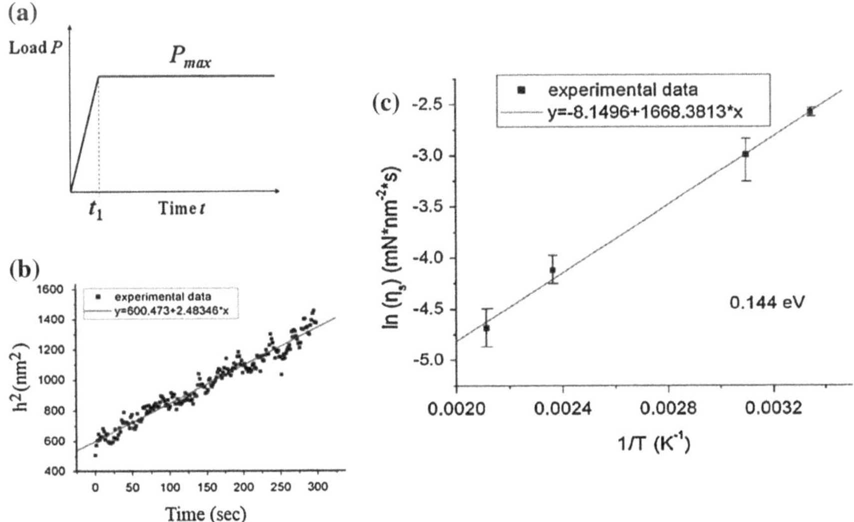

Fig. 5 Measurement of shear viscosity η_s in an amorphous Ge-Si alloy thin film. **a** Load schedule used. **b** h^2 versus t during load hold at P_{max}. **c** Arrhenius plot of η_s, exhibiting an activation energy of 0.14 eV. Data from (Xu 2008)

5 The Rate-Jump Protocol of Nanoindentation

The linear viscoelasticity analyses mentioned above often yield storage (elastic) or loss (viscous) coefficients which are supposed to be material properties on one hand, but on the other hand, are also strongly dependent of the test frequency, or rate of deformation in general. In the rheology literature, the rate dependence of such properties is accepted as "intrinsic" to the material itself, but in any case, this would still contradict the use of "springs" and "dashpots", supposedly with constant coefficients, in the constitutive law assumed. In an atomic model, the spring and dashpot elements in the correct constitutive law should correspond to the conservative stretching and permanent slippage or other dissipative events of the interatomic bonds in the solid, respectively. The spring elements of the viscoelastic network should, therefore, be characteristic of the nature and architecture of the atomic bonds in the solid and truly material constants independent of the rate of deformation and other extrinsic factors.

A recent "rate-jump" protocol for carrying out mechanical tests in general has been proven to be capable of returning an intrinsic elastic modulus that is independent of the test conditions from viscoelastic materials (Ngan and Tang 2009). The key assumption of the constitutive law for the material is very mild—a network of any arrangement of (in general) non-linear viscous dashpots and linear elastic springs, as shown in Fig. 6a,

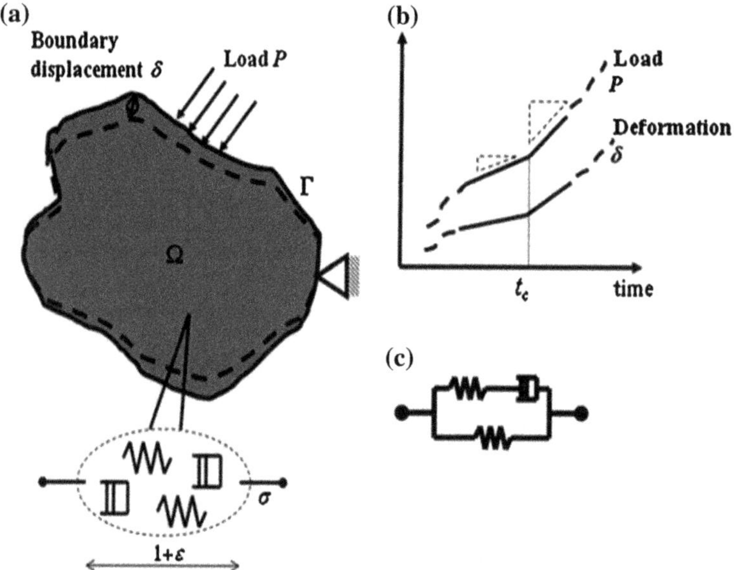

Fig. 6 **a** A general linear-elastic, nonlinear-viscous solid subjected to boundary load P undergoing deformation with boundary displacement δ. **b** Schematic of a rate jump in P and δ at time t_c. **c** An example of the viscoelastic network model of the material: the standard-linear-solid model, but any other network is admissible

is assumed to hold within a very short time window $[t_c^-, t_c^+]$ about time t_c, at which a sudden step change in either the loading rate or the displacement rate, depending on whether the test is load- or displacement-controlled, is applied on the sample (Fig. 6b). The dashpots and springs here are described respectively by relations of the form

$$\dot{\varepsilon}_{ij}(\text{dashpot}) = \dot{\varepsilon}_{ij}(\sigma_{kl}), \tag{22}$$

$$\varepsilon_{ij}(\text{spring}) = s_{ijkl}\sigma_{kl} \tag{23}$$

where ε_{ij} and σ_{kl} are strain and stress tensors, and a dot above denotes time rate. Note that the $\dot{\varepsilon}_{ij}$ versus σ_{kl} relation in Eq. (22) is not necessarily linear (i.e. nonlinear viscosity is anticipated), and also, any arrangement of the springs and dashpots in the constitutive model is admissible. The latter two points are important differences with the linear viscoelasticity analyses described in the section above.

A key point to note about the nonlinear dashpots is that, by virtue of Eq. (22), a step change $\Delta\dot{\sigma}_{kl}$ in the stress rate field at t_c, arising from the step change in loading rate in Fig. 6b, does not result in any non-zero change in the strain rate field $\dot{\varepsilon}_{ij}$ across the dashpots, because although the stress-rate $\dot{\sigma}_{kl}$ suffers a step jump $\Delta\dot{\sigma}_{kl}$, the stress itself must still be continuous across t_c. Thus, $\Delta\sigma_{kl} = 0$ across t_c, and from Eq. (22), $\Delta\dot{\varepsilon}_{ij}(\text{dashpot}) = 0$, i.e. the dashpots, whether they are linear or nonlinear and irrespective of their whereabouts in the constitutive network with respect to the springs, do not react to the rate-jump across t_c. Only the elastic springs react to the rate-jump according to Eq. (23), viz

$$\Delta\dot{\varepsilon}_{ij} = s_{ijkl}\Delta\dot{\sigma}_{kl}, \tag{24}$$

and $\Delta\dot{\varepsilon}_{ij}$ and $\Delta\dot{\sigma}_{kl}$ are also the overall strain-rate and stress-rate changes of the sample across t_c. Equation (24) says that the fields $\Delta\dot{\sigma}_{ij}$ and $\Delta\dot{\varepsilon}_{ij}$ can be solved as a *linear elastic* problem, with the *same* elastic spring elements in the original viscoelastic model of the material while the dashpot elements are ignored.

The solution of Eq. (24), for a given test geometry, would be a linear $\Delta\dot{P} \sim \Delta\dot{\delta}$ relation between the step changes in the load and displacement rates across t_c, with the linking proportionality constant being a lumped value of the elastic constants in the original viscoelastic model after removing all the dashpots. Fitting such a relation to experimental results allows this lumped value to be measured as an intrinsic elastic modulus of the material. As a simplest illustration, Fig. 7 shows results from macroscopic tensile tests performed on high-density polyethylene bars with such a rate-jump protocol applied (Chan and Ngan 2010). Figure 7a shows a typical load schedule of nominal stress versus time, where a step change in the stress rate is imposed at time marked as 0. The corresponding nominal-strain response is shown in Fig. 7b. From Eq. (24), an effective elastic modulus of the material is given by

$$E = \frac{\Delta\dot{\sigma}}{\Delta\dot{\varepsilon}} = \frac{\dot{\sigma}_+ - \dot{\sigma}_-}{\dot{\varepsilon}_+ - \dot{\varepsilon}_-} \tag{25}$$

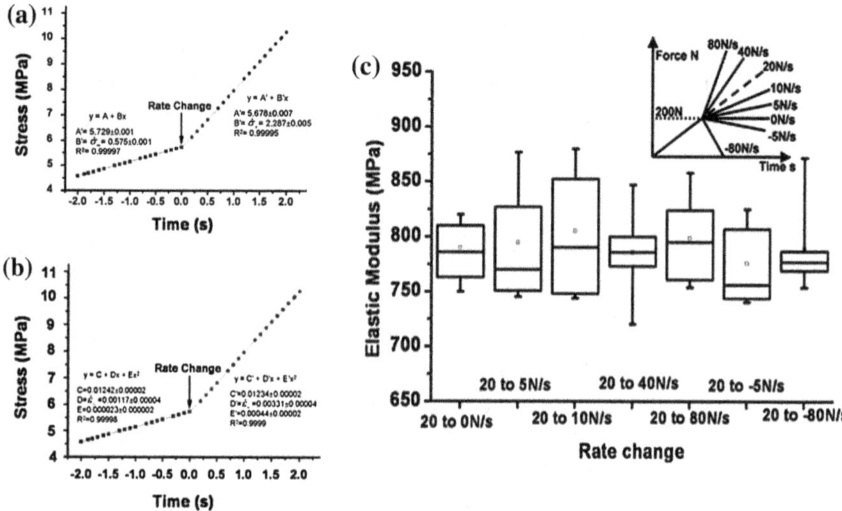

Fig. 7 Invariant effective modulus measured using the rate-jump protocol from tensile tests on high-density polyethylene. **a** Typical load schedule of nominal stress versus time with a rate jump imposed. **b** Response of nominal strain to the load schedule in (**a**). **c** E values calculated using Eq. (25) from different experiments with different magnitudes of load-rate jump. Inset in (**c**) shows the load schedules used. Data from (Chan and Ngan 2010)

where $\dot{\sigma}_{+/-}$ and $\dot{\varepsilon}_{+/-}$ are the stress and strain rates, i.e. the slopes of the graphs in Fig. 7a, b, measured before and after the rate jump and extrapolated to the latter. According to the above argument, E should be an intrinsic material constant independent of the test conditions. Figure 7c shows the values of E measured using different magnitudes of load-rate jump, and it can be seen that E is indeed invariant.

In any test platform, Eq. (24) says that, provided that the load–displacement ($P \sim \delta$) relation for a *linear elastic* specimen during load-ramp is known, the corresponding $\Delta\dot{P} \sim \Delta\dot{\delta}$ relation for a viscoelastic sample can be obtained by the following simple substitutions:

$$\sigma \to \Delta\dot{\sigma}; \quad \varepsilon \to \Delta\dot{\varepsilon}; \quad P \to \Delta\dot{P}; \quad \delta \to \Delta\dot{\delta}; \quad \text{etc.} \tag{26}$$

In the following, we analyze nanoindentation as carried out in commercial nanoindenters, as well as in the AFM.

5.1 Rate-Jump Method in Depth-Sensing Nanoindentation

In depth-sensing nanoindentation using the Oliver-Pharr protocol, the elastic modulus and hardness are evaluated at the onset of an unloading stage following a load-hold stage (e.g. see Fig. 3a). The onset point of unloading is, therefore, a rate-jump point pertinent to the above analysis. If the sample is purely elastic, the load–displacement relation is given by Eq. (6), i.e.

$$\frac{dh}{dP} = \frac{\sqrt{\pi}}{2E_r\sqrt{A_c}}. \tag{27}$$

Now, for a viscoelastic sample at the onset of unload, carrying out the substitutions in Eq. (26) in Eq. (27) gives

$$\frac{\Delta\dot{h}}{\Delta\dot{P}} = \frac{\dot{h}_h - \dot{h}_u}{\dot{P}_h - \dot{P}_u} = \frac{\sqrt{\pi}}{2E_r\sqrt{A_c}}, \tag{28}$$

where $\dot{h}_h$ and $\dot{h}_u$ are tip speeds just before and just after the unload onset point, and $\dot{P}_h$ and $\dot{P}_u$ are the load rates just before and after unload onset. The apparent contact stiffness, measurable as the slope of the P–h curve, is $S = \dot{P}_u/\dot{h}_u$, and writing $2E_r\sqrt{A_c}/\pi$ as S_{corr} the "corrected" elastic stiffness, then a relation between S_{corr} and S can be obtained from Eq. (28) as

$$\frac{1}{S_{corr}} = \left(\frac{1}{S} - \frac{\dot{h}_h}{\dot{P}_u}\right) \times \frac{1}{\left(1 - \dot{P}_h/\dot{P}_u\right)}. \tag{29}$$

Equation (29) then serves as a correction formula for the viscous effects on the elastic contact stiffness (Feng and Ngan 2002; Ngan et al. 2005); the quantities needed for the correction include the creep displacement rate $\dot{h}_h$ and load drop rate $\dot{P}_h$ (due to the suspending springs in the transducer) at the end of the load hold prior to the unload, and the unloading rate $\dot{P}_u$ (<0), in addition to the apparent stiffness $S = dP/dh$ at the onset of unload. If the load-drop due to the suspending springs is negligible, then, with the creep factor defined in Eq. (10), the following can be obtained from Eq. (29):

$$\frac{S_{corr}}{S} \approx 1 - C \tag{30}$$

which further indicates the effects of viscous deformation on the contact stiffness. In extreme cases (e.g. Fig. 3), C can be larger than unity, and then the apparent stiffness S will become negative. The corrected stiffness S_{corr} from Eq. (29) can be used to replace S in Eqs. (6) and (9) to obtain the reduced modulus E_r and hardness as in the original Oliver-Pharr protocol (Feng and Ngan 2002; Ngan et al. 2005). Equation (29) also says that if the unloading rate $\dot{P}_u$ is very fast, or the hold before unload is very long so that $\dot{h}_h$ and $\dot{P}_h$ become very small, then correction is not necessary. However, knowing these conditions a priori would be difficult for very soft samples.

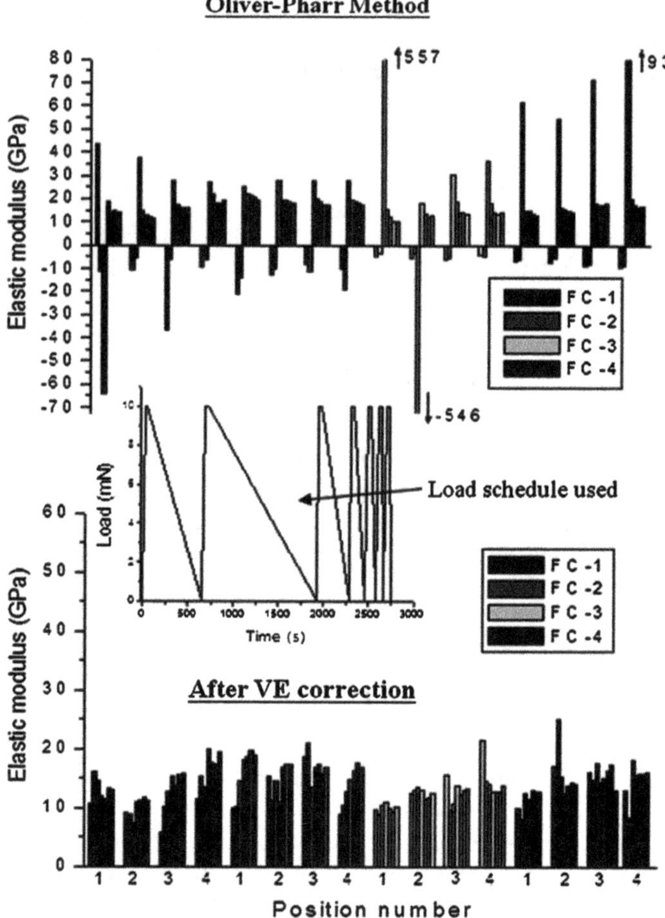

Fig. 8 The elastic modulus of mice cortical bone analyzed with the Oliver-Pharr method and rate-jump method. The inset shows the identical multi-cycle loading schedule for all the tests, in which the elastic modulus was calculated at the onset of each unloading portion. Data from (Tang 2005) and (Tang et al. 2006)

Figure 8 compares the elastic modulus of mice cortical bone calculated with the rate-jump and the Oliver-Pharr protocols (Tang 2005; Tang et al. 2006). Here, a multi-cycle loading schedule was used to evaluate the elastic modulus at the onset of each unloading cycle. In the earlier cycles, the elastic modulus obtained by the Oliver-Pharr method becomes negative due to very severe viscoelastic effects, which lead to the "nose" phenomenon in the $P–h$ curve (c.f. Fig. 3 for selenium). The rate-jump method is able to turn these cases back to normal with positive and rather consistent modulus values.

5.2 Rate-Jump Method in AFM Nanoindentation

As mentioned earlier, nanoindentation is also routinely carried out in commercial AFMs, especially for testing biological tissues and nano-scale objects such as nano-filaments or wires. Compared with using a commercial depth-sensing nanoindentation machine, one major challenge arises in the AFM, namely, unlike the diamond Berkovich tips usually used in depth-sensing nanoindentation which are quite durable, AFM tips are much sharper and more fragile, and may not survive the many indentations required in obtaining the tip-shape function $A_c = f(h_c)$ (Fig. 2b) from a calibrating sample. In fact, the uncertain shape of AFM tips is one major disadvantage involved in the Hertzian fit protocol, since the tip-end radius R has to be known when applying Eq. (5). To avoid the damaging tip-shape calibration, flat-ended tips are recommended for AFM nanoindentation work (Fig. 9), and these can be easily made from commercial AFM tips by focused-ion-beam milling. The tip-sample contact size a (c.f. Fig. 1) then remains constant for different indentation depths, and a can be obtained easily by electron microscope imaging of the tip. Accurate determination of the initial contact point would also not be necessary since the contact size a is constant.

As discussed above, the usual Hertzian fit method also gives rise to rate-dependent results in general, as illustrated in Fig. 4a. The rate-jump protocol is definitely more attractive, as this should return an intrinsic elastic modulus of the sample. Figure 10a, b shows two usual designs of commercial AFMs, one in which the tip-cantilever clamp is fixed and the sample moves up by displacement δ by piezoelectrics (Fig. 10a), and the other where the sample sits on a fixed platform while the cantilever clamp moves down by δ (Fig. 10b). In either case, the cantilever deflects with a displacement δ' at its free end, which is measured by a photo-diode. The cantilever deflection is related to the photo-diode signal D via a sensitivity constant A, i.e. $\delta' = AD$.

For the first situation in Fig. 10a, when the sample is purely elastic with tip-sample contact stiffness $S = 2E_r a$ (c.f. Eq. (6)), the indentation force is given by

Fig. 9 Flat-ended tip for AFM nanoindentation, made by cutting a commercial tip by FIB milling (side view on *left* and *top* view on *right*; courtesy B. Tang)

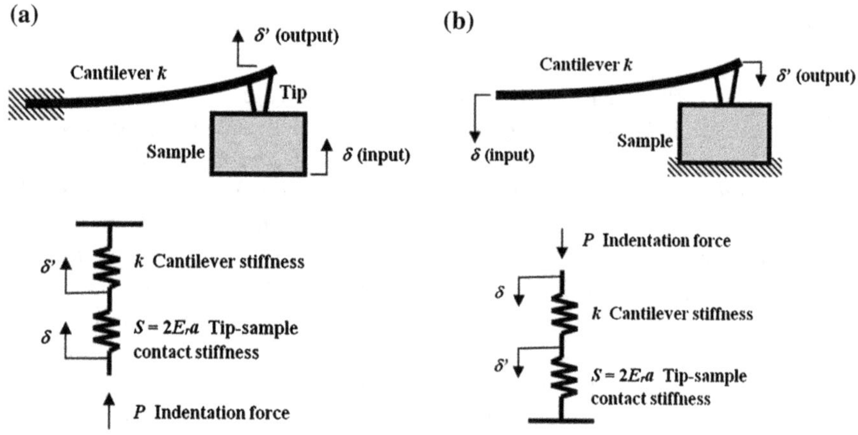

Fig. 10 Two designs of AFM. **a** Cantilever clamp fixed, sample moves up. **b** Cantilever clamp moves down, sample fixed

$P = k\delta' = S(\delta - \delta')$, where k is the force constant of the cantilever. This can be rearranged as $\delta/\delta' = (1 + \alpha/E_r)$, where $\alpha = k/2a$. For a viscoelastic sample under a rate-jump protocol, applying the substitutions in Eq. (26) gives

$$\frac{\Delta\dot{\delta}}{\Delta\dot{D}} = A\left(1 + \frac{\alpha}{E_r}\right), \quad \text{(cantilever fixed, sample moves)} \tag{31}$$

where $\Delta\dot{\delta}$ is an imposed step change in the movement rate of the sample base (i.e. the input), and $\Delta\dot{D}$ is the resultant step change in the rate of the photodiode signal D (the output).

For the second situation in Fig. 10b where the cantilever clamp moves, a similar analysis gives the following relation instead:

$$\frac{\Delta\dot{\delta}}{\Delta\dot{D}} = A\left(1 + \frac{E_r}{\alpha}\right) \quad \text{(sample fixed, cantilever moves)}. \tag{32}$$

In either situation, the rate-jump relation Eqs. (31) or (32) involves two machine constants, A the photo-diode sensitivity (i.e. cantilever deflection per unit sensor voltage or current generated), and $\alpha = k/2a$ which is a cantilever-tip constant, since both k and a are properties of the cantilever-tip. A also depends on the cantilever since different cantilevers will have different reflectivity for the laser. Thus, for a given cantilever-tip, if both A and α are pre-calibrated, Eqs. (31) or (32) can be used to evaluate the E_r of an unknown specimen, by measuring the $\Delta\dot{D}$ for a step change $\Delta\dot{\delta}$ imposed at some point during the load schedule. To calibrate A and α, single indentations can be performed on two samples with known E_r values (e.g. one hard one soft), and A and α can then be obtained by solving two simultaneous Equations of either (31) or (32) (Tang and Ngan 2011). This amount of calibration involving two single indentation tests should be the minimum required to achieve quantitative measurements by AFM nanoindentation, and tip damage can be minimized this way.

Figure 4b shows the elastic modulus values measured from the same batch of oral cancer cells as in Fig. 4a, using the above rate-jump protocol with three different rate-jump values of $\Delta\dot{\delta}$ (Zhou et al. 2012). The measured modulus does not exhibit any dependence on the magnitude of the $\Delta\dot{\delta}$ used, and so this is evidently an intrinsic constant of the cell line.

Referring back to Fig. 6b, while the constitutive law involving Eqs. (22) and (23) is expected to hold for a very short time span across the rate-jump time point t_c, for time- or strain-dependent materials, the constitutive law itself may evolve with time. For such materials, the effective elastic modulus measured from the rate-jump protocol would then be the material constant at t_c. Successive rate-jumps can be imposed along a load schedule to measure a series of E_r values along the strain path, and these should represent the evolution of the constitutive law or structure of the material over time or strain.

Although the rate-jump method is useful in returning an intrinsic elastic modulus of any viscoelastic sample, the viscous (dashpot) component of the deformation is subtracted out. The viscous component can only be obtained from linear viscoelasticity analysis of load-relaxation or creep response of the sample using an assumed constitutive model (see Eq. (21) and Fig. 5), or as the loss modulus by means of dynamic nanoindentation.

6 Discrete Yield Events in Soft Materials

Apart from smooth viscoelasticity or creep deformation, discrete plasticity events with time-dependent characteristics are also frequently observed in metals and polymers of small volumes. The first type of such discrete plasticity is delayed onset of

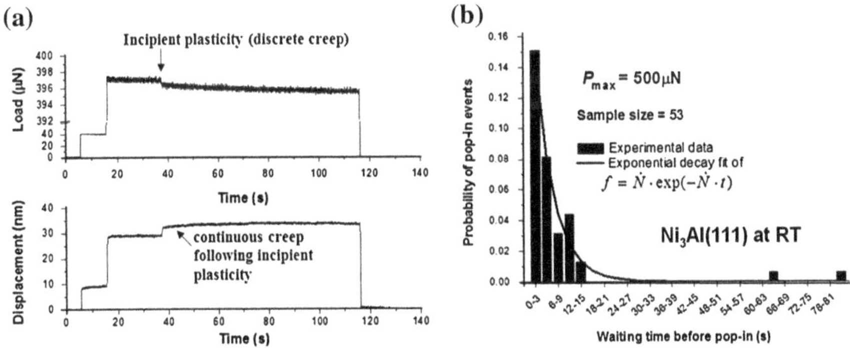

Fig. 11 Delayed incipient plasticity in Ni$_3$Al (111) during load-hold nanoindentation at room temperature. **a** Load and displacement versus time, showing the sudden occurrence of a discrete yield event during the load-hold. Creep follows immediately after the discrete event. **b** Statistical distribution of the waiting time for the discrete yield to occur at 500 μN. Data from (Wo et al. 2005)

yielding in well-annealed metals. It is well-known that annealed metals subjected to nanoindentation often exhibit a discrete yield point, which marks the onset of plastic deformation. When the load is held at a value slightly below the yield point, no yielding or creep occurs initially as expected, but with prolonged application of load, a discrete yield event may occur suddenly after some waiting time, as shown by the example in Fig. 11. This waiting time is shorter as the load increases (Chiu and Ngan 2002), and for a fixed load, it exhibits a stochastic distribution (Fig. 11b). This delayed yielding behavior is thought to be due to thermally agitated nucleation of incipient dislocations within the stressed volume (Ngan et al. 2006).

Discrete plasticity is not confined to the onset of yielding, but can occur intermittently after first yield. Figure 12a shows a type of creep deformation with discrete yield events in single-crystalline aluminum micro-pillars under uniform compression at room temperature (Ng and Ngan 2007). Figure 12b shows a similar type of discrete creep, superimposed on smooth creep deformation, observed during constant-load nanoindentation on high-density polyethylene (Li and Ngan 2010). The occurrence frequency of the discrete creep events in polyethylene was found to increase with crystallinity, suggesting that the discrete creep behavior is due to the crystal phases in the polymer. The discrete type of creep relaxation events represents an interesting contrast to the conventional smooth viscoelastic deformation, and is worthy of more investigations in the future.

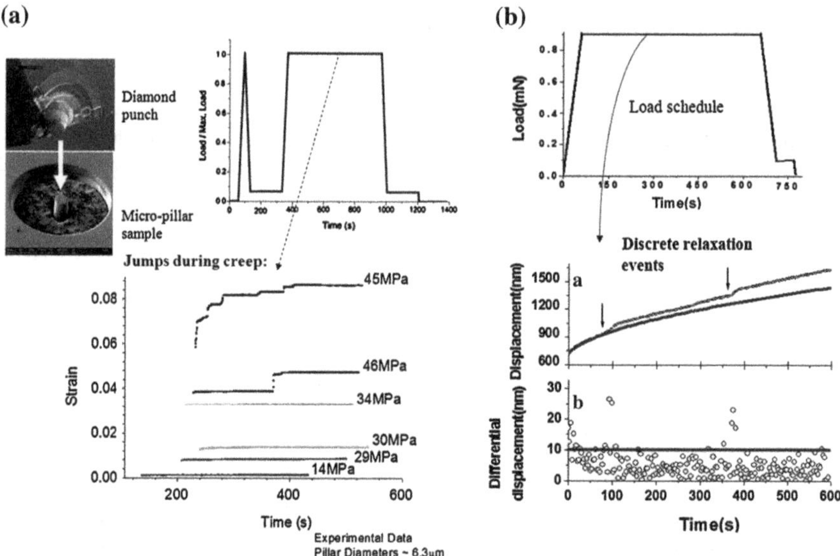

Fig. 12 Successive discrete yield events during load-hold experiments. **a** Aluminum single-crystalline micro-pillars subjected to uniform compression at constant load at room temperature exhibit a type of creep deformation with discrete jumps (Ng and Ngan 2007). **b** High-density polyethylene subject to Berkovich nanoindentation at constant load at room temperature exhibits similar creep behavior with discrete jumps (Li and Ngan 2010)

7 Concluding Remarks

In this chapter, we have reviewed selected behavior and critical issues when soft materials are subjected to nanoindentation-type testing. Soft samples such as polymers, biological specimens, glasses approaching glass transition, and so on, often exhibit time-dependent viscoelastic deformation during nanoindentation testing. Data analysis protocols based on Hertzian-type theories are inadequate for such materials, often returning erroneous results. Linear viscoelasticity analyses are straightforward to carry out, but the choice of the suitable constitutive law is not an easy step, and such analyses often return rate-dependent storage and loss coefficients of the model, which is not ideal. A rate-jump protocol can return an intrinsic elastic modulus for any general linear-elastic, nonlinear-viscous solid, but no information can be obtained concerning the viscous component of the constitutive law. In addition to continuous viscoelasticity, a wide range of materials also exhibit a discrete mode of creep deformation during nanomechanical testing.

Acknowledgments This review covers the work of previous group members, including B. Tang, G. Feng, P. C. Wo, Y. L. Chan, K. S. Ng, Z. W. Xu, J. Y. Li and Z. L. Zhou, to whom thanks are given. Some of the work was also supported by grants from the Research Grants Council (Project No. 7159/10E) as well as from the University Grants Committee (Project No. SEG-HKU06) of the Hong Kong Special Administrative Region.

References

Chan YL, Ngan AHW (2010) Invariant elastic modulus of viscoelastic materials measured by rate-jump tests. Polym Test 29:558–564

Cheng YT, Cheng CM (2005) Relationships between initial unloading slope, contact depth, and mechanical properties for spherical indentation in linear viscoelastic solids. Mater Sci Eng A 409:93–99

Chiu YL, Ngan AHW (2002) Time-dependent characteristics of incipient plasticity in nanoindentation of Ni_3Al single crystal. Acta Mater 50:1599–1611

Cross SE, Jin YS, Rao J, Gimzewski JK (2007) Nanomechanical analysis of cells from cancer patients. Nat Nano 2:780–783

Feng G, Ngan AHW (2002) Effects of creep and thermal drift on modulus measurement using depth-sensing Indentation. J Mater Res 17:660–668

Herbert EG, Oliver WC, Pharr GM (2008) Nanoindentation and the dynamic characterization of viscoelastic solids. J Phys D Appl Phys 41:074021

Hertz H (1882) Über die Berührung fester elastischer Körper. J Reine Angew Math 92:156–171

Johnson KL (1999) Contact mechanics. Cambridge University Press, Cambridge

Lekka M, Lekki J, Marszalek M, Golonka P, Stachura P, Cleff B, Hrynkiewicz AZ (1999) Local elastic properties of cells studied by SFM. Appl Surf Sci 141:345–350

Li JY, Ngan AHW (2010) Nano-scale fast relaxation events in polyethylene. Scripta Mater 62:488–491

Li QS, Lee GYH, Ong CN, Lim CT (2008) AFM indentation study of breast cancer cells. Biochem Biophys Res Comm 374:609–613

Ng KS, Ngan AHW (2007) Creep of micron-sized aluminum columns. Phil Mag Lett 87:967–977

Ngan AHW, Tang B (2009) Response of power-law-viscoelastic and time-dependent materials to rate jumps. J Mater Res 24:853–862

Ngan AHW, Wang HT, Tang B, Sze KY (2005) Correcting power-law viscoelastic effects in elastic modulus measurement using depth-sensing indentation. Int J Solids Strut 42:1831–1846

Ngan AHW, Zuo L, Wo PC (2006) Size dependence and stochastic nature of yield strength of micron-sized crystals: a case study on Ni_3Al. Proc Roy Soc Lond A 462:1661–1681

Oliver WC, Pharr GM (1992) An improved technique for determining hardness and elastic modulus using load and displacement sensing indentation experiments. J Mater Res 7:1564–1583

Oyen ML (2006) Analytical techniques for indentation of viscoelastic materials. Phil Mag 86:5625–5641

Radok JRM (1957) Visco-elastic stress analysis. Q App Math 15:198–202

Rosenbluth MJ, Lam WA, Fletcher DA (2006) Force microscopy of nonadherent cells: a comparison of leukemia cell deformability. Biophys J 90:2994–3003

Sakai M (2002) Time-dependent viscoelastic relation between load and penetration for an axisymmetric intenter. Phil Mag A 82:1841–1849

Sneddon IN (1965) The relation between load and penetration in the axisymmetric Boussinesq problem for a punch of arbitrary profile. Int J Eng Sci 3:47–57

Tang B (2005) Nanoindentation of viscoelastic materials. PhD thesis, University of Hong Kong, Hong kong

Tang B, Ngan AHW (2005) Investigation of viscoelastic properties of amorphous selenium near glass transition using depth-sensing indentation. Soft Mater 2:125–144

Tang B, Ngan AHW (2011) Nanoindentation using an atomic force microscope. Phil Mag 91:1329–1338

Tang B, Ngan AHW, Lu WW (2006) Viscoelastic effects during depth-sensing indentation of cortical bone tissues. Phil Mag 86:5653–5666

Wo PC, Zuo L, Ngan AHW (2005) Time-dependent incipient plasticity in Ni_3Al as observed in nanoindentation. J Mater Res 20:489–495

Xu ZW (2008) Phase transformation and properties of magnetron sputtered GeSi thin films. PhD thesis, University of Hong Kong, Hong Kong

Zhou ZL, Ngan AHW, Tang B, Wang AX (2012) Reliable measurement of elastic modulus of cells by nanoindentation in an atomic force microscope. J Mech Behav Biomed Mater 8:134–142

Nanoindentation Applied to Closed-Cell Aluminium Foams

Jiří Němeček

Abstract The chapter is devoted to the assessment of effective elastic properties of an aluminium alloy appearing on cell walls of a closed-cell foam system Alporas. The methodology used for this purpose is based on a bottom-up approach which includes identification of mechanically distinct material phases by means of combination of several analyses. Electron microscopy and image analyses are employed at first to identify miscrostructural and chemical entities. Mechanical properties of the distinct phases are studied by grid nanoindentation. Phase separation is performed using statistical deconvolution. Microstructural information is then used for the assessment of effective cell wall stiffness. Several analytical and numerical tools are tested and compared for this purpose. Good mutual agreement is achieved between the methods due to the close-to-isotropic nature of the phase dispersion within the cell wall volume.

1 Introduction

Metal foams are advanced lightweight materials used in applications ranging from automotive and aerospace industries to construction engineering. Depending on their specific density and a source material used for production they are used as various structural elements, e.g. bumpers, car body sills, filters, damping layers of motorcycle helmets or sandwich insulation panels (e.g. Banhart 2001). Macroscopically, the foams can be characterized by attractive mechanical and physical properties such as high stiffness and strength in conjunction with very low weight, excellent impact energy absorption, high damping capacity and good sound absorption capability. If aluminium alloy is used as a base material, other advantages like low density ($\sim$2,700 kg/m^3),

J. Němeček (✉)
Faculty of Civil Engineering, Department of Mechanics, Czech Technical University,
Thakurova 7, 16629 Prague 6, Czech Republic
e-mail: jiri.nemecek@fsv.cvut.cz

A. Tiwari (ed.), *Nanomechanical Analysis of High Performance Materials*,
Solid Mechanics and Its Applications 203, DOI: 10.1007/978-94-007-6919-9_9,
© Springer Science+Business Media Dordrecht 2014

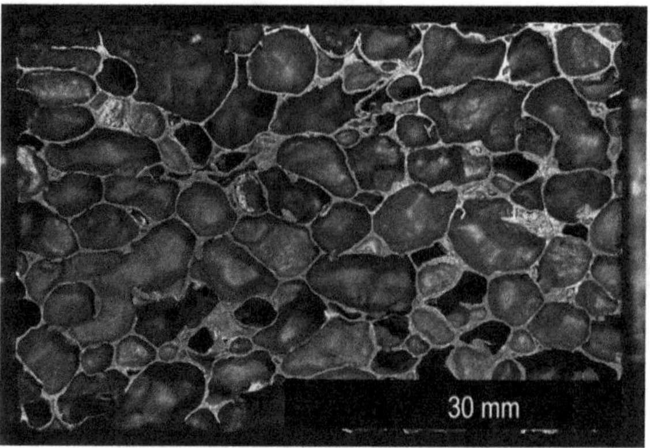

Fig. 1 Optical image of a typical foam cross section

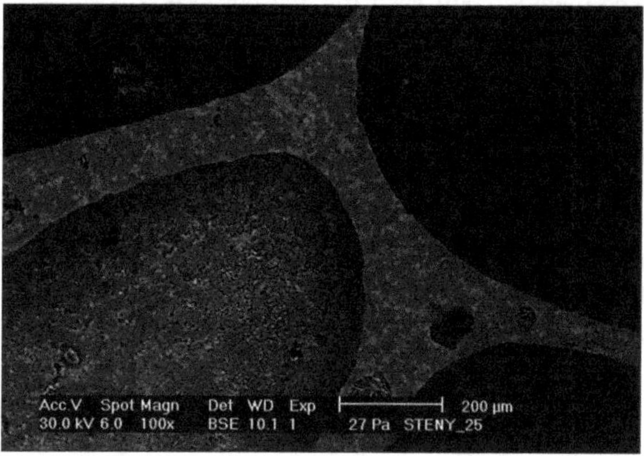

Fig. 2 SEM-BSE images showing the detail of a cell wall. Grey levels correspond to heterogeneous material phases

low melting point (~660 °C), non-flammability, possibility of recycling and excellent corrosion resistance are obtained.

Based on a particular purpose and production technology metal foams can be produced in an open-cell or closed-cell form. Our attention in this contribution is focused on a commercially available closed-cell foam system Alporas® developed by Shinko Wire Company, Ltd. (Miyoshi et al. 1998). Alporas is characterized with a hierarchical system of pores having different cell morphologies (in shape and size) in dependence on the foam density and inhomogeneous material properties of the cell walls (Hasan et al. 2008). The overall sample porosity can be found in the range from 60 to 90 % (Ashby et al. 2000). A typical cross section of the foam is shown in Fig. 1 in which large pores (having typically 1–13 mm in diameter) can be seen. Detailed view on thin cell walls is depicted in Fig. 2.

Traditionally, mechanical properties of metal foams are measured by conventional testing methods, e.g. by uniaxial compression tests (Papadopoulos et al. 2004; Jeon et al. 2009; Sugimura et al. 1997; Yongliang et al. 2010; De Giorgi et al. 2010) or microindentation tests (Idris et al. 2009). In principle, such measurements can give only overall response of this microscopically heterogeneous material and are experimentally difficult. It is mainly due to small wall dimensions, low local bearing capacity and stability problems caused by local wall yielding and bending. It is also uneasy to reach well defined geometry and boundary conditions for a miniaturized test on a single wall. Therefore, other means like nanoindentation are suitable techniques for exploring microscopically heterogeneous properties of the cell walls.

The material heterogeneity takes place at several length scales. In this paper we focus our attention on investigation of phase properties that appear at the level of cell walls (e.g. at the scale of 100 μm). At first, we use the information from nanoindentation (Fischer-Cripps 2002) for prediction of overall elastic properties of the cell wall which is difficult to measure by other means. In this task, massive grid indentation is utilized together with statistical evaluation and deconvolution of phase properties (Constantinides et al. 2006; Němeček et al. 2011b). Finally, we apply analytical and numerical homogenization schemes (Zaoui 2002; Moulinec and Suquet 1994; Michel et al. 1999) to predict overall elastic properties of the wall.

2 Experimental Part

2.1 Test Samples

Commercial aluminium foam samples with high porosity (Alporas, Shinko Wire Company, Ltd) were used in this study. The foam production process covers melting of aluminium at 680 °C and mixing it with blowing (1.6 wt% of TiH_2) and thickening (1.5 wt% of Ca) agents (Miyoshi et al. 1998). Hydrogen is released from TiH_2 which forms large bubbles in the molten whereas calcium increases the molten viscosity and stabilizes cell walls. Finally, the molten is cooled down. The resulting internal structure of the aluminium foam (Fig. 1) shows wide distribution of pores and high overall sample porosity. On the other hand, an Al/Ca/Ti alloy which develops in the thin cell walls during hardening process (Fig. 2) exhibits high degree of heterogeneity.

2.2 Sample Preparation

Firstly, an Alporas foam block was cut into pieces. Each piece was cut into slices (~5 mm thick) and embedded into epoxy resin to fill large pores. The sample surface was mechanically grinded and polished to reach minimum surface roughness suitable for micromechanical testing and image analyses. Very low roughness $R_q \approx 10$ nm on 10×10 μm area (ISO 4287-1997) was achieved on cell walls. The sample was investigated with scanning electron microscopy (SEM) at first. Acquired images were segmented to binary ones and used further in image analyses. Subsequently, nanoindentation measurements were performed on cell walls situated perpendicularly to the surface (as checked with optical microscope).

2.3 SEM and Image Analyses

Scanning electron microscope equipped with secondary electron (SE), back-scattered electron (BSE) and energy dispersive X-ray (EDX) analyses was used to study the cell wall heterogeneity. It was confirmed that a significant inhomogeneity of micro-strutural material phases exists on the level of tens of micrometers (Figs. 2 and 3). Two distinct phases, that exhibit different color in BSE images, can be distinguished. The chemical composition of the two phases was checked with EDX element analysis in SEM. It was found that the majority of the sample volume (dark zone in Figs. 2 and 3) consists of aluminium (~67 wt %), oxygen (~32 wt %) and other trace elements (Mg, Ti, Fe, Co, Ni, Cu, Si < 2 wt %). Lighter zones in Figs. 2 and 3 consist of Al (~60 wt %), O (~30 wt %), Ca (~5 wt %), Ti (~5 wt %) and other elements (<1 wt %). As expected, the majority of the volume (dark zone) is composed of aluminium and aluminium oxide Al_2O_3. This phase is further denoted as Al-rich area.

It follows also from other studies (Simone and Gibson 1998) that Ca/Ti-rich discrete precipitates and diffuse Al_4Ca areas develop in the Al-matrix during hardening. These areas that appear as lighter zones in Figs. 2 and 3 are further denoted as Ca/Ti-rich phase.

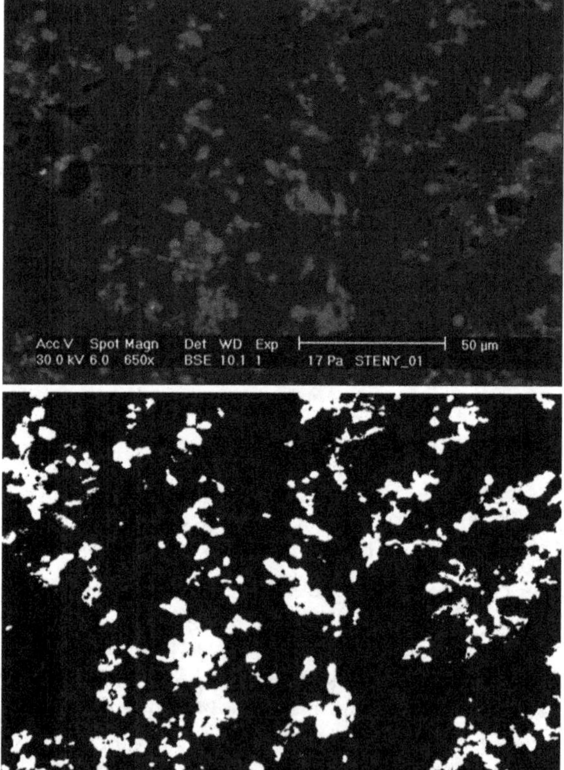

Fig. 3 Detail of the cell wall microstructure in SEM-BSE image (*top*) and segmented image in which *black* = Al-rich, *white* = Ca/Ti-rich areas (*bottom*)

It is possible to estimate the volume fraction of the two phases in the sample by an image analysis performed on BSE images. For this purpose, ten arbitrary BSE images were taken and segmented to two phases using common grey threshold. The Ca/Ti-rich area was estimated to cover 22 ± 4 % of the whole image area. It is shown later in the text, that this volume fraction does not correspond to the volume of mechanically distinct phase which is larger than the portion estimated by the pixel colors in BSE images.

2.4 Nanoindentation

Nanoindentation tests on cell walls were performed using Hysitron Tribolab system® at the Czech Technical University in Prague. The system is equipped with quasi-static and dynamic measuring modes and with in situ SPM imaging for scanning the sample surface with the same tip which is used for the sample penetration. Pyramidal diamond tip (Berkovich) was used for all measurements. Several distant locations were chosen on the sample to capture its heterogeneity. Each location was covered by a series of indents with 10 μm spacing (Fig. 4). The total amount of 200 was performed which was considered to give sufficiently large statistical set of data.

Since we are dealing with a two phase system, the mutual phase influence needs to be experimentally minimized to receive their intrinsic properties. The size of an indent (h) must be kept small enough to be within a phase (d). As a rule of thumb, the indent's size h < d/10 is usually used to access material properties of individual constituents without any dependence on the length scale (Durst et al. 2004). For our case, the minor Ca/Ti-rich phase has a characteristic dimension of 4 μm (from the image analysis). Therefore, the indent's size should be kept within 400 nm in our experiments.

Fig. 4 Indentation matrix showing individual indents with 10 μm spacing scanned by in situ imaging

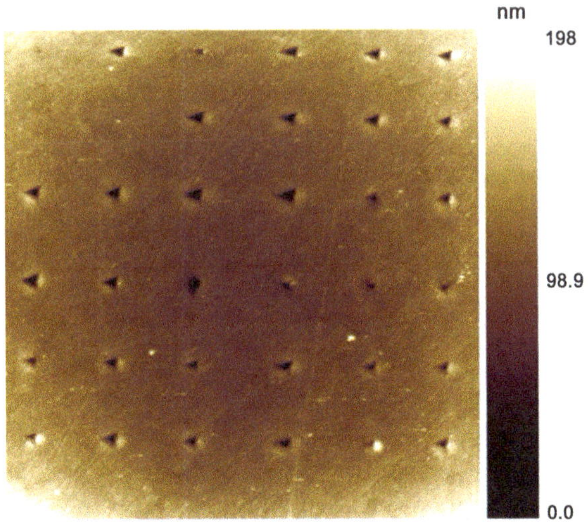

Image Scan Size: 55.000 μm

Standard load controlled test was prescribed. The loading history consisted of three segments: loading, holding at the peak and unloading. Loading and unloading of this trapezoidal loading function lasted for 5 s. The holding part lasted for 10 s and was included to minimize any time-dependent material effects (Fischer-Cripps 2002; Němeček 2009). Maximum applied load was 1 mN. Maximum indentation depths were ranging between 100 and 300 nm depending on the stiffness of the indented phase.

Elastic modulus was evaluated for individual indents using standard Oliver and Pharr methodology (Oliver and Pharr 1992) which accounts for elasto-plastic contact of a conical indenter with an isotropic half-space as in Eq. 1:

$$E_r = \frac{1}{2\beta} \frac{\sqrt{\pi}}{\sqrt{A}} \frac{dP}{dh}$$

(1)

in which E_r is the reduced (combined) modulus measured in an experiment, A is the projected contact area of the indenter at the peak load, β is a geometrical factor accounting for the indenter shape ($\beta = 1.034$ for the used Berkovich tip) and $\frac{dP}{dh}$ is a slope of the unloading branch evaluated at the peak. Elastic modulus E of the measured sample can be found using basic contact mechanics from Eq. 2:

$$\frac{1}{E_r} = \frac{\left(1 - \nu^2\right)}{E} + \frac{\left(1 - \nu_i^2\right)}{E_i}$$

(2)

in which ν is the Poisson's ratio of the tested material, E_i and ν_i are known elastic modulus and Poisson's ratio of the indenter ($E_i = 1{,}140$ GPa, $\nu_i = 0.07$). Note that the choice of Poisson's ratio from a reasonable range of similar metals and alloys (0.3–0.4) does not have a significant influence on values of sample elastic moduli evaluated from measured reduced moduli. Therefore, the sample Poisson's ratio was taken $\nu = 0.35$ as an estimate for all indents (even if not measured for individual phases).

3 Numerical Part

3.1 Statistical Deconvolution

Since the cell walls show high degree of heterogeneity, it is effective to use a statistical approach for the analysis of the measured set of mechanical data. Quite frequently, it is not possible to distinguish between the mechanically different phases with a sufficient resolution according to optical images that are coupled with the indentation system. Even using SEM images does not give a clue how to match different color in BSE images with the true stiffness of the phase. Such color interpretation could lead to erroneous conclusions (as demonstrated later in indentation results). Therefore, statistical analysis which was formerly proved to be a reliable technique for the analysis of microscopically heterogeneous structural materials, e.g. for cementitious composites (Ulm et al. 2007; Němeček et al. 2013), alkali-activated fly ash or gypsum (Němeček et al. 2011b; Němeček 2012), high-performance concretes (Sorelli et al. 2008; Němeček et al. 2011a) and also for metal alloys (Němeček and Králík 2012) has been applied.

The statistical technique is based on the phase deconvolution that seeks for parameters of individual phase distributions included in overall results. It searches for r distributions (Gaussian fits are assumed) in an experimental histogram of a chosen mechanical property. Random seed and minimizing criteria of the differences between the experimental and theoretical overall histograms (particularly quadratic norm of the differences) are computed in the algorithm to find the best fit (Fig. 6).

The analysis begins with the generation of experimental Probability Density Function (PDF) or Cumulative Distribution Function (CDF) for the data set. Using of PDF is more physically intuitive since significant peaks associated with mechanically distinct phases can be often distinguished in the graph. On the other hand, the construction of PDF requires the choice of a bin size. Application of CDF (Ulm et al. 2007) is more straightforward (does not require the choice of a bin size) and is more appropriate for cases where no clear peaks occur in the property histogram. The deconvolution algorithms based on PDF of CDF are analogous and thus the procedure will be demonstrated for the case of PDF.

Experimental PDF is firstly constructed from all measurements whose number is N^{exp}, using equally spaced N_{bins} bins of the size b. Each bin is assigned a frequency of occurrence f_i^{exp} that can be normalized with respect to the overall number of measurements as $\frac{f_i^{exp}}{N^{exp}}$. From that, one can compute the experimental probability density function (PDF) as a set of discrete values (as in Eq. 3):

$$p_i^{exp} = \frac{f_i^{exp}}{N^{exp}} \frac{1}{b} \quad i = 1,\ldots,N_{bins} \tag{3}$$

The task of deconvolution into M phases represents finding of $r = 1,\ldots,M$ individual distributions related to single material phase. Assuming normal (Gauss) distributions, the single phase PDF can be written as Eq. 4:

$$p_r(x) = \frac{1}{\sqrt{2\pi s_r^2}} \exp \frac{-(x - \mu_r)^2}{2s_r^2} \tag{4}$$

in which μ_r and s_r are the mean value and standard deviation of the r-th phase computed from n_r values as Eq. 5:

$$\mu_r = \frac{1}{n_r} \sum_{k=1}^{n_r} x_k \quad s_r^2 = \frac{1}{n_r - 1} \sum_{k=1}^{n_r} (x_k - \mu_r)^2 \tag{5}$$

and x is the approximated quantity (i.e. elastic modulus or hardness). The overall PDF constructed from M phases is then (Eq. 6):

$$C(x) = \sum_{r=1}^{M} f_r p_r(x) \tag{6}$$

where f_r is the volume fraction of a single phase defined as Eq. 7:

$$f_r = \frac{n_r}{N^{\exp}} \tag{7}$$

Individual distributions can be found by minimizing the following error function (Eq. 8):

$$\min \sum_{i=1}^{N^{bins}} [(P_i^{\exp} - C(x_i))P_i^{\exp}]^2 \tag{8}$$

in which quadratic deviations between experimental and theoretical PDFs are computed in a set of discrete points. The function is weighted by the experimental probability in order to put emphasis on the measurements with a higher occurrence. The minimization in Eq. 8 can be based on the random Monte Carlo generation of M probability density functions satisfying the condition (Eq. 9):

$$\sum_{r=1}^{M} f_r = 1 \tag{9}$$

As mentioned above, the bin size needs to be chosen prior the computation. Also, it is beneficial to fix the number of mechanically distinct phases M in advance to minimize the computational burden and to stabilize the ill-posed problem (Němeček et al. 2011b). Such knowledge can be supplied by some independent analyses (chemical composition, SEM or image analyses). In our studied case, a two-phase system (one dominant Al-rich phase and one minor Ca/Ti-rich phase) was assumed in the statistical deconvolution.

3.2 Effective Material Properties

Once the phase properties and volume fractions are known it is necessary to predict the overall (effective) properties for the whole material level (i.e. the cell wall level). Several approaches can be employed for this task using a variety of analytical or numerical homogenization methods. We decided to study several approaches.

Continuum micromechanics serve as a fundamental tool in our assessment of effective material properties. A material is considered as macroscopically homogeneous with microscopically inhomogeneous phases that fill a representative volume element (RVE) with characteristic dimension l. The scale separation condition requires to be $d \ll l \ll D$, where d stands for a size of the largest micro-level inhomogeneity in the RVE (e.g. particles or phases), l is the RVE size and D stands for structural dimension of a macroscopically homogeneous material which can be continuously built from the RVE units. The characteristic structural dimension is usually at least 4–5 times larger than the RVE size (Drugan and Willis 1996). In our case, the characteristic level size is determined by the cell wall thickness whose average dimension is 60 μm (from the image analysis in Němeček and Králík 2012). Microscopic heterogeneities within RVE take the form of Ca/Ti-rich precipitates with a characteristic dimension of 4 μm.

3.3 Analytical Homogenization

As mentioned earlier, the heterogeneous microstructure of materials is taken into account by representative volume elements (RVE). The RVE with substantially smaller dimensions than the macroscale body allows imposing homogeneous boundary conditions over the RVE (Hill 1963, 1965; Hashin 1983). Then, continuum micromechanics provide a framework, in which elastic properties of heterogeneous microscale phases are homogenized to give overall effective properties of the upper scale (Zaoui 2002). A significant group of analytical homogenization methods relies on the Eshelby's solution (Eshelby 1957) that uses an assumption of the uniform stress field in an ellipsoidal inclusion embedded in an infinite body. Effective elastic properties are then obtained through averaging over the local contributions. The methods are bounded by rough estimates based on the mixture laws of Voigt (parallel configuration of phases with perfect bonding) and Reuss (serial configuration of phases). The bonds are usually quite distant so that more precise estimates need to be used. Very often, the Mori-Tanaka method (Mori and Tanaka 1973) is used for the homogenization of composites with continuous matrix (reference medium) reinforced with spherical inclusions. In this method, the effective bulk and shear moduli of the composite are computed as follows (Eq. 10):

$$k_{eff} = \frac{\sum_r f_r k_r \left(1+\alpha_0\left(\frac{k_r}{k_0}-1\right)\right)^{-1}}{\sum_r f_r \left(1+\alpha_0\left(\frac{k_r}{k_0}-1\right)\right)^{-1}} \qquad \mu_{eff} = \frac{\sum_r f_r \mu_r \left(1+\beta_0\left(\frac{\mu_r}{\mu_0}-1\right)\right)^{-1}}{\sum_r f_r \left(1+\beta_0\left(\frac{\mu_r}{\mu_0}-1\right)\right)^{-1}} \tag{10}$$

$$\alpha_0 = \frac{3k_0}{3k_0+4\mu_0}, \beta_0 = \frac{6k_0+12\mu_0}{15k_0+20\mu_0}$$

where f_r is the volume fraction of the rth phase, k_r is its bulk modulus, μ_r is its shear modulus, and the coefficients α_0 and β_0 describe bulk and shear properties of the 0-th phase, i.e. the reference medium (Mori and Tanaka 1973). The bulk and shear moduli can be directly linked with Young's modulus E and Poisson's ratio ν used in engineering computations as (Eq. 11):

$$E = \frac{9k\mu}{3k+\mu} \qquad \nu = \frac{3k-2\mu}{6k+2\mu} \tag{11}$$

The reference medium is principally chosen as the prevailing (matrix) phase of the composite. Also in our case, dominant Al-rich area was prescribed as the reference medium. Materials with no preference of matrix phase (i.e. polycrystal-line metals) are usually modeled with the self-consistent scheme (Zaoui 2002). It is an implicit scheme, similar to Mori-Tanaka method, in which the reference medium points back to the homogenized medium itself. For comparison, such approach was also included to our study.

3.4 Numerical Homogenization Based on FFT

The homogenization problem, i.e. finding the link between microscopically inhomo-geneous strains and stresses and overall behavior of a RVE can be solved e.g. by finite element calculations or by applying advanced numerical schemes that solve the problem using fast Fourier transformation (FFT), for example. The later was found to be numerically efficient in connection with grid indentation that serves as a source of local stiffness parameters in equidistant discretization points. The behavior of any heterogeneous material consisting of periodically repeating RVE occupying domain Ω can be described with differential equations with periodic boundary conditions and prescribed macroscopic strain (ε^0) as (Eqs. 12 and 13):

$$\sigma(\mathbf{x}) = \mathbf{L}(\mathbf{x}) : \varepsilon(\mathbf{x}) \quad div\sigma(\mathbf{x}) = \mathbf{0} \quad \mathbf{x} \in \Omega \tag{12}$$

$$\langle \varepsilon \rangle := \frac{1}{|\Omega|} \int_\Omega \varepsilon(\mathbf{x})d\mathbf{x} = \varepsilon^0 \tag{13}$$

where $\sigma(\mathbf{x})$ denotes second order stress tensor, $\varepsilon(\mathbf{x})$ second order strain tensor and $\mathbf{L}(\mathbf{x})$ the fourth order tensor of elastic stiffness at individual locations $\mathbf{x}$. The effec-tive (homogenized) material tensor $\mathbf{L}_{\mathrm{eff}}$ is such a tensor satisfying (Eq. 14):

$$\langle \sigma \rangle = \mathbf{L}_{\mathrm{eff}}\langle \varepsilon \rangle \tag{14}$$

Local strain tensor can be decomposed to homogeneous (macroscopic) and fluctuation parts which leads to the formulation of an integral (Lippmann–Schwinger type) Eq. 15:

$$\varepsilon(\mathbf{x}) = \varepsilon^0 - \int_\Omega \Gamma^0(\mathbf{x} - \mathbf{y}) : (\mathbf{L}(\mathbf{y}) - \mathbf{L}^0) : \varepsilon(\mathbf{y})d\mathbf{y} \tag{15}$$

where Γ^0 stands for a periodic Green operator associated with the reference elasticity tensor $\mathbf{L}^0$ which is a parameter of the method (Moulinec and Suquet 1998). The problem is further discretized using trigonometric collocation method (Saranen and Vainikko 2002) which leads to the assemblage of a non-symmetrical linear system of equations. The system can be resolved e.g. by the conjugate gradient method as proposed in Zeman et al. (2010). Elastic constants received from grid nanoindentation have been used as input parameters for this FFT homogenization with the assumption of plane strain conditions.

The resulting homogenized stiffness matrix is generally anisotropic (but must be symmetric and positive definite). The degree of anisotropy of the matrix depends on topology of inclusions (phases) in the solved volume (called periodic unit cell) regardless of the fact that the individual points are treated as locally isotropic. Note also, that the FFT homogenization takes no assumptions on the morphology of the phases as in case of analytical schemes and is, therefore, more general. It works only with the stiffness coefficients distributed within the periodic unit cell and the accuracy of the method depends only on the density of grid points.

4 Results

4.1 Nanoindentation

Results from grid nanoindentation proved the existence of mechanically different phases within the cell walls. Examples of loading diagrams received from nanoindentation at different locations are shown in Fig. 5. Results from Al-rich areas exhibit higher penetration depths with more compliant response whereas Ca/Ti-rich area is characterized by smaller penetration depths and increased stiffness. An average maximum depth of penetration reached by the indenter was around ~180 nm. Higher values for more compliant Al-rich zone were reached (~190 nm) whereas the indentation depths to harder but less frequent Ca/Ti-rich areas were around 100 nm.

Elastic properties evaluated for each individual indent were subsequently merged for the purpose of statistical analysis. The deconvolution algorithm was applied to the experimental histogram (PDF) of elastic moduli with assumption of a two-phase system (one dominant Al-rich phase and one minor Ca/Ti-rich phase) and a fixed bin size of elastic modulus used for the construction of PDF equal to 1 GPa. It can be seen in Fig. 6 that a significant peak appears around 62 GPa. This value can be considered as a dominant characteristic of the prevailing Al-rich phase. The distribution of the second phase (i.e. minor Ca/Ti-rich zone) shows much larger scatter which is likely due to the inhomogeneity of the precipitates.

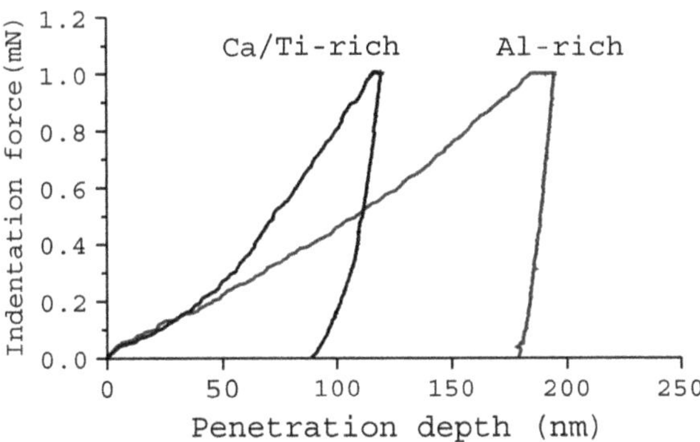

Fig. 5 Typical loading diagrams for Al-rich and Ca/Ti-rich zones

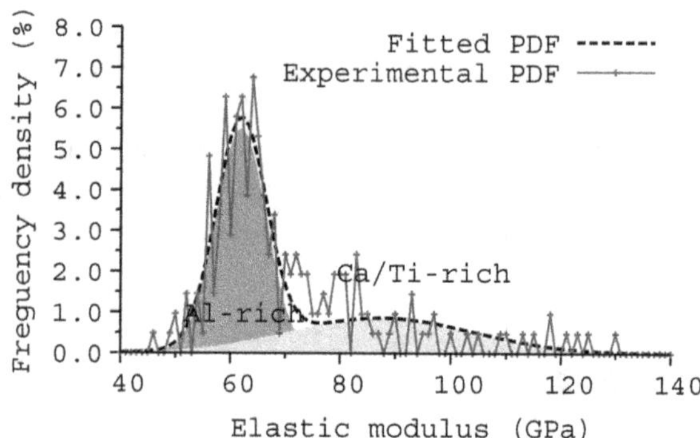

Fig. 6 Statistical deconvolution applied to the experimental histogram of elastic modulus

Table 1 Elastic moduli and volume fractions received from deconvolution

Phase	Mean (GPa)	St. dev. (GPa)	Volume fraction (−)
Al-rich	61.88	4.6	0.638
Ca/Ti-rich	87.40	16.7	0.362

Table 1 contains numerical results from the deconvolution with the estimated volume fractions of the phases.

4.2 Results of Elastic Homogenization

Results from statistical deconvolution were further applied in analytical homogenization schemes. The homogenized elastic modulus for the two considered microscale phases in the cell wall is summarized in Table 2 for individual homogenization techniques. Very close bounds and insignificant differences in elastic moduli estimated by the schemes were found. In the following, we take results from the Mori-Tanaka method ($E = 70.083$ GPa, $v = 0.35$) as a product of analytical homogenization.

It is possible to construct an elastic stiffness matrix of an isotropic media from this result (in Mandel's notation) as (Eq. 16):

$$L_{\text{eff}}^{A} = \frac{E}{(1+v)(1-2v)} \begin{bmatrix} 1-v & v & 0 \\ v & 1-v & 0 \\ 0 & 0 & 1-2v \end{bmatrix} = \begin{bmatrix} 112.479 & 60.566 & 0 \\ 60.566 & 112.479 & 0 \\ 0 & 0 & 51.913 \end{bmatrix}. \tag{16}$$

Numerical FFT-based technique was further applied. Elastic moduli from grid indentation served as input values for the method. Resulting elastic stiffness matrix (in Mandel's notation) was (Eq. 17):

$$L_{\text{eff}}^{\text{FFT}} = \begin{bmatrix} 117.1300 & 62.7413 & -0.1625 \\ 62.7413 & 117.1060 & -0.1430 \\ -0.1625 & -0.1430 & 54.3132 \end{bmatrix} \text{(GPa)}. \tag{17}$$

Table 2 Effective Young's modulus computed by different homogenization schemes

Scheme	Mori-Tanaka	Self-consist. scheme	Voigt bound	Reuss bound
E (GPa)	70.083	70.135	71.118	69.195

The stiffness matrices in Eqs. 16 and 17 contain similar values on relevant positions, for the first sight. To compare the matrices more rigorously, one can apply different measures. Here, a matrix error norm is computed as (Eq. 18):

$$\delta = \sqrt{\frac{\left(L_{eff}^{FFT} - L_{eff}^{A}\right) :: \left(L_{eff}^{FFT} - L_{eff}^{A}\right)}{\left(L_{eff}^{FFT} :: L_{eff}^{FFT}\right)}} = 0.04 \tag{18}$$

Relatively low value ($\delta = 4\ \%$) indicates the matrix similarity.

The homogenized matrix in Eq. 17 is symmetric and positive definite. Some of the matrix components are negative which has no physical meaning. It just shows a certain (low) level of anisotropy that comes from the distribution of phases within the calculated periodic unit cell. To assess the degree of material anisotropy received in the FFT method, again different measures can be used. One of the choices is to compute a matrix error norm with a particular choice of a reference (isotropic) matrix such that (Němeček et al. 2013) (Eqs. 19–21):

$$\delta_{ISO}^{FFT} = \sqrt{\frac{\left(L_{eff}^{FFT} - L_{ISO}^{FFT}\right) :: \left(L_{eff}^{FFT} - L_{ISO}^{FFT}\right)}{\left(L_{eff}^{FFT} :: L_{ISO}^{FFT}\right)}} \tag{19}$$

$$L_{ISO}^{FFT} = \begin{bmatrix} k_{ISO} + \frac{4}{3}\mu_{ISO} & k_{ISO} - \frac{2}{3}\mu_{ISO} & 0 \\ k_{ISO} - \frac{2}{3}\mu_{ISO} & k_{ISO} + \frac{4}{3}\mu_{ISO} & 0 \\ 0 & 0 & 2\mu_{ISO} \end{bmatrix} \tag{20}$$

$$\mu_{ISO} = \frac{L_{eff,33}^{FFT}}{2}, \quad k_{ISO} = \frac{L_{eff,11}^{FFT} + L_{eff,22}^{FFT}}{2} - \frac{4}{3}\mu_{ISO} \tag{21}$$

Supplying results from Eq. 17 to Eqs. 19–21 yields $\delta_{ISO} = 0.0016$. Close to zero value shows that the degree of anisotropy is small in our case and the material has randomly distributed phases compared to materials with directionally dependent microstructures. Therefore, we conclude that the material is characterized with a close-to-isotropic behavior.

5 Discussion on Results

Results from nanoindentation proved the existence of two mechanically distinct phases in the cell walls that correspond to their chemical counterparts (Al-rich and Ca/Ti-rich zones). The characteristic elastic modulus ~62 GPa obtained for the

Al-rich phase is lower than that for a pure aluminium (~70 GPa, e.g. Webelements on-line library). The lower value obtained from nanoindentation suggests that probably some small-scale porosity or impurities (Ca) added to the molten are intrinsically included in the results of this mechanically dominant phase. On the other hand, the determined elastic modulus value of Al-rich zone is in excellent agreement with the value 61.7 GPa measured by Jeon et al. (2009) on melted Al-1.5 wt % Ca alloy.

Further, results from mechanical analysis (grid nanoindentation) revealed that simple image analysis does not allow estimation of volume fractions of the mechanically distinct phases on the walls. Based on the color in SEM images the area occupied with Ca/Ti-rich phase is estimated as 22 ± 4 % by image analysis. Results from statistical nanoindentation (36.2 %) suggest that a substantially larger part of the matrix is mechanically influenced by the Ca/Ti addition and a higher fraction of the volume belongs to this mechanically distinct phase.

Analytical homogenizations provided almost the same results regardless on the scheme used which is likely due to not so high stiffness contrast between the phases and relatively high volume fraction of Ca/Ti-rich inclusions in the composite. Numerical FFT-based scheme fully respects the spatial distribution of mechanical results from an indentation grid. However, similar stiffness matrix was obtained compared to analytical results (4 % difference). This outcome suggests that the material shows close-to-isotropic nature and the microstructural inhomogeneities are relatively uniformly dispersed in its RVE. Consequently, this finding also justifies the usage of analytical methods producing isotropic effective properties for the studied case.

6 Concluding Remarks

Micromechanical properties of an Al/Ca/Ti alloy appearing on cell walls of a closed-cell foam system Alporas have been studied in this contribution. Effective elastic properties of the walls have been assessed with the aid of several experimental and numerical techniques.

Electron microscopy and image analyses provided the first insight to the multiphase microstructural system. Two characteristic zones named as Al-rich and Ca/Ti-rich phases have been identified. However, results from the image analysis did not correspond to the mechanical ones in terms of phase volume fractions. Therefore, separate measurements performed on the phases and using volume estimates taken just from the image analysis would lead to erroneous results if used for computation of effective wall properties.

On the other hand, it has been demonstrated that the use of grid nanoindentation in conjunction with statistical deconvolution method provides an effective tool for the assessment of mechanical properties of the studied two-phase system. Results of elastic stiffness and volume fractions for both the dominant and minor phases have been obtained and used for subsequent homogenization.

Application of analytical and FFT-based numerical approach led to the assessment of the effective elastic stiffness matrix. Differences only up to 4 % have been found between the methods. Relatively good dispersion of differently stiff regions in the tested material contributed to close-to-isotropic nature of the alloy and low differences obtained from various homogenization methods.

Results of the lower level material characterization described in this contribution can be further used in a multiscale analysis of the whole foam system taking into account pore shapes and distribution of cell walls (Němeček and Králík 2012).

Acknowledgments Support of the Czech Science Foundation (P105/12/0824) is gratefully acknowledged.

References

Ashby MF, Evans AG, Fleck NA, Gibson LJ, Hutchinson JW, Wadley HNG (2000) Metal foams: a design guide. Butterworth-Heinemann, UK

Banhart J (2001) Manufacture, characterisation and application of cellular metals and metal foams. Prog Mater Sci 46:559–632

Constantinides G, Chandran FR, Ulm F-J, Vliet KV (2006) Grid indentation analysis of composite microstructure and mechanics: principles and validation. Mater Sci Eng A 430(1–2):189–202

De Giorgi M, Carofalo A, Dattoma V, Nobile R, Palano F (2010) Aluminium foams structural modelling. Comput Struct 88:25–35

Drugan WR, Willis JR (1996) A micromechanics-based nonlocal constitutive equation and estimates of representative volume element size for elastic composites. J Mech Phys Solids 44(4):497–524

Durst K, Göken M, Vehoff H (2004) Finite element study for nanoindentation measurements on two-phase materials. J Mater Res 19:85–93

Eshelby JD (1957) The determination of the elastic field of an ellipsoidal inclusion and related problem. Proc Roy Soc London A 241:376–396

Fischer-Cripps AC (2002) Nanoindentation. Springer Verlag, ISBN 0-387-95394-9

Hasan MA, Kim A, Lee H-J (2008) Measuring the cell wall mechanical properties of Al-alloy foams using the nanoindentation method. Compos Struct 83:180–188

Hashin Z (1983) Analysis of composite materials—a survey. ASME J Appl Mech 50:481–505

Hill R (1963) Elastic properties of reinforced solids—some theoretical principles. J Mech Phys Solids 11:357–372

Hill R (1965) Continuum micro-mechanics of elastoplastic polycrystals. J Mech Phys Solids 13:89–101

Idris MI, Vodenitcharova Hoffman M (2009) Mechanical behaviour and energy absorption of closed-cell aluminium foam panels in uniaxial compression. Mater Sci Eng A 517:37–45

ISO 4287-(1997) geometrical product specifications (GPS)—surface texture: profile method—terms, definitions and surface texture parameters

Jeon I, Katou K, Sonoda T, Asahina T, Kang K-J (2009) Cell wall mechanical properties of closed-cell Al foam. Mech Mater 41:60–73

Michel JC, Moulinec H, Suquet P (1999) Effective properties of composite materials with periodic microstructure: a computational approach. Comput Methods Appl Mech Eng 172:109–143

Miyoshi T, Itoh M, Akiyama S, Kitahara A (1998) Aluminium foam "ALPORAS": the production process, properties and application. Materials research society symp. proc, p 521

Mori T, Tanaka K (1973) Average stress in matrix and average elastic energy of materials with misfitting inclusions. Acta Metall 21(5):571–574

Moulinec H, Suquet P (1994) A fast numerical method for computing the linear and non-linear properties of composites. Comptes-Rendus de l'Académie des Sciences série II 318:1417–1423

Moulinec H, Suquet P (1998) A numerical method for computing the overall response of nonlinear composites with complex microstructure. Comput Methods Appl Mech Eng 157(1–2):69–94

Němeček J (2009) Creep effects in nanoindentation of hydrated phases of cement pastes. Mater Charact 60(9):1028–1034. doi:10.1016/j.matchar.2009.04.008

Němeček J (2012) Nanoindentation based analysis of heterogeneous structural materials. In: Němeček J (ed) Nanoindentation in materials science, intech, pp 89–108. doi:10.5772/50968

Němeček J, Králík V (2012) A two-scale micromechanical model for closed-cell aluminium foams. In: Proceedings of the thirteenth international conference on civil, structural and environmental engineering computing. Civil-Comp Press, Edinburgh, paper 259, pp 1–12. doi:10.4203/ccp.99.259

Němeček J, Lehmann C, Fontana P (2011a) Nanoindentation on ultra high performance concrete system. Chem Listy 105(17):656–659

Němeček J, Šmilauer V, Kopecký L (2011b) Nanoindentation characteristics of alkali-activated aluminosilicate materials. Cement Concr Compos 33(2):163–170. doi:10.1016/j.cemconc omp.2010.10.005

Němeček J, Králík V, Vondřejc J (2013) Micromechanical analysis of heterogeneous structural materials. Cement and Concrete Composites 36:85–92. doi:10.1016/j.cemconc omp.2012.06.015

Oliver W, Pharr GM (1992) An improved technique for determining hardness and elastic modulus using load and displacement sensing indentation experiments. J Mater Res 7(6):1564–1583

Papadopoulos DP, Konstantinidis IC, Papanastasiou N, Skolianos S, Lefakis H, Tsipas DN (2004) Mechanical properties of Al metal foams. Mater Lett 58(21):2574–2578

Saranen J, Vainikko G (2002) Periodic integral and pseudo differential equations with numerical approximation. Springer, Berlin

Simone AE, Gibson LJ (1998) Aluminum foams produced by liquid-state processes. Acta Mater 46(9):3109–3123

Sorelli L, Constantinides G, Ulm F-J, Toutlemonde F (2008) The nano-mechanical signature of ultra high performance concrete by statistical nanoindentation techniques. Cem Concr Res 38:1447–1456

Sugimura Y, Meyer J, He MY, Bart-Smith H, Grenstedt J, Evans AG (1997) On the mechanical performance of closed cell Al alloy foams. Acta Mater 45:5245–5259

Ulm F-J, Vandamme M, Bobko C, Ortega JA (2007) Statistical indentation techniques for hydrated nanocomposites: concrete, bone, and shale. J Am Ceram Soc 90(9):2677–2692

Webelements on-line library. http://www.webelements.com/aluminium/physics.html

Yongliang M, Guangchun Y, Hongjie L (2010) Effect of cell shape anisotropy on the compressive behavior of closed-cell aluminum foams. Mater Des 31:1567–1569

Zaoui A (2002) Continuum micromechanics: survey. J Eng Mech 128(8):808–816

Zeman J, Vondřejc J, Novák J, Marek I (2010) Accelerating a FFT-based solver for numerical homogenization of periodic media by conjugate gradients. J Comput Phys 229(21):8065–8071. doi:10.1016/j.jcp.2010.07.010

Mechanical Properties of Biomaterials Determined by Nano-Indentation and Nano-Scratch Tests

A. Karimzadeh and M. R. Ayatollahi

Abstract Nano-indentation and nano-scratch tests are appropriate methods for measuring the mechanical and tribological properties of bulk samples, thin films and coatings. Using nano-scale tests, numerous properties like modulus of elasticity, hardness, fracture toughness, elastic–plastic behavior and wear resistance can be obtained. Macro-scale tests are conventional methods for determining the mechanical properties of biomaterials, but they are often expensive and require rather large test samples. In this article, the mechanical and tribological properties of hand-mixed and vacuum-mixed bone cements and a dental nano-composite determined by using nano-indentation and nano-scratch tests will be described and discussed. It is shown that bone cement mixed using the vacuum mixing exhibits significantly improved mechanical and tribological properties compared with the hand-mixed bone cement. Moreover, it is demonstrated that thermocycling affect the mechanical properties of dental nano-composite considerably.

1 Introduction

The mechanical properties of biomaterials are often obtained using the macro-scale or micro-scale experiments. For instance, fracture toughness measurements for biomaterials are sometimes performed on various test specimens such as single edge notch, chevron notched short rod or by using the Vickers indentation technique. However, such methods might have a number of disadvantages like:

A. Karimzadeh · M. R. Ayatollahi (✉)
Fatigue and Fracture Research Laboratory, School of Mechanical Engineering,
Iran University of Science and Technology, Tehran, Iran
e-mail: m.ayat@iust.ac.ir

A. Tiwari (ed.), *Nanomechanical Analysis of High Performance Materials*,
Solid Mechanics and Its Applications 203, DOI: 10.1007/978-94-007-6919-9_10,
© Springer Science+Business Media Dordrecht 2014

1. The specimen preparation in the first two methods is costly and highly time consuming.
2. It is often difficult to create sharp pre-cracks in brittle and quasi-brittle materials (including many biomaterials), without catastrophic damage in the test sample.
3. Fracture toughness data obtained from the notched specimens sometimes give erroneously high values (Munz et al. 1980). Therefore, some researchers have preferred to investigate fracture toughness of biomaterials like bone cement by direct measurements of cracks created using a sharp diamond indenter (Lawn et al. 1980; Anstis et al. 1981; Fett 2002, Fett et al. 2005).
4. In the Vickers indentation method, it is common to observe other undesirable cracks such as lateral cracks or cone cracks, in addition to the main crack. These additional cracks disturb the stress field of the main crack and hence underestimate the toughness value (Cook and Pharr 1990; Kruzic and Ritchie 2003).

In the nano-indentation test, the threshold load to induce cracking is significantly lower than the Vickers indentation test (Kruzica et al. 2009) and thus the risk of generating lateral cracks or cone cracks is reduced. The highly controlled applied loads and positioning of the nano-indentation system can provide sensitive measurement of the load and the initiation of fracture. Similar advantages can be found for measuring other mechanical properties of materials using the nano-indentation technique. Therefore, nano-indentation can be considered as a reliable and convenient method for obtaining fracture toughness of brittle materials in addition to their other mechanical properties.

For reasons such as the accuracy of load control and scan size and also the absence of electrical and chemical fields the nano-scratch test too can be a good alternative to the conventional wear tests for measuring the tribological properties of materials.

In this article, the nano-indentation and nano-scratch tests are employed to investigate the procedure for determination of the mechanical and tribological properties of hand-mixed and vacuum-mixed bone cements and a dental nano-composite.

2 Test Procedure

2.1 Sample Preparation

Sample preparation in the nano-indentation and nano-scratch tests is based on the ISO 14577 standard (parts 1 and 4). According to this standard, the nano-indentation and nano-scratch tests are independent of the specimen geometry and depend only on the specimen thickness which should be large enough relative to the indentation depth such that the test result is not influenced by the test piece support. Accordingly, the specimen thickness should be at least 10 times the indentation depth or 3 times the indentation diameter. In the case of coated samples, the coating thickness should be considered as the test piece thickness (ISO-14577-1 2002; ISO-14577-4 2007). Meanwhile, the specimen dimension must be small enough to be fixed firmly in the test piece holder of the test instrument.

The contact area between the test piece and the indenter shall be free of fluids or lubricants except where this is essential for the performance of the test. Care shall be taken that the contact area is free from extraneous materials (e.g. dust particles). Surface finish has a significant influence on the test results. The preparation of the test surface shall be carried out in such a way that any likely alteration in the surface hardness (e.g. due to heat or cold-working) is minimized.

A polishing process that is suitable for the test material shall be used. The test surface shall be normal to the test force direction with a tilting angle typically no more than 1° (ISO-14577-1 2002; ISO-14577-4 2007).

2.2 Nano-Indentation Test

The nano-indentation test uses an established method in which an indenter tip is pressed into specific sites of the test material, by applying an increasing normal load. When the penetration depth of the indenter tip reaches a pre-set maximum value, the normal load is reduced until partial or complete relaxation occurs. During the test, a high-precision instrument records the values of load and displacement, continuously.

Previous studies demonstrated that the penetration depth of the indenter affects the measured mechanical properties of materials and, in most cases, after a depth of 200 nm the measured properties are almost stable (Hu et al. 2006; Liu et al. 2004). Indeed, the indentation depth should be deep enough to minimize the surface effect. Meanwhile, the indentation depth should also be less than 10 % of the film thickness when the sample is mounted on a hard substance; otherwise the measured value is usually larger than it should be due to the effect of the support (Jee and Lee 2010).

In the nano-indentation test, the tip of the testing instrument is calibrated by the Oliver-Pharr method, and the same method is used for analyzing the experimental data. The Oliver-Pharr method depends on the unloading segment of the load–displacement curve and assumes that only the elastic displacements are recovered. However, with this assumption, an error occurs in determination of the mechanical properties of polymers due to their time and rate dependent behavior. Therefore, to eliminate the error in viscous materials, it is common to hold the indenter at the maximum load for a period of time (Chudoba and Richter 2001; Ngan et al. 2005).

Fig. 1 Schematic illustration of the indenter direction in the nano-scratch test

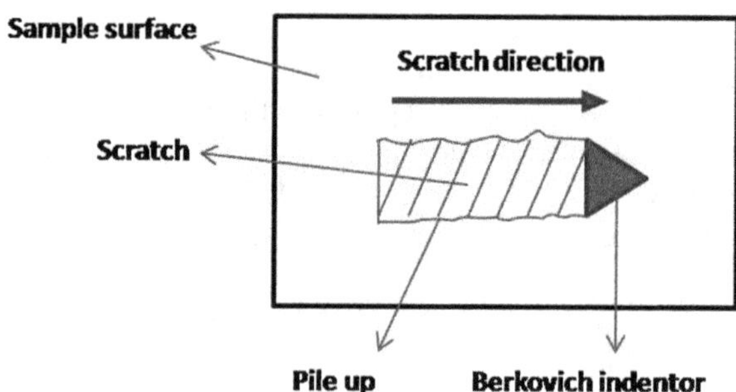

2.3 Nano-Scratch Test

During this test, a scratch is made on the sample surface with an indenter tip which is drawn at a constant speed across the sample under a constant load or, more commonly, a progressive load of fixed loading rate.

The direction of indenter tip in a nano-scratch test affects the scratch results. Thus, in order to consider the viscous behavior in the elastic and plastic parts of deformation, the indenter is often drawn in the direction which creates high plastic deformation in the specimens (Sinha et al. 2009). A schematic illustration of the indenter direction in the nano-scratch test is shown in Fig. 1.

3 Mechanical Properties of Bone Cement

Bone cements have numerous applications in orthopedics such as making hip prostheses and packing the defects caused by bone tumor. The base of these cements is often polymethylmethacrylate (PMMA).

The bone cement is the weakest element in prosthesis and the mechanical failure in this cement is the main reason for prosthesis failure. Cement failure due to the loads imposed under working conditions occasionally forces the prosthetic replacement surgery to be repeated. Moreover, the abrasion of bone cement may result in small cement debris in the contact surface between the bone and the cement, causing the bone to degrade more rapidly. Thus, the assessment of the mechanical and tribological properties of bone cements, such as hardness, modulus of elasticity and wear resistance is very important when using these cements.

Polymer powder and liquid monomer are two components of bone cements, which can be mixed in different ways. Hand mixing is one of the common methods for mixing the components of bone cement but it can produce considerable porosity in the cement, while the person mixing the cement can also be exposed to harmful methyl methacrylate vapors. Another common method for mixing the bone cement components is vacuum mixing in which the polymer powder and the liquid monomer are mixed under vacuum. The porosity of bone cement is expected to reduce in vacuum mixing (Dunne and Orr 2001). The mechanical and tribological properties of bone cement recently studied by Ayatollahi and Karimzadeh using the nano-indentation and nano-scratch tests (Karimzadeh and Ayatollahi 2012; Ayatollahi and Karimzadeh 2012) are describe in this section.

3.1 Bone Cement Sample Preparation

CEMEX RX (TECRES Company, Italy) was used for preparing the bone cement samples. Monomer liquid and polymer powder were mixed according to the manufacturer's instructions. Two methods of mixing were employed. In the first method, the components of bone cement were mixed by hand in air at a temperature of 23 °C for 60 s. In the second method, vacuum mixing was used where powder and liquid were poured into the vacuum mixing machine and held for 15 s under vacuum pressure of 0.7 bars at a temperature of 23 °C. The mixture was mixed for 30 s in the conditions described above and then held for a further 15 s at 0.7 bar vacuum pressure and 23 °C. After mixing the cement components by either of hand mixing and vacuum mixing methods, the mixture was injected separately into cubic molds of the size $10 \times 10 \times 5$ mm^3. The cements were allowed to cure for 15 min in air and the specimens were subsequently removed from the molds.

In order to obtain a smooth surface for performing nano-indentation test, all the samples were ground with 400–2,500 grit sandpaper and then polished with an alumina suspension. The roughness values of samples were checked using atomic force microscopy (AFM). The highly polished specimens were kept for 2 months in the ambient conditions at 23 °C before conducting the nano-indentation and nano-scratch tests.

3.2 Experiments on Bone Cement

The nano indentation and nano scratch tests were performed using a Berkovich indenter and a Triboscope system (Hysitron Inc., USA) based on ISO 14577. The test setup and Berkovich indenter tip are shown in Fig. 2. The Berkovich indenter has an average curvature radius of about 150 nm and is primarily used for bulk materials and thin films of greater than 100 nm thicknesses.

3.3 Nano-Indentation Test of Bone Cement

An indentation load of 450 mN with constant rate of 15 mN s^{-1} was applied. The maximum indentation depth in the experiments was limited to 210 nm. A holding time of 15 s was selected to minimize the likely error due to the time dependent behavior of the material.

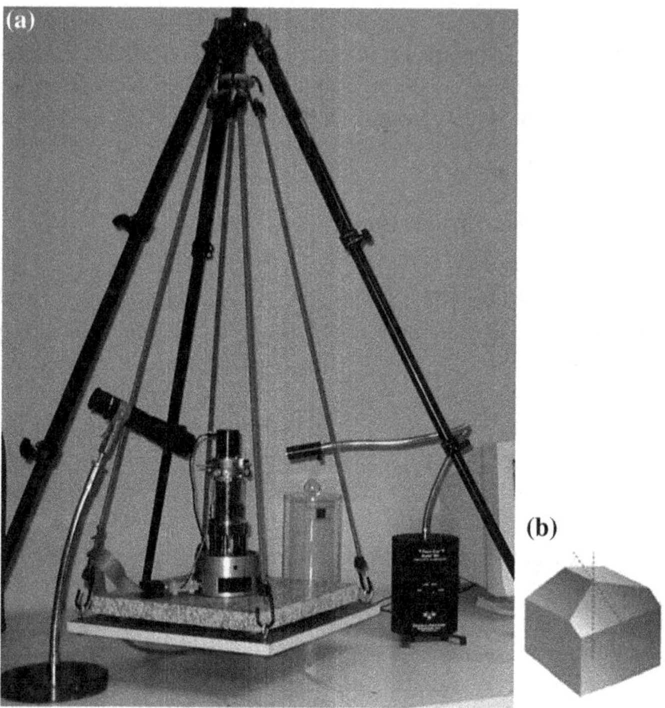

Fig. 2 **a** Test device. **b** Berkovich indenter tip

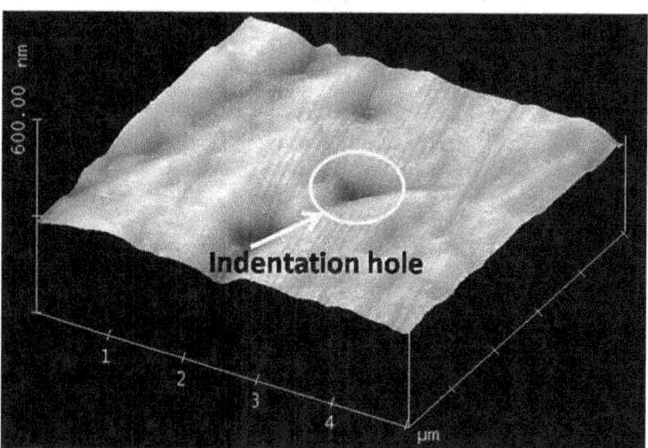

Fig. 3 AFM Image of an indentation hole on the bone cement sample

According to the described method, at least five indentations on the randomly selected sites of the samples were performed at a temperature of 23 °C. The AFM images taken before and after the indentations were used for the subsequent analyses. AFM image of an indentation hole on the bone cement is shown in Fig. 3.

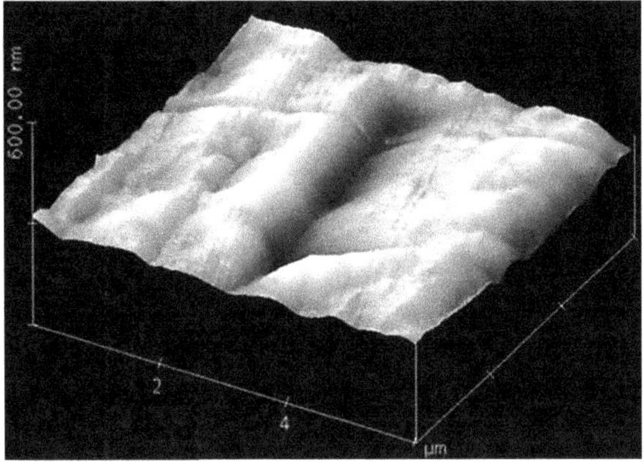

Fig. 4 AFM Image of a scratch on the bone cement sample

3.4 Nano-Scratch Test of Bone Cement

An indenter with penetration load of 500 mN and constant scratch speed of 0.13 mN s^{-1} was used. The scratch length was 4 mm and the indenter penetration depth was 300 nm. The AFM images were taken before and after the test for analyzing the sample deformation at the scratch site. Figure 4 shows the AFM image of an indentation hole on the bone cement.

3.5 Modulus of Elasticity of Bone Cement

Based on the continuous load–displacement data obtained from a complete cycle of loading and unloading during the nano-indentation test and by using the Oliver-Pharr method (Oliver and Pharr 2004), the effective modulus of elasticity is calculated. A sample load–displacement curve obtained from the nano-indentation test of bone cement is shown in Fig. 5.

Using the Sneddon relationship, the elasticity modulus of bone cement samples can be calculated from (Sneddon 1965):

$$\frac{1}{E_{eff}} = \frac{1 - v^2}{E} - \frac{1 - v_i^2}{E_i} \tag{1}$$

where E and v are the elasticity modulus and Poisson's ratio of the test samples, E_i and v_i are the elasticity modulus and the Poisson's ratio of indenter tip and E_{eff} is the effective elasticity modulus of material obtained from the nano-indentation test.

The AFM images of all indentation tests were studied and the elasticity moduli of the hand-mixed and vacuum mixed samples were calculated by using Eq. 1 and considering $v = 0.3$ (Murphy and Prendergast 1999), $v_i = 0.07$ and

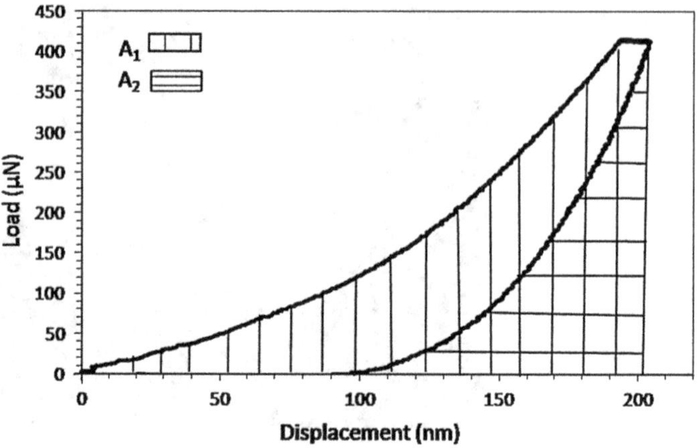

Fig. 5 Load–displacement curve in nano-indentation test of the bone cement sample

Table 1 Mean value and standard deviation of elastic modulus obtained for the hand-mixed and vacuum-mixed cements

Samples	Mean elastic modulus (GPa)	Standard deviation
Hand-mixed cement	5.56	0.17
Vacuum-mixed cement	6.00	0.10

$E_i = 1{,}140$ GPa, but only for those tests where no redial crack around the indentation hole were observed. The values used for v_i and E_i are based on the technical data available for the related Triboscope system. It is noteworthy that the load–displacement curves of each test sample obtained from the nano-indentation technique demonstrated good repeatability for the bone cement.

The mean values obtained for the elasticity modulus of the hand-mixed and vacuum-mixed CEMEX RX cements and their corresponding standard deviations are given in Table 1.

The results indicate that the elasticity modulus of vacuum-mixed cement is higher than that of the hand-mixed samples. The comparison between the values of hand-mixed and vacuum-mixed moduli of elasticity with t test indicates that 0.44 difference between the two values is statistically significant (p-value $= 0.01$). Thus, the vacuum mixing method makes the cement stiffer compared with the hand mixing method.

According to previous studies (Lidgren et al. 1987; Wang et al. 1996; Smeds et al. 1997), the percent volume of porosity in the bone cements produced using the vacuum mixing method decreases significantly in comparison with the hand-mixed cements. The reduction in the percent porosity increases the connection between the polymer chains leading to enhanced material stiffness. Therefore, the vacuum-mixed CEMEX RX cements are expected to be stiffer than the hand-mixed ones.

3.6 Hardness of Bone Cement

Material hardness is defined as the material resistance against surface deformation caused by an external load. It can be calculated by dividing the normal load by the projected area of the surface on which it was imposed.

As suggested by (Briscoe et al. 1996), the hardness or the normal hardness (H_n) can be determined in the nano-indentation test from:

$$H_n = \frac{F'_N}{A} \tag{2}$$

where F'_N is the maximum load for a given indentation and A is the projected area of the contact surface between the specimen and the indenter. The load–displacement curve obtained from the nano-indentation test and also the Oliver-Pharr method were used (Oliver and Pharr 2004) in this study in order to calculate F'_N and A in Eq. 2.

The mean values obtained for the normal hardness (H_n) of the hand-mixed and vacuum-mixed CEMEX RX cements and their corresponding standard deviations are given in Table 2.

Table 2 shows that normal hardness of the bone cement increases when the vacuum mixing method is used instead of hand mixing. The use of the t-test indicates that 0.03 difference between the two values is statistically significant (p-value $= 0.03$).

Scratch hardness (H_s) is one of the parameters in the nano-scratch test which is used for determining the resistance of material to the scratch. If the Berkovich indenter is used in the nano-scratch test, the scratch hardness is calculated from (Williams 1996):

$$H_s = 2.31 \frac{F_N}{d^2} \tag{3}$$

where F_N is the maximum normal load applied on the indenter in the nano-scratch test and d is the residual width of the scratch. According to the definition of hardness mentioned earlier, one may suggest that the use of the residual scratch width in Eq. 3 imposes some errors in the calculation of scratch hardness. (Briscoe and Sinha 2003) showed that in polymers, the visco-elastic recovery has only a slight effect on the scratch width, thus it is reasonable to use the residual scratch width for the calculation of the scratch hardness.

The AFM pictures taken from the scratch sites (see Fig. 6) were used to measure the values of residual scratch width i.e. d in Eq. 3. The material pile up around the scratches caused some variations in the measured scratch widths. Therefore, at

Table 2 Mean value and standard deviation of normal hardness obtained for the hand-mixed and vacuum-mixed cements

Samples	Mean normal hardness (GPa)	Standard deviation
Hand-mixed cement	0.29	0.010
Vacuum-mixed cement	0.32	0.015

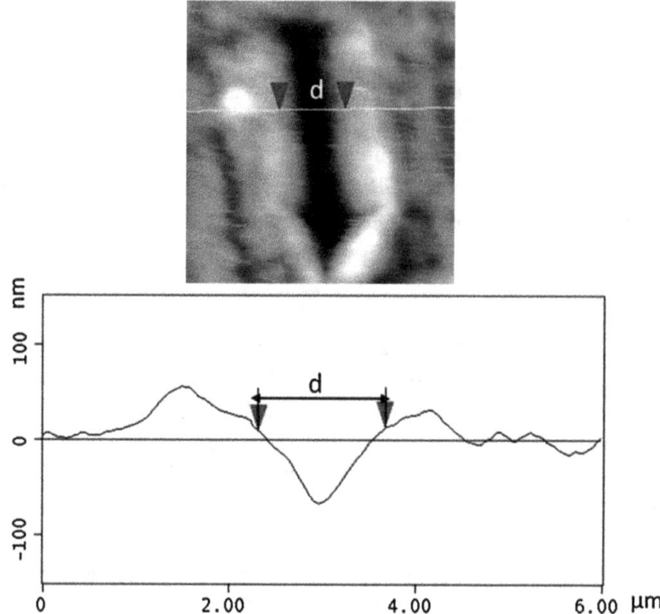

Fig. 6 An AFM image of scratch and its associated cross section

Table 3 Mean values obtained for the scratch hardness of the hand-mixed and vacuum-mixed cements

Samples	Mean scratch hardness (GPa)	Standard deviation
Hand-mixed cement	0.48	0.010
Vacuum-mixed cement	0.62	0.053

least 10 measurements of the scratch widths were taken and the average value for each scratch was used in Eq. 3.

It was found from the AFM profiles of the scratches that under the same load, the residual widths of the scratch for the vacuum-mixed samples were less than the residual widths of the scratch for the hand-mixed ones. This indicates that the values of scratch hardness for the vacuum-mixed cement increase in comparison with those of the hand-mixed cement.

Table 3 present the mean values obtained for the scratch hardness (H_s) of the hand-mixed and vacuum-mixed CEMEX RX cements and their corresponding standard deviations.

Independent-samples t-test indicates that 0.14 difference between the two values of scratch hardness is statistically significant (P-value $= 0.009$). Therefore, one can suggest that the vacuum-mixed bone cement is more resistant to deformation caused by scratching than the hand-mixed bone cement (Karimzadeh and Ayatollahi 2012).

If hardness is defined as the resistance of the material against surface deformation, the same values are expected to be obtained for the normal hardness and the scratch hardness. Although the trends for the values of normal and scratch hardness obtained in this study are similar for the hand-mixed and vacuum-mixed cements, the ratio of the scratch hardness to the normal hardness (i.e. H_s/H_n) is in the range of 1.6–1.9. This difference can be due to the various mechanisms involved in the two test methods. The normal hardness is the resistance of material against local deformation created by vertical penetration of an indenter (near static condition) (Duncan 1999), whereas the scratch hardness is ploughing deformation caused by the interfacial friction between the indenter and the material (Dunne and Orr 2001).

During the dynamic deformation of the surface which occurs in the nano-scratch process, the indenter moves laterally and the net energy for causing deformation on the material surface is calculated from the scratch force or the tangential component of the load (Briscoe et al. 1996). The scratch force depends on the mechanism of material deformation and also on the interfacial friction between the indenter and the material. The attack angle during the scratch is another factor which influences the scratch force, because the mode of material deformation may change with this angle (Sinha et al. 2009).

3.7 Fracture Toughness of Bone Cement

The nano-indentation test can be used for determining fracture toughness of brittle materials. Lawn et al. (1980) proposed Eq. 4 for calculating fracture toughness of ceramics from an indentation test.

$$K_c = \alpha \sqrt{\frac{E}{H}} \frac{P}{c^{\frac{3}{2}}} \tag{4}$$

where P is the applied load, E is the elasticity modulus, H is the hardness, and c is the length of the surface radial crack measured from the center of the indentation hole. α is an empirically determined calibration constant which can be considered 0.04 for sharp indentation geometries, such as Berkovich (Pharr 1998).

The nano-indentation technique was employed for measuring fracture toughness for both hand-mixed and vacuum-mixed bone cements. The AFM images of the nano-indentation holes and the corresponding surface radial cracks are shown in Fig. 7a and b for the hand-mixed and the vacuum-mixed cements, respectively. The length of surface radial crack (c in Eq. 4) is measured from these AFM images.

The values of elasticity modulus and hardness of bone cement obtained earlier from the nano-indentation test were used in Eq. 4 and the values of fracture toughness (K_c) were obtained 0.39 MPa $\sqrt{\text{m}}$ for the hand-mixed cement and 0.62 MPa $\sqrt{\text{m}}$ for the vacuum-mixed cement. The results indicate that fracture toughness of bone cement increases more than 50 % when vacuum mixing is used instead of hand mixing (Ayatollahi and Karimzadeh 2012).

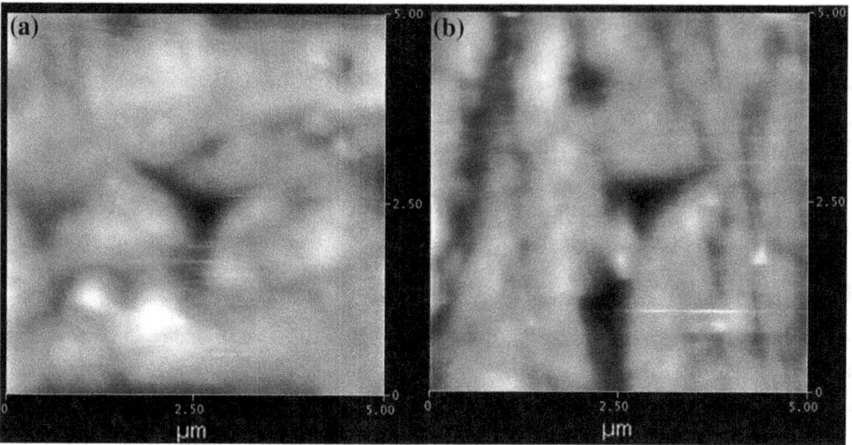

Fig. 7 The surface radial crack around the nano-indentation hole for (**a**) hand-mixed, and (**b**) vacuum-mixed cements

As reported by several researchers (Lidgren et al. 1987; Smeds et al. 1997; Wang et al. 1996), the percent volume of porosity in the bone cements produced using the vacuum mixing method decreases significantly in comparison with the hand-mixed cements. Once the percent porosity is reduced, the connection between the polymer chains and hence the material resistance against crack propagation increase. In addition, local stress concentrations are reduced in the samples having lower porosity. Therefore, fracture toughness of the vacuum-mixed bone cement is expected to be higher than that of the hand-mixed one.

According to the AFM images of the nano-indentation tests, no lateral crack has been formed and cracks are observed only at the three corners of the indentation holes, among which the longest crack was considered for calculating fracture toughness of the bone cement samples from Eq. 4. Due to the relatively large angles between these three cracks, the interference between their crack tip stress fields is negligible.

Conventional fracture toughness tests are often performed on rather large test samples which impose undesirable extra costs. Since the nano-indentation technique needs less test material and decreases sample preparation costs in comparison with the macro or micro-scale test methods (e.g. Vickers indentation), it can be considered as a suitable alternative for determining fracture toughness of brittle or quasi-brittle materials like bone cements.

3.8 Elastic–Plastic Behavior of Bone Cement

The elastic–plastic behavior of the material can be characterized by using the nano-indentation test. The plasticity index (ψ) and the recovery resistance parameter (R_s) are two important parameters which can be used for such characterizations.

The plasticity index of a solid describes the elastic–plastic response of the material under external stresses and strains. In other words, this variable determines the relation between the elastic and the non-elastic (either plastic or visco-elastic) response of a material.

If the load–displacement data are recorded accurately during the nano-indentation test, the plasticity index of the material can be estimated from the test results using Eq. 5.

$$\psi = \frac{A_1 - A_2}{A_1} \tag{5}$$

where A_1 is the area under the loading section of the load–displacement curve (i.e. the total work during the nano-indentation test) and A_2 is the area under the unloading section (i.e. the reversible work due to the visco-elastic behavior during the nano-indentation test), as shown in Fig. 5. According to the descriptions, the difference between A_1 and A_2 indicates the irreversible or plastic work during the nano-indentation test.

The plasticity index (ψ) can vary between 0 and 1, where $\psi = 0$ indicates the fully-elastic behavior and $\psi = 1$ represents fully-plastic behavior of the material.

Since in this study the same indenter is used for all samples, the plasticity index (ψ) can also be calculated by Eq. 6:

$$\psi = \frac{h_m - h_e}{h_m} \tag{6}$$

where h_m is the maximum indentation depth and h_e is the elastic reversible indentation depth which are measured from the load–displacement curve obtained from the nano-indentation test. It should be mentioned that $h_m - h_e$ in Eq. 6 shows the residual depth in the nano-indentation test.

Recovery resistance parameter (R_s) is another quantitative variable which is sometimes used for characterization of the elastic–plastic behavior of materials. This parameter determines the energy dissipation during one cycle of loading and unloading in a nano-indentation test and is estimated by the following equation (Bao et al. 2004)

$$R_s = 2.263 \frac{E_{eff}^2}{H} \tag{7}$$

where E_{eff} and H are the elasticity modulus and hardness of the material, respectively, obtained from the nano-indentation test.

Table 4 shows the values of plasticity index (ψ) and recovery resistance parameter (R_s) calculated by Eqs. 6 and 7.

The reduction in the plasticity index (ψ) and the increase in the recovery resistance parameter (R_s) for the vacuum-mixed cement compared to the hand-mixed cement indicate that the portion related to the elastic work is enhanced, and hence the elastic recovery of the surface is improved. Therefore, the residual depth in the vacuum-mixed cement after the nano-indentation test is lower than the hand-mixed one. The reduction in the residual depth causes an increase in the hardness of the vacuum-mixed cement.

Table 4 The values of plasticity index and recovery resistance parameter obtained for the bone cements prepared by hand-mixing and vacuum-mixing

Samples	Plasticity index (ψ)	Recovery resistance parameter (R_s)
Hand-mixed cement	0.78	241.23
Vacuum-mixed cement	0.76	254.59

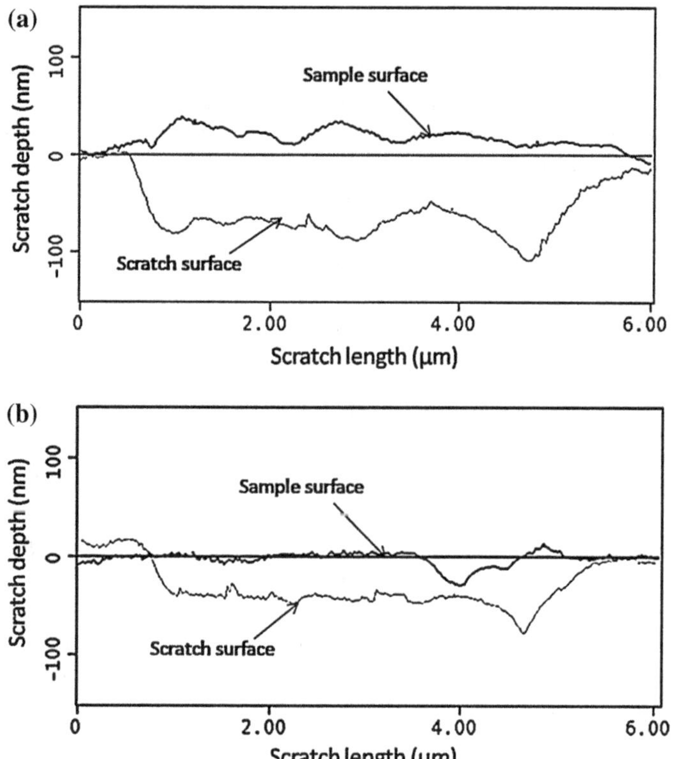

Fig. 8 The longitudinal scratch sections of (**a**) hand-mixed, and (**b**) vacuum-mixed samples

3.9 Wear Resistance of Bone Cement

The residual depth described earlier in the nano-scratch test is a good indicator for investigating the material deformation. Figure 8 displays the longitudinal scratch sections of the hand-mixed and vacuum-mixed samples. A comparison between these curves shows that the residual depth of the vacuum-mixed samples is less than that of the hand-mixed ones. Based on the scratch profile, the mean value of the residual depth for the hand-mixed cements is 80 nm while for the vacuum-mixed cements this value is 45 nm, indicating that the vacuum-mixed samples have higher resistance than the hand-mixed ones (Karimzadeh and Ayatollahi 2012).

One may suggest that the increase in the elasticity modulus and the decrease in the plasticity index of the vacuum-mixed bone cement (as shown in this study) play an important role in the improvement of the elastic recovery, and thus decrease the residual depth of the scratch.

4 Mechanical Properties of a Dental Nano-Composite

Dental nano-composites are relatively new category of restorative composites which contain fillers with nano-meter dimensions (20–75 nm). These nano-particles can improve the physical and mechanical properties of dental composites.

Oral conditions such as moisture and thermal conditions affect the mechanical properties of dental materials. Thus, the investigation of the mechanical properties of dental materials in oral conditions is very important when using these materials (Montes-G and Draughn 1986). In this section, the mechanical behavior of a dental nano-composite in moist media and its thermocycling effects are studied by nano-indentation test.

4.1 Dental Nano-Composite Sample Preparation

Some disk specimens each of the diameter 10 mm and the thickness 4 mm were prepared from Filtek Z350 XT (3 M ESPE, Germany) nano-composite, according to the manufacturer's instructions. Half of the specimens were stored in distilled water at room temperature (24 °C) and the other half were thermo-cycled in distilled water for 1,000 cycles between 5 and 55 °C according to Gale and Darvell (1999). The surfaces of all specimens were polished by diamond pastes with meshes of 1 and 0.5 μm. After the sample preparation, the nano-indentation test was performed at room temperature (24 °C) on both non-thermocycled and thermocycled specimens.

4.2 Nano-Indentation Test on Dental Nano-Composite

The indenter and test setup for the nano-indentation test of dental nano-composite were the same as those of the bone cement experiment. An indentation load of 750 mN with a constant rate of 15 mN s^{-1} was selected for testing the dental nano-composite. The maximum indentation depth in these experiments was limited to 250 nm. In this study, a holding time of 10 s was selected to minimize the likely error due to the time dependent behavior of the material.

Fig. 9 AFM image of an indentation hole on the dental nano-composite sample

According to the described method, at least five indentations on the randomly selected sites of the samples were performed at a temperature of 24 °C. The AFM images were used for the analysis of results before and after the indentations. The AFM image of an indentation hole obtained from this experiment is shown in Fig. 9.

4.3 Modulus of Elasticity of Dental Nano-Composite

The modulus of elasticity was calculated based on the continuous load–displacement data obtained from a complete cycle of loading and unloading during the nano-indentation test, by using the Oliver-Pharr's method and Sneddon relationship (Sneddon 1965). Table 5 displays the mean values and the standard deviations for the modulus of elasticity obtained for the dental nano-composite samples. A sample load–displacement curve obtained from the nano-indentation test is shown in Fig. 10.

4.4 Hardness of Dental Nano-Composite

Similar to the bone cement samples, hardness was calculated by dividing the normal load by the projected area of the surface where load was imposed on. The mean values and the standard deviations of hardness of the dental nano-composite specimens are presented in Table 5. It is seen that the modulus of elasticity of the test samples decreases when the dental nano-composite is subjected to thermocycling, but its hardness increases.

A comparison between the mechanical properties obtained from the non-thermocycled and thermocycled samples made by using the independent-samples t-test indicates that the decrease of the elasticity modulus and the increase of hardness for the two groups are statistically significant (p-value < 0.05). Humid

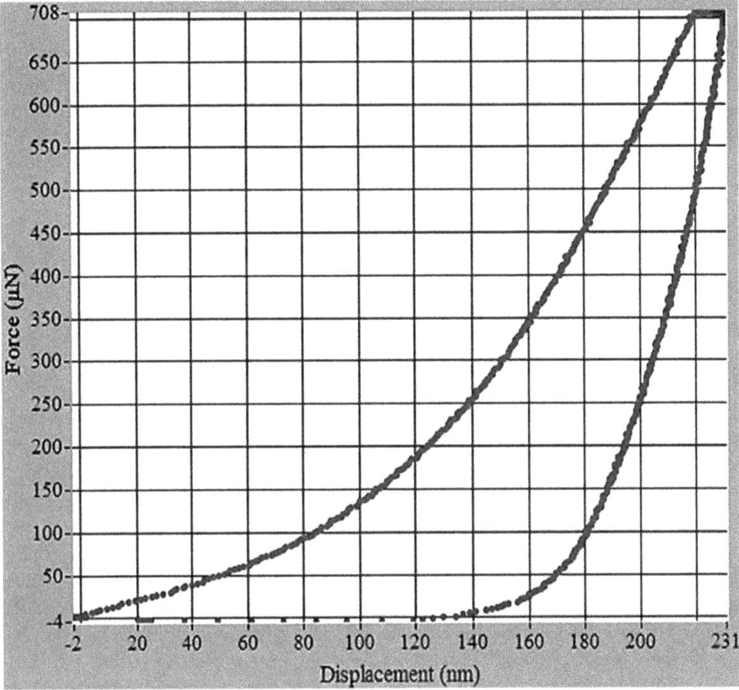

Fig. 10 Load–displacement curve in nano-indentation test of dental nano-composite

Table 5 The mean values and standard deviations for modulus of elasticity and hardness of the dental nano-composite specimens

Samples	Modulus of elasticity (GPa)	Hardness (GPa)
Non-thermocycled nano-composite	15.07 ± 0.91	0.47 ± 0.05
Thermocycled nano-composite	12.07 ± 0.15	0.61 ± 0.01

environment and thermal stresses generated between different components of the nano-composite can be the main reasons for the change in the mechanical properties of the thermocycled dental nano-composites.

5 Concluding Remarks

Nano-indentation and nano-scratch tests were used successfully for measuring the mechanical and tribological properties of hand-mixed and vacuum-mixed bone cements and also the thermocycled and non-thermocycled dental nano-composites.

Using these tests, the elasticity modulus, hardness, fracture toughness, elastic–plastic response and wear resistance can be obtained for brittle and quasi-brittle materials including many biomaterials. The experimental results showed that the mixing method and thermocycling have significant effects on the mechanical properties of bone cement and dental nano-composite, respectively.

References

Anstis GR, Chantikul P, Lawn BR, Marshall DB (1981) A critical evaluation of indentation techniques for measuring fracture toughness. I. Direct crack measurements. J Am Ceram Soc 64(9):533–538

Ayatollahi MR, Karimzadeh A (2012) Determination of fracture toughness of bone cement by nano-indentation test. Int J Fract 175:193–198

Bao YW, Wang W, Zhou YC (2004) Investigation of the relationship between elastic modulus and hardness based on depth-sensing indentation measurements. Acta Mater 52:5397

Briscoe BJ, Evans PD, Pelillo E, Sinha SK (1996) Scratching maps for polymers. Wear 200:137

Briscoe BJ, Sinha SK (2003) Scratch resistance and localised damage characteristics of polymer surfaces – a review. Materialwissenschaft und Werkstofftechnik 34(10–11):989–1002

Chudoba T, Richter F (2001) Investigation of creep behaviour under load during indentation experiments and its influence on hardness and modulus results. Surf Coat Technol 148:191

Cook RF, Pharr GM (1990) Direct observation and analysis of indentation cracking in glasses and ceramics. J Am Ceram Soc 73(4):787–817

Duncan JC (1999) Mechanical properties and testing of polymers. Kluwer Academic Publishers, Dordrecht, p 43 12

Dunne NJ, Orr JF (2001) Influence of mixing techniques on the physical properties of acrylic bone cement. Biomaterials 22:1819

Fett T (2002) Computation of the crack opening displacements for Vickers indentation cracks. Forschungszentrum Karlsruhe, Report FZKA 6757, Karlsruhe, Germany

Fett T, Njiwa ABK, Rodel J (2005) Crack opening displacements of Vickers indentation cracks. Eng Fract Mech 72(5):647–659

Gale MS, Darvell BW (1999) Thermal cycling procedures for laboratory testing of dental restorations. J Dent 27:89–99

Hu Y, Shen L, Yang H, Wang M, Liu T, Liang T, Zhang J (2006) Nanoindentation studies on nylon 11/clay nanocomposites. Polym Test 25:492

ISO-14577-1 (2002) Metallic materials: instrumented indentation test for hardness and materials parameters part 1: test method. Switzerland, Geneva

ISO-14577-4 (2007) Metallic materials: instrumented indentation test for hardness and materials parameters part 4-test method for metallic and non-metallic coatings. Switzerland, Geneva

Jee A, Lee M (2010) Comparative analysis on the nanoindentation of polymers using atomic force microscopy. Polym Test 29:95

Karimzadeh A, Ayatollahi MR (2012) Investigation of mechanical and tribological properties of bone cement by nano-indentation and nano-scratch experiments. Polym Test 31:828–833

Kruzic JJ, Ritchie RO (2003) Determining the toughness of ceramics from Vickers indentations using the crack-opening displacements: an experimental study. J Am Ceram Soc 86(8):1433–1436

Kruzica JJ, Kimb DK, Koesterc KJ, Ritchiec RO (2009) Indentation techniques for evaluating the fracture toughness of biomaterials and hard tissues. J Mech Behav Biomed Mater 2:384–395

Lawn BR, Evans AG, Marshall DB (1980) Elastic/plastic indentation damage in ceramics: the median/radial crack system. J Am Ceram Soc 63(910):574–581

Lidgren L, Bodelind B, Moller J (1987) Bone cement improved by vacuum mixing and chilling. Acta Orthop Scand 57:27

Liu TX, Phang IY, Shen L, Chow SY, Zhang W-D (2004) Morphology and mechanical properties of multiwalled carbon nanotubes reinforced nylon-6 composites. Macromolecules 37:7214

Montes-G GM, Draughn RA (1986) In vitro surface degradation of composites by water and thermal cycling. Dent Mater 2:193–197

Munz D, Bubsey RT, Shannon J (1980) Fracture toughness determination of A12O3 using four-point-bend specimens with straight-through and chevron notches. J Am Ceram Soc 63(5–6):300–305

Murphy BP, Prendergast PJ (1999) Measurement of non-linear microcrack accumulation rates in polymethylmethacrylate bone cement under cyclic loading. J Mater Sci 10:779

Ngan AHW, Wang HT, Tang B, Sze KY (2005) Correcting power-law viscoelastic effects in elastic modulus measurement using depth-sensing indentation. Int J Solid Struct 42:1831

Oliver WC, Pharr GM (2004) Measurement of hardness and elastic modulus by instrumented indentation: advances in understanding and refinements to methodology. J Mater Res 19:3–20

Pharr GM (1998) Measurement of mechanical properties by ultra-low load indentation. Mater Sci Eng A253:151–159

Sinha SK, Song T, Wan X, Tong Y (2009) Scratch and normal hardness characteristics of polyamide 6/nano-clay composite. Wear 266:814–821

Smeds S, Goertzen D, Ivarsson I (1997) Influence of temperature and vacuum mixing on bone cement properties. Clin Orthop Relat Res 334:326

Sneddon LN (1965) The relation between load and penetration in the axisymmetric Boussinesq problem for a punch of arbitrary profile. Int J Eng Sci 3:47

Wang JS, Toksvig-Larsen S, Müller-Wille P, Franzén H (1996) Is there any difference between vacuum mixing systems in reducing bone cement porosity? J Biomed Mater Res Appl Biomater 33:115

Williams JA (1996) Analytical models of scratch hardness. Tribol Int 29:675

Nanomechanical Characterization of Brittle Rocks

Annalisa Bandini, Paolo Berry, Edoardo Bemporad and Marco Sebastiani

Abstract After an introduction describing the indentation techniques traditionally applied to the study of micromechanical properties of minerals and rocks, phenomena induced by the diamond tip's penetration into crystalline rocks are analyzed. Crystalline rocks are characterized by low values of the critical breakage load, i.e. the threshold load corresponding to the transition from a ductile to a brittle behavior. As a consequence, it seems more convenient to examine the mechanical behavior of crystalline rocks by using instrumented nanoindentations. Above the critical load, ranging from rock to rock, fractures occur, affecting the indentation results and thus invalidating the values of the rock mechanical properties obtained by indentation data processing. In order to determine the correct values of the hardness and elastic modulus of brittle rocks, an innovative measurement modality for rocks, i.e. Continuous Stiffness Measurement mode, is proposed. By providing the continuous evolution of the hardness and of the elastic modulus as a function of the indentation depth, it has proven particularly suited to analyze the effects of induced fracturing on the load versus displacement curve.

1 Introduction

Crystalline rocks consist of crystals of the same mineral or of aggregates of several mineral grains and contain discontinuities of different nature. Simmons and Richter (1976) and Kranz (1983) classified such microdiscontinuities into four types:

- grain boundaries (following the contacts between the grains);
- intragranular (lying inside the grains and not contacting the boundaries);

A. Bandini (✉) · P. Berry
Department of Civil, Chemical, Environmental and Materials Engineering (DICAM),
University of Bologna, via Terracini, 28, 40131 Bologna, Italy
e-mail: annalisa.bandini4@unibo.it

E. Bemporad · M. Sebastiani
Engineering Department, Roma Tre University, via della Vasca Navale 79, 00146 Rome, Italy

A. Tiwari (ed.), *Nanomechanical Analysis of High Performance Materials*,
Solid Mechanics and Its Applications 203, DOI: 10.1007/978-94-007-6919-9_11,
© Springer Science+Business Media Dordrecht 2014

- intergranular (intersecting grain boundaries);
- multigranular (the longest cracks, crossing several grains and their boundaries).

At the scale of the laboratory specimen, crystalline rocks are considered mostly continuous and homogeneous media and, when there is a marked iso-orientation of the grains, anisotropic. For a detailed analysis at the microscopic scale, they are revealed as discontinuous solids, affected by a dense network of cracks and grain boundary cavities. Figure 1 shows some examples of microdiscontinuities in a granite's sample. More generally, the discontinuous nature and heterogeneity of rocks at the grain scale are well described in Blair and Cook (1998) and in Lan et al. (2010).

Depending on their distribution, nature and statistically prevailing orientation more than grains iso-orientation, the defects affect the rock's mechanical behavior at a varying extent.

The microdiscontinuities are surfaces along which the failure processes of the rock samples under critical state of stresses originate (Tapponnier and Brace 1976; Kranz 1979; Wong 1982). Consequently, to properly understand the mechanisms leading to the rock's macroscopic failure, analyses should be performed even at the micro and nanoscale (discontinuities scale), where the failure begins.

The micro and nano scale analysis on the rock's mechanical response highlights the clear influence of the above mentioned discontinuities on the indentations performed with the different methods available today. Indeed, unlikely the observations in metals, for instance, the impression's shape in rock samples completely differs from the geometry provided by the interpretative models, as it is generally characterized by an intense fracturing, by detachments of chips, wedges, that may give rise to unformed craters around the contact area between the sample and the indenter.

Despite the deviation, in terms of indent's shape, from experimentation and theoretical models, the mechanical properties (hardness and elastic modulus) are calculated assuming the rock as a continuous, isotropic or anisotropic medium. As a consequence, those values can be affected by significant errors, which can be due to the fact that the method does not consider the phenomena induced by the tip

Fig. 1 Polished surface of a granite sample. Aggregates of quartz, feldspar and biotite can be recognized. In quartz the microcracks net clearly stands out, while in feldspars grains the cleavage planes traces are more obvious. In biotite some microcracks can be seen even at naked eye

penetration into the rock sample during loading and unloading, because the micro-discontinuities act as stress concentrators.

In order to investigate on the influence of the defects network on the mechanical behavior under indentation of rocks, instrumented Berkovich nanoindentations in Continuous Stiffness Measurement (CSM) mode were carried out. This modality, developed by Oliver and Pharr in 1992 to evaluate the elastic modulus and the hardness of continuous, isotropic and elasto-plastic materials, has been adopted for the first time for the study of rock materials by Bandini et al. (2012) to examine the relationships between the cracks net and the micromechanical parameters of brittle and discontinuous media, such as hard rocks.

In the rock mechanics literature, before the work of Bandini et al. (2012), instrumented indentation techniques were limited to the determination of micromechanical properties of rocks adopting the conventional quasi-static method, not in CSM modality (Broz et al. 2006; Whitney et al. 2007; Zhu et al. 2007; Mahabadi et al. 2012).

Compared to the conventional method, the CSM allows analyzing dynamically the mechanical behavior (hardness and elastic modulus) and the fracture mechanisms of materials during indentation, as a function of penetration depth with a spatial resolution of few nanometers, and consequently it is particularly suited to study the failure mechanisms of rock materials.

1.1 Conventional Indentation Methods for Rocks Characterization

In rock mechanics the most widespread indentation techniques are the not instrumented ones (Delgado et al. 2005; Xie and Tamaki 2007; Yilmaz 2011).

Not instrumented indentation is a static method involving applying and removing a specified load onto the sample surface by a diamond probe. The indentation measurements are taken after the indentation event.

The Knoop diamond tip is usually utilized in rocks, since, compared to the other indenters, has the advantage to produce an elongated and therefore easily measurable impression, also when the imprint tends to undergo cleavage deformation to be charged to the material brittleness, as occurs in many minerals and rocks.

The reference surface commonly used to calculate the Knoop hardness is the projected contact area (impression section), A_{PCA}, and the Knoop microhardness is determined with the Eq. (1):

$$HK = \frac{P}{A_{PCA}} = \frac{P}{L^2} \cdot \frac{2}{\cot\left(\frac{\alpha}{2}\right) \cdot \tan\left(\frac{\beta}{2}\right)} = 14.229 \cdot \frac{P}{L^2} \qquad (1)$$

where $\alpha = 172°\ 30'$ e $\beta = 130°$ are the two characteristic angles of the Knoop indenter tip, and P and L the load and the long diagonal's length of the impression.

The Vickers microhardness, instead, is conventionally calculated by considering, as reference surface, the true contact area, A_{TCA}, represented by the side surface of the pyramid, in accordance with the following equation:

$$ HV = \frac{P}{A_{TCA}} = \frac{P}{\frac{d^2}{2 \cdot \sin(\phi/2)}} = 1.8544 \cdot \frac{P}{d^2} \qquad (2) $$

where $\phi = 136°$ is the characteristic angle of the indenter, and P and d are the load and the diagonal length of the squared section impression.

In both Eqs. (1) and (2), hardness is determined purely from the measurements of the diagonals length of the permanent impression under a given load, irrespective of the impression's shape.

The impressions in rocks are rarely regular, sometimes fully developed, often partly developed and occasionally they are simply a cross mark. In performing Vickers indentations on coal, Das (1974) observed several times that impressions having nearly identical length of diagonals and consequently the same hardness according to Vickers formula had different shape of the impressions.

These remarks are also confirmed by the results obtained with other testing methods, including scratching and impact tests on rock samples (Berry et al. 1989; Beste et al. 2004).

The impressions shape in rocks are more complex than the models' geometry (Hughes 1986; Lindqvist and Hai-Hui 1983), valid for homogeneous and continuous media. Such complexity is due to the microstructural parameters, which play an important role: position, orientation according to the stress direction, frequency, size and type of discontinuities (cracks across mineral grains, mineral contacts, cracks crossing different species, cleavage planes), discontinuities' alteration degree, grains' mineralogical composition.

The theoretical models that describe the behavior of rocks under point loads are limited to consider the brittleness of homogeneous materials and then assume symmetrical shapes of the permanent impression, whereas they do not take into account the inhomogeneities also at the scale of the indenter's tip, which might cause distortions or asymmetries.

These models predict the formation of regular and symmetrical chips around the indenter (Paul and Sikarskie 1965; Pariseau and Fairhurst 1967; Lundberg 1974; Hughes 1986) and according to Paul and Sikarskie (1965), a zone of intense comminution is also created below the tip.

The models lead to conclude that for brittle materials, such as rocks, the contact area between the tip and the sample remains approximately constant from the beginning to the end of the test. It follows that in brittle rocks the impression is never the "negative" portion of the penetrated indenter ("positive").

The Eq. (1) and (2) give the hardness in terms of ratio between the applied load and the contact surface between the tip and the material under observation, assuming that the entire portion of the penetrated tip is in contact (hypothesis that is verified for ductile materials, but not for brittle rocks).

Thus, it may be reasonably assumed that the rocks' hardness values calculated according to Eqs. (1) and (2) are overestimated and each value deviates from the "true" value, in first approximation, proportionally to the induced effects (microcracking, craters, etc.), which vary point to point, as a function of the fabric conditions around the tip.

1.2 Rock's Indentation: Induced Phenomena

From examining the residual impressions on rock samples, it appears that the states of stress due to a shaped tip's penetration generate effects in an area much larger than the indenter's section, depending upon the rock's heterogeneity degree (pores and microcracks).

Cracks within each crystalline element, contact surfaces between grains of the same mineral or different species, schistosity surfaces and cleavage planes act as "weakness points" in the rock structure. As the microstructure's conditions of the rock change in the contact area between the rock sample and the indenter, various phenomena can be noted.

We may observe rock's plasticization, as pointed by the mark on the left side of the impression in Fig. 2a. The Fig. 2b shows a bifurcation of the impression, as well as another mark. The pictures c and d (Fig. 2) are characterized by more or less symmetrical marks of various extents. In the last two pictures of Fig. 2(e and f) two craters were generated next to the indent. Similar phenomena are visible even in Fig. 3 relating to indentations on feldspar grains of the same granite sample of Fig. 2. The two impressions in Fig. 3a are affected by intense streaks. Moreover, the Fig. 3 (b, c, d, e) gives examples of craters of varying shape and extension. The Fig. 3e displays an intense network of induced microfractures at the ends of the long diagonal of the impression.

The Fig. 4, referring to impressions on biotite grains, highlights the influence on the microstructure on the shape and extension of the induced effects. The indents with the longest diagonal parallel to the cleavage planes (Fig. 4a) are accompanied by a set of cracks parallel to the longest diagonal. It means that the biotite lamellae are widened under indentation. The indentations perpendicular to cleavage (Fig. 4b) induce the rock deformation up to cause a dense network of microcracks parallel to the short diagonal of the rhombus, giving to the indent a squatter shape. The impression produced along a direction slightly inclined (Fig. 4d) with respect to the configuration of Fig. 4a is markedly asymmetrical. It seems more pronounced toward the left side and such extension is clearly due to the lamellae orientation, which opens under stress.

In the same rock sample we can often find both regular and shapeless indents, surrounded by craters that make the diagonal's length difficult to measure and therefore alter the impression's shape. Figure 5 refers to the same Carrara marble sample. In the first case (Fig. 5a) a close net of well-marked microfractures extends both on the right and on the left side, but anyway the impression is very

Fig. 2 Optical micrographs of Knoop residual impressions ($P = 1.9$ N) on quartz grains of a granite sample. The first four pictures (**a, b, c, d**) show signs of coaction or induced yielding. The pictures (**e** and **f**) depict the craters produced by the penetration of the indenter tip, from the edges of the impression

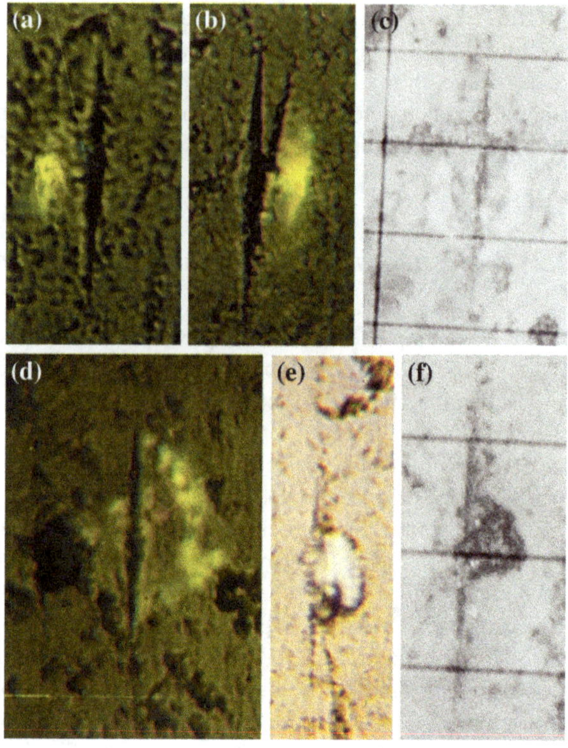

clear and visibly contrasts into the background. In the second case (Fig. 5b), it is not easy to define the exact contour of the impression since the detachment of a large piece of material occurred.

According to Brace (1961), the impression's shape should be clearer and more regular in extremely fine-grained rocks because the network of microcracks has a mesh less than or comparable with the tip size and the structural weakness is homogenously spread throughout the rock.

Generally, the effects of each indentation extend well beyond the contact area between the sample and the indenter tip. The failure does not occur with radial cracks, as assumed by the theoretical models and observed in other materials (Oliver and Pharr 1992). Fractures arise along preexisting microdiscontinuities. For instance, Wong and Bradt (1992) noted that indentations inside calcite crystals always induce cleavage cracks and twins, regardless of the tip's orientation with reference to the cleavage planes directions (Carter et al. 1993).

The indentation effects also depend on the mineralogical composition of the grains. The grooves produced by scratching within single grains are characterized by different shapes depending on the involved mineral species (Berry et al. 1989; Beste et al. 2004). For example, in Berry et al. (1989) the minerals of granite respond differently to the shear stresses. In the feldspar the groove appears to be

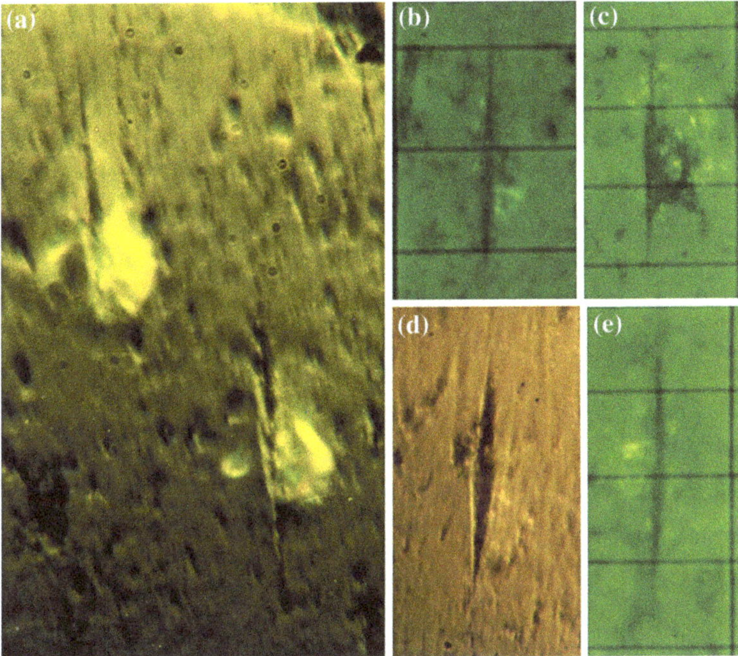

Fig. 3 Optical micrographs of Knoop residual impressions ($P = 1.9$ N) on feldspar grains of a granite sample. The first picture (**a**) shows marks of coaction or induced yielding (two impressions). The other four pictures (**b, c, d, e**) highlight microfracturing and micro-craters beyond the yielding signs

controlled by the induced cracks and it mainly develops along the cleavage planes. Sometimes the groove is barely detectable. In quartz the boundary is more irregular than in the feldspar, consisting of the microcracks produced by the cutter and by the preexisting ones. Moreover, along the grooves' boundaries a band of discontinuities can be noticed. In the biotite, the groove is contained inside the mineral and marked effects beyond the streak are not visible.

At the contacts between the various rock minerals, the effects vary in dependence of the species of neighboring minerals, of the size of the crystals elements, of the joint strength and of its orientation with respect to the groove. If the orientation is favorable compared to the stress direction, stresses concentrate in the flaws edges determining the propagation and coalescence along the preexisting discontinuities.

The extent and the intensity of the produced phenomena depend on the value of the applied load, in addition to the rock fabric. For example, a load of 2 N induces cracks in all the Mohs minerals (Broz et al. 2006). Each Mohs mineral responds differently to indentation, as is evident by the length, number, direction and type of cracks produced by Vickers microindentation (Broz et al. 2006).

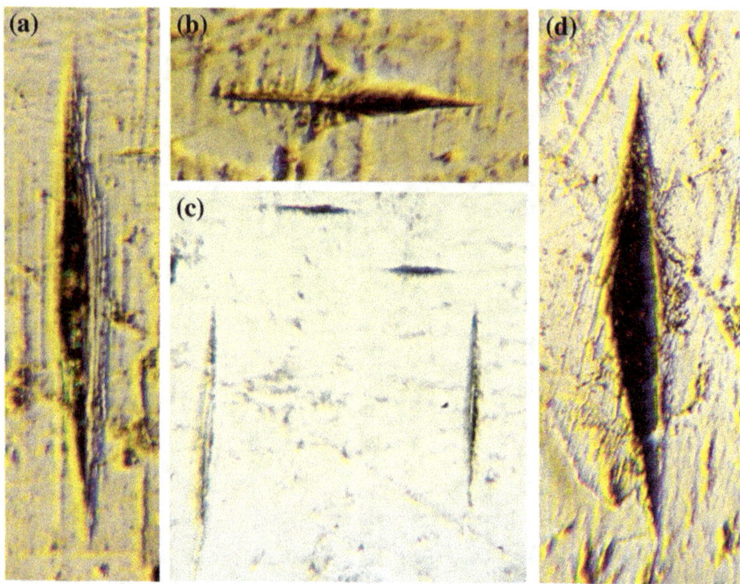

Fig. 4 Optical micrographs of Knoop residual impressions ($P = 1.9$ N) on biotite grains of a granite sample. Cleavage markedly affects the impression's shape and sizes. The first picture (**a**) shows effects induced by an indentation along the cleavage planes direction. The second one (**b**) represents the produced effects when the tip is rotated through 90° relative to the cleavage planes direction. The third one (**c**) compares the previous configurations. The last one (**d**) shows an impression produced by rotating the tip's longest diagonal by 18° with respect to the cleavage.

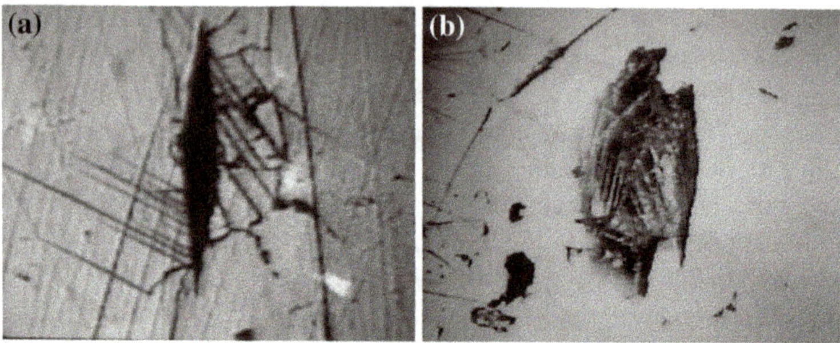

Fig. 5 Optical micrographs of two Knoop residual impressions ($P = 1.9$ N) on Carrara marble

In crystalline rocks induced cracking is evident also at lower loads. Figures 6 and 7 depict illustrative examples of Knoop and Vickers indents, by applying a load of 100 mN on a marble sample. Cleavage cracks and twins are clearly identifiable, despite the low applied load.

In Bandini et al. (2012) a load of a few mN is enough to have fracturing around the indents in marble. Likewise, also Skrzypczak et al. (2009) always noticed

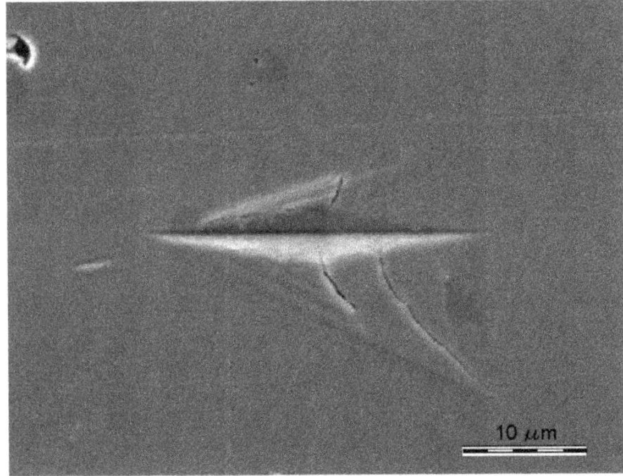

Fig. 6 SEM micrograph (2000x, SE) of a Knoop indent ($P = 100$ mN) into a calcite grain of a marble sample (Bandini et al. 2012)

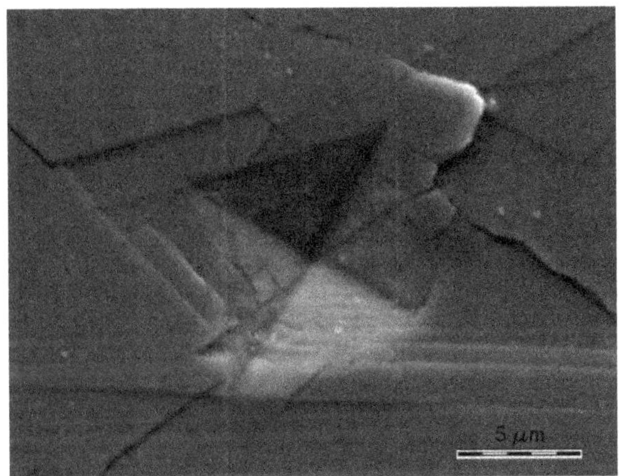

Fig. 7 SEM micrograph (5000x, SE) of a Vickers indent ($P = 100$ mN) into a calcite grain of a marble sample (Bandini et al. 2012)

cracks and were not able to experimentally observe the transition from plastic deformation to brittle fracture, because the range of applied loads (from 250 mN to 5 N) was higher than the threshold load between ductile and brittle response.

1.3 Critical Breakage Size

During an indentation, the energy given to the sample may be spent in two different ways. Fracturing occurs, as the energy allows creating new surfaces, or the material deforms plastically, the energy being dissipated by atoms rearrangement. In addition, a part of energy, which can be neglected, is dissipated as heat.

The one or the other of the two above mentioned phenomena prevails depending on the value of the applied load and, as a consequence, on the sizes of the indent produced by indentation.

The critical breakage size, a_{crit}, is defined as the size corresponding to the transition from a ductile to a brittle behavior. Consequently, it results from the energetic balance between plastic deformation and crack propagation.

In particular, the energy W_S needed to plastically deform a particle of volume a^3 can be expressed with the Eq. (3):

$$W_S = \frac{\sigma_P}{2 \cdot E} \cdot a^3 \tag{3}$$

where σ_P and E are, respectively, the yield stress and the Young's modulus of the material.

The energy W_P required to create a new surface a^2 is:

$$W_P = G \cdot a^2 \tag{4}$$

where G is the crack propagation energy.

It results:

$$a_{crit} = \frac{2 \cdot E \cdot G}{\sigma_P^2} \tag{5}$$

The energy balance becomes favorable to plastic deformation for sizes lower than the critical one, a_{crit}. Above this size, the behavior becomes brittle and fractures occur.

Determining accurately the critical breakage size with Eq. (5) is complex, since it requires to know σ_P and G. On the other hand, the heterogeneity of the rock materials leads to a high dispersion in the values of such parameters and then it is difficult to choose the more appropriate values.

Skrzypczak et al. (2009) introduced a procedure to experimentally determine the critical breakage size through Vickers indentation techniques. This procedure consists in characterizing the indents by the diagonal length d of the plastically deformed area and by the length $2c$ of the cracks around the indent (Fig. 8).

The diagonal length d of the permanent impression is related to the Vickers hardness HV through Eq. (2).

The cracks length $2c$ depends on the material resistance to crack propagation. The fracture toughness under the fracture mode I, K_{IC}, can be calculated with Eq. (6):

$$K_{IC} = 0.016 \cdot \left(\frac{E}{H}\right)^{\frac{1}{2}} \frac{P}{c^{\frac{3}{2}}} \tag{6}$$

Assuming that E, HV and K_{IC} are independent of the applied load P, the two Eqs. (2) and (6) are represented by two straight lines of slope 2 and 3/2 respectively in the plot of Fig. 8b. Thus the intersection point obtained by extrapolating the two straight lines defines the critical load P_{crit} and the corresponding size a_{crit}.

For loads higher than P_{crit}, brittle cracks appear, while, for load lower than P_{crit}, only plastic deformation is observed. The a_{crit} corresponds to the critical size

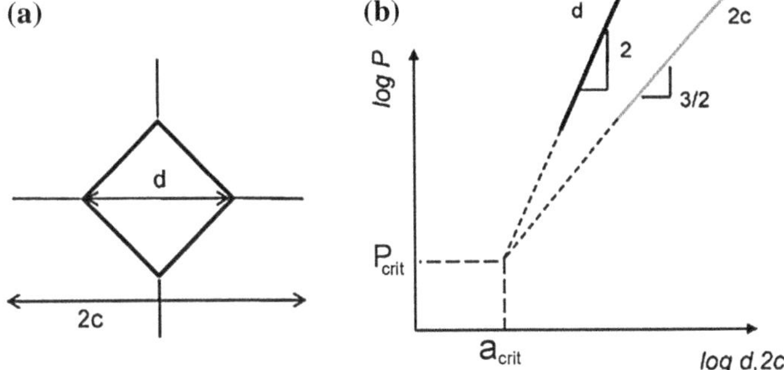

Fig. 8 **a** Scheme of a Vickers indent with radial cracks; **b** Theoretical evolution of the diagonal length d of the Vickers indentation and crack length $2c$ versus the applied load P (Skrzypczak et al. 2009)

below which no cracking occurs. By applying this methodology, Skrzypczak et al. (2009) obtained low value for critical load and size (4-5 mN e 1.9 μm, respectively) in Carrara marble, suggesting that microindentation of crystalline rocks results in cracking.

2 Instrumented Nanoindentation in Continuous Stiffness Measurement Modality in Rocks

Compared to the not instrumented one, the instrumented indentation testing has the advantage of continuously measuring the penetration depth of the indenter into the sample surface and the load applied to the sample.

Unlikely the materials science where it is largely used, the instrumented indentation has been rarely utilized to study the micromechanical properties of minerals and rocks and never in CSM mode before the work of Bandini et al. (2012).

Broz et al. (2006) used nanoindentation to determine the mechanical properties of Mohs scale minerals. Whitney et al. (2007) reported the elastic modulus and hardness by instrumented indentations for common metamorphic minerals. Zhu et al. (2007) attempted to correlate the hardness and the elastic modulus maps of rocks samples (quartzite and granodiorite), by Berkovich nanoindentation, to the mineralogy and shape of the rocks-forming minerals grains, resulting from analysis with optical and scanning electron microscopies.

Mahabadi et al. (2012) conducted grid micro-indentation tests, to determine the instrumented indentation modulus and fracture toughness of the constituent phases of a crystalline rock as input parameters for numerical modeling. Such research represents one of the first attempts to consider the rocks heterogeneity when modeling their mechanical behavior.

The instrumented indentation testing was developed to measure the mechanical properties of a material from the indentation load and indentation depth data obtained during on cycle of loading and unloading. A tip is pushed into the sample surface. It produces a load versus displacement curve from which the indentation modulus and the hardness of the tested material can be quantified by following the method developed by Oliver and Pharr (1992, 2004).

2.1 Oliver and Pharr's Method

During an instrumented indentation, the indenter displacement h, relative to the initial undeformed sample surface, is continuously registered under the controlled application of a normal load P. h_{max} represents the displacement at the peak load P_{max}, h_C is the contact depth and is defined as the depth of the indenter in contact with the sample under load, h_f is the final displacement after complete unloading and S is the slope of the unloading curve defined as contact stiffness (Fig. 9).

During loading, deformation is supposed to be both elastic and plastic, since the permanent impression forms. During unloading, it is assumed that only the elastic displacements are recovered. For this reason, the method does not apply to materials in which additional plastic deformation is observed during unloading path of the load versus displacement curve, a phenomenon that is usually referred as "reverse plasticity". However, such events are very unusual and have been observed only in a few cases and only in case of metals (Oliver and Pharr 1992, 2004), so they can be considered as negligible in the investigated rock and more generally in rock materials.

The elastic nature of the unloading curve facilitates the analysis and the contact process can be modeled by using the Sneddon's elastic solution (Sneddon 1965).

The reduced modulus E_R, which takes into account elastic displacements occur in both the specimen and the indenter, describes the elastic contact between the indented material and the indenter's tip. It is calculated from the contact stiffness S

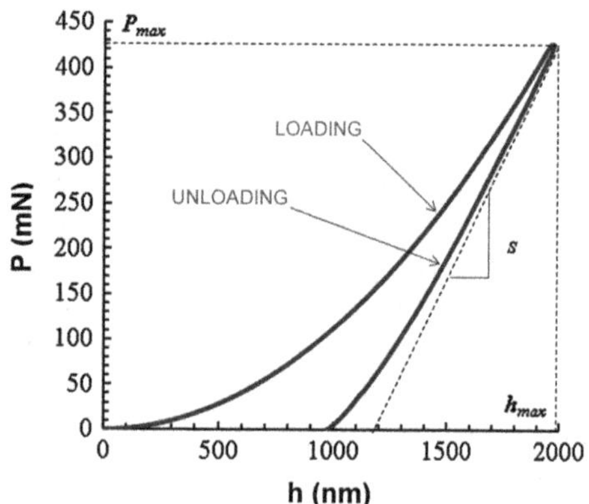

Fig. 9 A typical load P versus displacement h curve for a conical indenter

and the contact area A_C, by using the Eq. (7) derived from the Sneddon's solution (Sneddon 1965) for the elastic contact between a cone and a flat surface:

$$E_R = \frac{\sqrt{\pi}}{2 \cdot \beta} \cdot \frac{S}{\sqrt{A_C}} \tag{7}$$

where β is a constant depending on the indenter's geometry ($\beta = 1.034$ for a Berkovich indenter and $\beta = 1$ for a spherical tip).

Oliver and Pharr (1992) suggested the use of a power law to describe the unloading curve:

$$P = B \cdot (h - h_f)^m \tag{8}$$

where B and m are material constants determined by best fitting analysis ($1.2 \leq m \leq 1.6$).

The slope of the upper part of the unloading curve defines the contact stiffness S:

$$S = \left(\frac{dP}{dh}\right)_{h_{max}} = B \cdot m \cdot (h - h_f)^{m-1} \tag{9}$$

The contact area A_C can be calculated continuously from the depth of the indenter-sample contact h_C.

By applying the elastic model of Sneddon (1965), it follows:

$$h_C = h_{max} - \eta \cdot \frac{P_{max}}{S} \tag{10}$$

where the constant depends on the geometry of indentation and can be derived analytically from the Sneddon's solution ($\eta = 0.72$ for cones and $\eta = 0.75$ for spheres).

Therefore:

$$A_C = f(h_C) \tag{11}$$

where the function $f(h_C)$, known as area function, is calibrated using standard sample of amorphous silica (SiO_2). The indenter shape at the apex is complex and in nanoindentation the function area is well expressed by a polynomial relation:

$$A = a_0 \cdot h_C^2 + a_1 \cdot h_C + a_2 \cdot h_C^{1/4} + a_3 \cdot h_C^{1/3} + \dots \tag{12}$$

By knowing the Poisson's ratio, the elastic modulus of the material E can be determined from the reduced modulus E_R, by the Eq. (13):

$$\frac{1}{E_R} = \frac{1 - v^2}{E} + \frac{1 - v_i^2}{E_i} \tag{13}$$

where E, E_i, v e v_i are Young's moduli and Poisson's ratios for the specimen and the indenter.

The material hardness is expressed as follows:

$$H = \frac{P_{max}}{A} \tag{14}$$

where A is the projected contact area at the peak load P_{max}.

2.2 Continuous Stiffness Measurement Modality

The CSM mode lies in superimposing a small sinusoidal oscillation on the primary loading signal and analyzing the resulting response of the system (Oliver and Pharr 1992). The loading curve can be seen as the superposition of a series of small cycles of loading and unloading, from each of which the elastic modulus and hardness to that value of load may be determined. The hardness and elastic modulus can be recorded as a continuous function of the surface penetration depth by continuous measurement of the dynamic contact stiffness.

By providing the continuous evolution of the elastic modulus and hardness during the loading phase, the CSM allows to assess the critical load P_{crit} and, as a consequence, until which load value the elastic modulus and the hardness may be considered actually representative of the material.

2.3 Case Study

The experimentation has been carried out on a calcitic marble (Bandini et al. 2012; Bandini and Berry 2013), in which the granoblastic (with regular grains and triple points) and xenoblastic (with well interlocked grains of irregular shape) textures coexist. Such textures show a different response to stresses (Table 1) at the laboratory's scale even if the mineralogical composition is almost monomineralic and the grain size does not vary substantially as a function of microstructure.

Berkovich nanoindentations in CSM mode were performed inside calcite grains composing the two textures (xenoblastic and granoblastic) in order to examine the evolution of the indentation hardness, elastic modulus and fracture mechanisms of the two marble textures.

The experimental results show failure occurs from preexisting flaws, acting as stress concentrators in controlling the rock's strength.

Indentations cause the brittle failure of the calcite grains, whatever their shape (regular for granoblastic and irregular for xenoblastic one) and the induced fracturing is ruled by the cleavage planes (Fig. 10).

Such intragranular failures can be recognized by the load versus indentation depth curves and by the SEM micrographs of the indents (Fig. 11). These curve are never regular and in the loading part (Fig. 11a) often exhibit pop-in phenomena, i.e. sudden discrete increase of the penetration depth at an approximately

Table 1 Mechanical properties of the marble under investigation (Bandini et al. 2012)

		N	Min	Max	m	Δm^*(%)	v (%)
n [%]	X	6	3.2	2.9	3.5	\	6.3
	G	11	3.2	2.9	3.4		4.9
σ_c [MPa]	X	4	87.5	104.0	94.3	34.0	6.9
	G	4	60.8	72.0	66.9		6.3
σ_t [MPa]	X	3	7.3	8.8	8.1	33.1	7.5
	G	3	5.4	6.1	5.8		5.2
$V_{p,dry}$ [m/s]	X	6	5,446	5,296	5,616	60.5	1.9
	G	11	2,538	2,233	3,009		8.3
$V_{p,sat}$ [m/s]	X	6	6,136	6,108	6,175	7.8	0.4
	G	11	5,679	5,638	5,709		0.4

$$^*\Delta m = \frac{|mX-mG|}{(mX+mG)/2} \cdot 100$$

X xenoblastic; G granoblastic; n total porosity; σ_c UCS; σ_t Brazilian indirect tensile strength; $V_{p,dry}$ and $V_{p,sat}$ dry and saturated P-waves velocities; N number of tests; *min* minimum value; *max* maximum value; m mean value; Δm^* relative difference, in percent, between the mean values of xenoblastic and granoblastic parameters (mX, mG); v variation coefficient

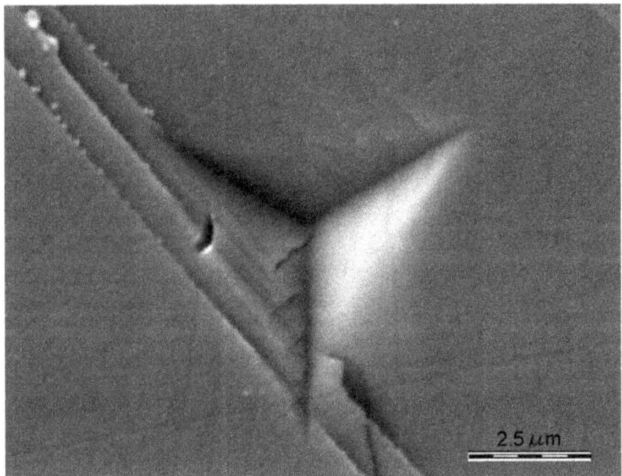

Fig. 10 SEM micrograph (10.000×) of a Berkovich indenter: fracturing induced by the tip penetration (Bandini et al. 2012)

constant load, as also noticed in other materials (Jian 2007; Saber-Samandari and Gross 2009; Tulliani et al. 2009). The pop-in is due to the formation of cracks, fractures around the indenter, as found by Presser et al. (2010), because of the rock inhomogeneity even on the scale of hundreds of nanometers.

Cracks are noted around each indent and the transition from plastic deformation to brittle fracture cannot be directly observed because the range of the maximum applied loads appears higher than the plastic-brittle threshold of calcite.

The fracturing induced by the indenter's penetration into the sample causes the progressive decrease of the elastic modulus and of the hardness, continuously recorded during loading (CSM mode) (Fig. 12).

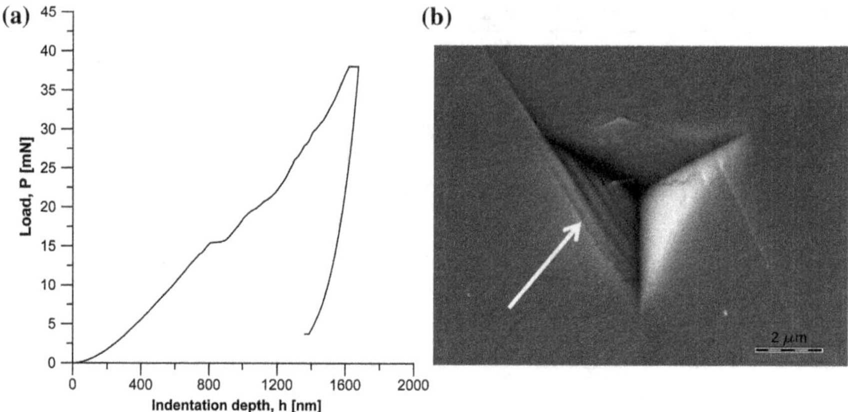

Fig. 11 Berkovich nanoindentations. **a** Load P versus indentation depth h; **b** SEM micrograph ($10.000\times$) of the impression generated by the indentation, where fracturing is indicated by an arrow (Bandini et al. 2012)

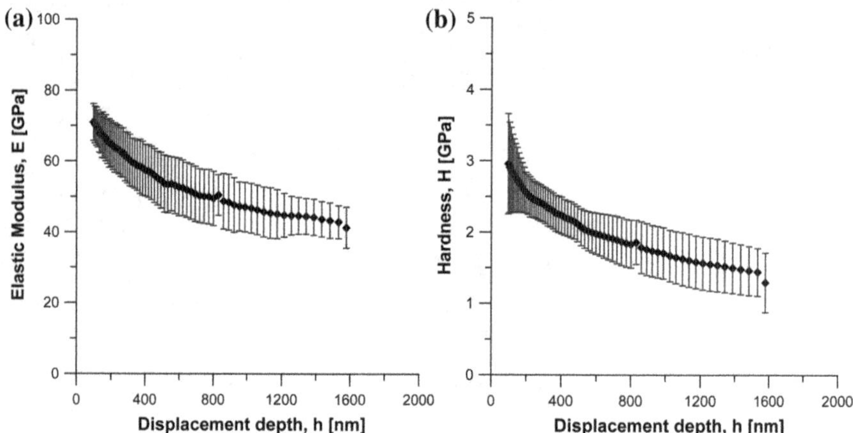

Fig. 12 Elastic modulus E (**a**) and Berkovich hardness H (**b**) variations during an indentation (nanoindentations in displacement control, CSM mode)

Actually, they should be considered as apparent elastic modulus and hardness, since such reduction appears due not to actual variations of the calcite grains properties but to failure and sliding along the cleavage planes of calcite grains.

The elastic modulus and the hardness obtained by the conventional instrumented indentation (without applying the CSM mode) correspond to the values at the deepest penetration depth, i.e. E_{max} (Fig. 13) and H_{max}.

For the extensive cracking during loading which reduces the apparent elastic modulus and hardness during indentation, E_{max} and H_{max} would be affected by a significant error and consequently the conventional instrumented indentation would give a wrong description of the rock properties.

On the other hand, the CSM mode, giving the complete elastic modulus and hardness versus indentation depth profiles, allows to correctly estimating the rock properties.

As cracking is not observed for low applied loads (in this case, lower than 3 mN), the correct values of the elastic modulus and of the hardness can be achieved by extrapolating these profiles to zero depth (E^* and H^*) (Fig. 13), for

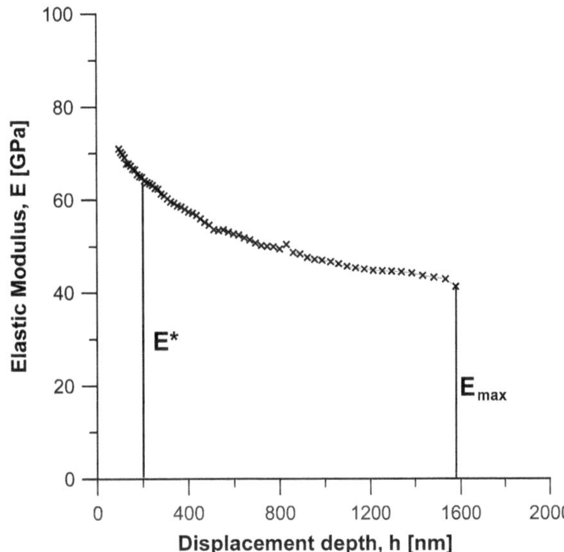

Fig. 13 Berkovich nanoindentations in CSM mode: elastic moduli extrapolated at zero depth (E^*) and to the maximum applied load (E_{max})

Table 2 Berkovich nanoindentations in CSM mode on xenoblastic marble: elastic modulus and hardness values extrapolated at zero depth (E^* and H^*) and at the maximum applied load (E_{max} and H_{max}) (Bandini et al. 2012)

	E* [GPa]	H* [GPa]	E_max [GPa]	H_max [GPa]
m [GPa]	75.0	3.4	42.3	1.4
v (%)	6.0	9.1	10.4	28.4

instance, by fitting the profiles with a polynomial function. As can be seen, such values differ from the mechanical properties at the maximum load, E_{max} and H_{max} (Tables 2 and 3).

It results the calcite grains forming the two textures have the same intrinsic properties, for instance an elastic modulus of around 70 GPa (Fig. 14) in agreement with literature data for calcite (Broz et al. 2006; Presser et al. 2010), as we can see by the comparison of the curves of the elastic modulus versus

Table 3 Berkovich nanoindentations in CSM mode on granoblastic marble: elastic modulus and hardness values extrapolated at zero depth (E^* and H^*) and at the maximum applied load (E_{max} and H_{max}) (Bandini et al. 2012)

	E* [GPa]	H* [GPa]	E_{max} [GPa]	H_{max} [GPa]
m [GPa]	73.9	3.7	57.3	1.7
ν (%)	9.7	19.4	7.3	11.1

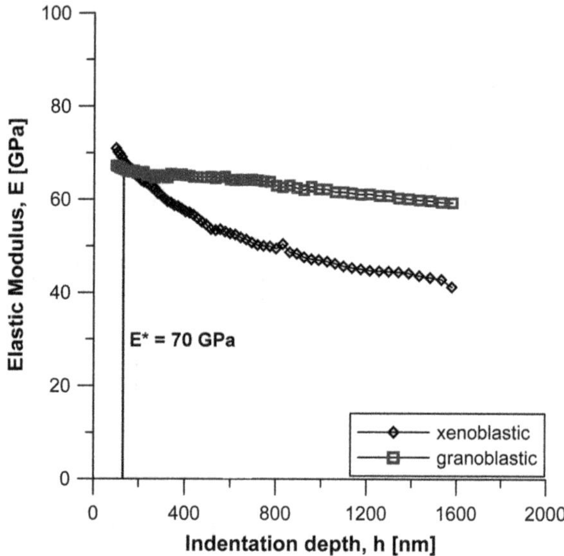

Fig. 14 Elastic modulus E versus indentation depth h: comparison between two calcite grains of xenoblastic and granoblastic marble

indentation depth (Fig. 14), which show no significant variation for depth lower than 200 nm.

Moreover, the CSM Berkovich nanoindentations picked out a more marked tendency to cleavage fracture of the xenoblastic marble grains, since the elastic modulus decreases more significantly with increasing the indentation depth (Fig. 14).

The calcite grains of the xenoblastic marble, which shows higher strengths at the scale of the laboratory sample (Table 1), appear more brittle than the calcite grains of the granoblastic texture and fracturing is already noticeable to a load of 5 mN, critical threshold load for the transition between plastic deformation and brittle fracture, similar to the value obtained by Skrzypczak et al. (2009) in Carrara marble.

With increasing the indentation depth, the elastic modulus of the calcite grains decreases more markedly in the xenoblastic marble, for a more significant tendency to cleavage fracture. It explains why we obtained lower elastic moduli at the maximum applied load, E_{max}, in the xenoblastic marble (Tables 2 and 3).

The same considerations can be made for the hardness. Similarly, the Berkovich hardness decreases more significantly in the calcite grains of the xenoblastic texture, resulting in lower apparent hardness (Tables 2 and 3), for the same reason.

3 Discussion

The study clearly shows that reliable values of the elastic modulus and hardness of rock materials can be obtained only after a careful analysis of the load versus displacement curve and of profiles of the elastic modulus and hardness during indentation (CSM mode). On the other hand, the conventional instrumented indentation does not take into account the tendency to brittle failure of the rock material. As a consequence, it can significantly underestimate the rock's mechanical properties, since the induced cracking determines a gradual decrease of the indentation modulus and hardness. The CSM allows to evaluate to which load value only plastic deformation occurs and the rock's hardness can be defined. Above the critical load, from rock to rock, fractures occur, affecting the indentation results and thus invalidating the values of the rock mechanical properties. The correct values of the elastic modulus and hardness can be achieved only for loads lower than the critical load. In the case study, indentantion results in cracking also at loads of few mN. As a consequence, the correct values of the elastic modulus and hardness can be achieved by extrapolating their profiles to zero depth.

4 Concluding Remarks

In rock mechanics the Knoop and Vickers not instrumented indentations are traditionally adopted. Because of the discontinuous nature of rocks, the residual impression produced by microindentation is almost never perfectly regular and several phenomena (fracturing, chips and wedges detachment, craters formation) are induced by the tip's penetration into a rock sample. Consequently, the calculated hardness deviates from the "true" value, proportionally to the induced effects.

The critical breakage load of crystalline rocks, corresponding to the transition from a ductile to a brittle behavior, is low (order of few mN). In other words, a load of few mN is enough to generate fractures around the impression. Therefore, it is more convenient to examine the mechanical behavior of such rocks by using instrumented nanoindentations. Above the critical load, ranging from rock to rock, fracturing occurs, affecting the indentation results, thus invalidating the values of the elastic modulus and hardness by instrumented indentation data processing. The research work proposed an innovative approach in rock mechanics, consisting in nanoindentation in continuous stiffness measurement (CSM) mode, to determine the correct elastic modulus and hardness of the rock.

By using the CSM modality, which provides the continuous evolution of the hardness and of the elastic modulus as a function of the indentation depth, the correct mechanical properties of the rock can be deduced by analyzing the effects of the induced fracturing on the load versus displacement curve. In the investigated rock, the hardness and elastic modulus may be obtained by extrapolating their profiles (CSM) during indentation, to zero depth. Moreover, the research shows that, on the other hand, the mechanical properties by the conventional instrumented indentation, without applying the CSM, neglecting the fracturing's effect, cannot be considered as representative of the rock (apparent elastic modulus and hardness).

In conclusions, the study should be seen as a preliminary investigation aimed at evaluating the research opportunities offered by the application of the Oliver and Pharr's method to the rocks and, in particular, of the CSM mode for determining the mechanical properties of rock materials.

The theme deserves further investigations aimed at studying the micromechanical properties of soft rocks through instrumented indentation technique in CSM modality.

References

Bandini A, Berry P (2013) Influence of marble's texture on its mechanical behavior. Rock Mech Rock Eng 46(4):785–799

Bandini A, Berry P, Bemporad E, Sebastiani M (2012) Effects of intra-crystalline microcracks on the mechanical behavior of a marble under indentation. Int J Rock Mech Min Sci 54:47–55

Berry P, Bonifazi G, Fabbri S, Pinzari M (1989) Technological problems in cutting granite. In: MMIJ/IMM (ed) Proceedings of MMIJ/IMM joint symposium: today's technology for the mining and metallurgical industries, Kyoto, 2–4 October 1989, pp 215–227

Beste U, Lundvall A, Jacobson S (2004) Micro-scratch evaluation of rock types-a means to comprehend rock drill wear. Tribol Int 37(2):203–210

Blair SC, Cook NGW (1998) Analysis of compressive fracture in rock using statistical techniques: Part II. Effect of microscale heterogeneity on macroscopic deformation. Int J Rock Mech Min Sci 35(7):849–861

Brace WF (1961) Dependence of fracture strength of rocks on grain size. In: Proceedings 4th symposium on rock mechanics, Pennsylvania State University, pp. 99–103

Broz ME, Cook RF, Whitney DL (2006) Microhardness toughness, and modulus of Mohs scale minerals. Am Mineral 91:135–142

Carter GM, Henshall JL, Wakeman RJ (1993) Knoop hardness and fracture anisotropy of calcite. J Mater Sci Lett 12:407–410

Das B (1974) Vicker's hardness concept in the light of Vicker's impression. Int J Rock Mech Min Sci Geom Abstr 11:85–89

Delgado NS, Rodríguez-Rey A, Suárez del Río LM, Díez Sarriá I, Calleja L, Ruiz de Argandona VG (2005) The influence of rock microhardness on the sawability of Pink Porrino granite (Spain). Int J Rock Mech Min Sci 42:161–166

Hughes HM (1986) The relative cuttability of coal-measures stone. Min Sci Technol 3:95–109

Jian S-R (2007) Mechanical deformation induced in Si and GaN under Berkovich nanoindentation. Nanoscale Res Lett 3:6–13

Kranz RL (1979) Crack growth and development during creep of Barre granite. Int J Rock Mech Min Sci 18:23–35

Kranz RL (1983) Microcracks in rocks: a review. Tectonophysics 100:449–480

Lindqvist PA, Hai-Hui L (1983) Behaviour of the crushed zone in rock indentation. Rock Mech Rock Eng 16:199–207

Lan H, Martin CD, Hu B (2010) Effect of heterogeneity of brittle rock on micromechanical extensile behavior during compression loading. J Geophys Res 115:B01202

Lundberg B (1974) Penetration of rock by conical indenters. Int J Rock Mech Min Sci Geomech Abstr 11(6):209–214

Mahabadi OK, Randall NX, Zong Z, Grasselli G (2012) A novel approach for micro-scale characterization and modeling of geomaterials incorporating actual material heterogeneity. Geophys Res Lett 39:L01303

Oliver WC, Pharr GM (1992) An improved technique for determining hardness and elastic modulus using load and displacement sensing indentation experiments. J Mater Res 7:1564–1583

Oliver WC, Pharr GM (2004) Measurement of hardness and elastic modulus by instrumented indentation: advances in understanding and refinements to methodology. J Mater Res 19:3–20

Pariseau WG, Fairhurst C (1967) The force-penetration characteristic for wedge penetration into rock. Int J Rock Mech Min Sci 4(2):165–180

Paul B, Sikarskie DL (1965) A preliminary theory of static penetration by a rigid wedge into a brittle material. Trans Soc Mining Eng 232:372–383

Presser V, Gerlach K, Vohrer A, Nickel K, Dreher W (2010) Determination of the elastic modulus of highly porous samples by nanoindentation: a case study on sea urchin spines. J Mater Sci 45:2408–2418

Saber-Samandari S, Gross KA (2009) Micromechanical properties of single crystal hydroxyapatite by nanoindentation. Acta Biomater 5:2206–2212

Simmons G, Richter D (1976) Microcracks in rocks. In: Sterns RGJ (ed) The physics and chemistry of minerals and rocks. Wiley-Interscience, New York, pp 105–137

Skrzypczak M, Guerret-Piecourt C, Bec S, Loubet JL, Guerret O (2009) Use of a nanoindentation fatigue test to characterize the ductile–brittle transition. J Eur Ceram Soc 29:1021–1028

Sneddon IN (1965) The relation between load and penetration in the axisymmetric Boussinesq problem for a punch of arbitrary profile. Int J Eng Sci 3:47–57

Tapponnier P, Brace WF (1976) Development of stress-induced microcracks in Westerly Granite. Int J Rock Mech Min Sci Geomech Abstr 13(4):103–112

Tulliani JM, Bartuli C, Bemporad E, Naglieri V, Sebastiani M (2009) Preparation and mechanical characterization of dense and porous zirconia produced by gel casting with gelatin as a gelling agent. Ceram Int 6:2481–2491

Whitney DL, Broz M, Cook RF (2007) Hardness, toughness, and modulus of some common metamorphic minerals. Am Mineral 92:281–288

Wong TF (1982) Micromechanics of faulting in Westerly granite. Int J Rock Mech Min Sci 19:49–64

Wong TY, Bradt RC (1992) Microhardness anisotropy of single crystals of calcite, dolomite and magnesite on their cleavage planes. Mater Chem Phys 30:261–266

Xie J, Tamaki J (2007) Parameterization of micro-hardness distribution in granite related to abrasive machining performance. J Mater Process Technol 186:253–258

Yılmaz N (2011) Abrasivity assessment of granitic building stones in relation to diamond tool wear rate using mineralogy-based rock hardness indexes. Rock Mech Rock Eng 40:725–733

Zhu W, Hughes JJ, Bicanic N, Pearce CJ (2007) Nanoindentation mapping of mechanical properties of cement paste and natural rocks. Mater Charact 58:1189–1198

Examining Impact of Particle Deagglomeration Techniques on Microstructure and Properties of Oxide Materials Through Nanoindentation

Sharanabasappa B. Patil, Ajay Kumar Jena and Parag Bhargava

Abstract Microstructure and properties of sintered components produced from nanoparticulate materials are critically dependent on degree of deagglomeration of particulates prior to their consolidation. While all nanoparticulate materials have an inherent tendency to agglomerate owing to attractive Van der Waals forces the impact of agglomeration on sintering behavior/sintered density of powder compacts and associated properties is significant. Although a lot of work has been carried out on developing approaches to deagglomeration of nanopowders it is a challenging task to evaluate the extent of deagglomeration by examining powder compacts. Microscopy of powders (TEM) or of compacts (SEM) is unable to provide any clear distinction between powders with different degrees of particulate agglomeration/deagglomeration. The present chapter cites two case studies from processing of dye sensitized solar cells and synthesis of nanocrystalline yttria stabilized zirconia (YSZ) powders respectively to illustrate that nanoindentation can be an effective way of characterizing the impact of deagglomeration approaches and the consequent deagglomeration extent on powders and compact characteristics. Electrical characterization of the titania based dye-sensitized solar cells, characteristics of green and sintered compacts prepared from synthesized nano YSZ powders are supported by observations from nanoindentation studies.

1 Introduction: Dye Sensitized Solar Cells

Since 1991, following the demonstration of Dye Sensitized Solar Cells (DSSCs) for the first time by Professor Michael Gratzel at EPFL (O'Regan and Gratzel 1991), DSSCs have been attracting attention of both researchers and industries

S. B. Patil · A. K. Jena · P. Bhargava (✉)
Metallurgical Engineering and Materials Science, IIT Bombay, Mumbai, India
e-mail: pbhargava@iitb.ac.in; pbmatsc@yahoo.com

A. Tiwari (ed.), *Nanomechanical Analysis of High Performance Materials*,
Solid Mechanics and Its Applications 203, DOI: 10.1007/978-94-007-6919-9_12,

"

worldwide. Due to its low material cost, easy and inexpensive methods of fabrication and reasonably good power conversion efficiency (10–12 %) (Chiba et al. 2006; Ito et al. 2008), DSSCs are being considered to be a potential alternative to expensive conventional inorganic solar cells. Light harvesting in DSSCs is achieved through dye molecules that are chemisorbed on the available titania surface within the mesoporous titania films. Unlike silicon solar cells, electrons and holes in a DSSC are transported in two different phases; TiO_2 and the electrolyte respectively because of which, the chances of recombination in the cell become low. Hence, DSSCs do not require ultra high pure materials unlike inorganic solar cells. In addition, DSSCs have been proved to perform better than conventional solar cells, without significant change in the power conversion efficiency, in diffused light (Toyoda et al. 2004) and at moderate temperature—up to 50 °C (Berginic et al. 2007). Moreover, these cells can be made on flexible substrates (Pichot et al. 2000; Durr et al. 2005) and hence find wide range of applications, such as on clothes, bags, car tops, etc. Transparency of these cells and their invariant performance under diffused light make these cells more attractive for indoor applications.

A schematic of a dye sensitized solar cell is shown in Fig. 1. The dye molecule adsorbed on TiO_2 surface (photoanode) absorbs light (light harvest) and undergoes photoexcitation. Electron in the excited state of the dye is then injected into conduction band of TiO_2 and diffuses through the film (electron transport) to reach the substrate. From the substrate it goes through external circuit across a load to the counter electrode where it reduces oxidized species (tri-iodide) in the electrolyte. Since light harvest and electron transport are two key processes that affect photovoltaic

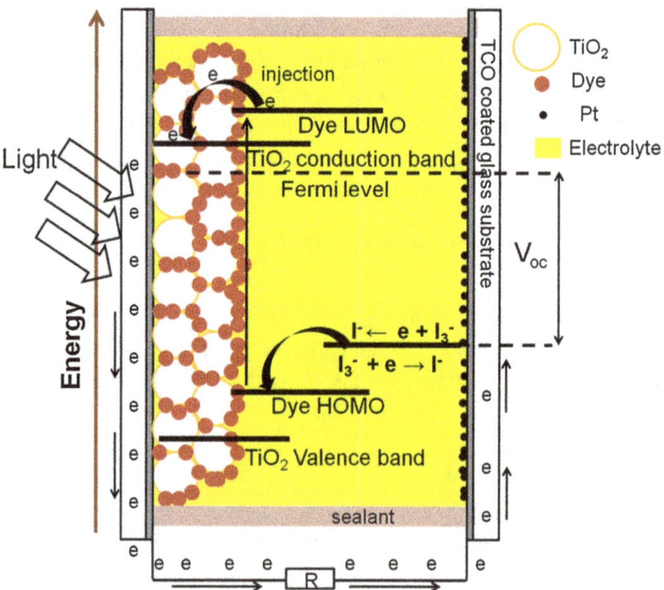

Fig. 1 Schematic diagram of a dye sensitized solar cell

properties of a DSSC, performance of the cell is known to be critically dependent on the titania photoanode characteristics. The photoanode is usually prepared by doctor blading or screen printing of titania slurries/pastes to form a film with thickness ranging from 3–10 μm on transparent and conductive substrates (glass or plastic coated with transparent conductive oxides). The extent of deagglomeration achieved in titania slurries/paste can have a significant impact on particle size, pore size, surface area of the film which eventually alter the cell performance.

1.1 Effect of size of TiO$_2$ Agglomerates on cell Performance

Studies on effect of particle/agglomerate size of titania on performance of DSSCs reported in literature have been carried out mainly from the view point of impact of surface area/pore size on dye loading. Smaller particle size, leading to larger internal surface area, is always expected to enhance the dye loading and hence the current density in the cell. But, contrary to that it has been reported in many studies that J_{sc} and overall efficiency of the cell increases with particle size (Saito et al. 2004; Chou et al. 2007). For instance, TiO$_2$ films of same thickness produced by screen printing of TiO$_2$ pastes prepared in three different dispersion media; PEG 600, mixture of water and ethanol, and terpineol showed highest J_{sc} and efficiency for the film having largest TiO$_2$ particles (Ingli et al. 2003). The films made from PEG-based TiO$_2$ paste had larger particles and pores compared to that of films made from water based paste and showed highest current density and efficiency. Bigger pores probably allowed better permeation of the dye into TiO$_2$ films made from PEG based paste and consequently, increased the dye loading despite lower surface area. Highest efficiency obtained for the film made from PEG based paste was attributed to its greater dye content and more scattering of light in the film due to larger particles. Although the surface area and pore size for film made from water based dispersion was small compared to that of the film made from terpineol based dispersion, the efficiency of the former was higher than the latter because of greater amount of dye adsorbed on the film made from water based paste.

A study on effect of TiO$_2$ particle size on performance of DSSC, undertaken by Chou et al. (2007), also showed increase in J_{sc} and efficiency with particle size. It was found that the TiO$_2$ nanoparticle film with smaller particles ($\sim$10 nm in diameter) resulted in a lower overall light conversion efficiency of 1.4 % with an open-circuit voltage of 730 mV, a short-circuit current density of 3.6 mA/cm^2, and a fill factor of 54 %. Larger particles ($\sim$23 nm in diameter) resulted in a higher efficiency of 5.2 % with an open-circuit voltage of 730 mV, a short-circuit current density of 12.2 mA/cm^2, and a fill factor of 58 %. The increase in short-circuit current density and overall light conversion efficiency with the increase in particle size might have been due to the better dye adsorption behavior of the larger TiO$_2$ nanoparticles, where larger particles allow for more dye adsorption through easier access of dyes. While the performance of DSSCs for different particle/agglomerate sizes has been analyzed in terms of dye loading none of the studies have examined impact of agglomerate size on physical interparticle connectivity and its correlation to DSSC performance.

1.2 Fabrication and Characterization of TiO₂ film

In order to compare the impact of TiO_2 agglomerate size on performance of DSSCs, slurries were prepared by roller milling (rpm = 90) and centrifugal ball milling (rpm = 1,100). The resulting slurries with different agglomerate size were used to make photoanodes. The slurries made by both methods contained TiO_2 powder (Degussa P25, BET specific surface area 47 m^2/g) in ethanol and PEG600 (polyethylene glycol). After adding the required amount of titania powder (6 vol. %) to ethanol and PEG (1 ml per g of titania), the mixture was milled on roller mill for 24 h (the slurry denoted as RM24) or ball milled on a centrifugal mill for 3 h (BM3), ball milled for 3 h followed by roller mixing for 24 h (BM3RM24) or 5 days (BM3RM5D). The particle size and its distribution in the slurries were measured by dynamic light scattering technique using particle size analyzer (Beckman Coulter, Delsa Nano C). TiO_2 films (1 cm × 1 cm) were made from all the prepared slurries by doctor blade technique on an FTO glass substrate (8Ω/square, Pilkington) by a glass rod. Nanoindentation (Load control mode, 500 μN/s, Turboindenter TI 900, Hysitron) on the sintered TiO_2 films was done to know the apparent hardness of the films which indirectly related to TiO_2 interparticle connectivity and particle packing in the film (Raicham et al. 2006).

DSSCs were fabricated using the above photoanodes and a platinized (sputter deposited) FTO substrate as counter electrode. Surlyn (25 μm thick, Solaronix) gasket of size 13 mm × 13 mm was used as spacer between the electrodes. The electrolyte used in the cells was composed of 0.1 M LiI (Merck), 0.05 M I_2 (Thomas Baker), 0.5 M TBP (Sigma Aldrich) in acetonitrile (Merck). A dye sensitized solar cell at different stages of preparation including an assembled cell are shown in Fig. 2. I–V measurements under illumination (100 mW/cm^2) and in dark were taken on the DSSCs with a Keithley 2,420 source meter. A Newport class A solar simulator was used to provide the 1 sun condition (100 mW/cm^2) during the I–V measurement.

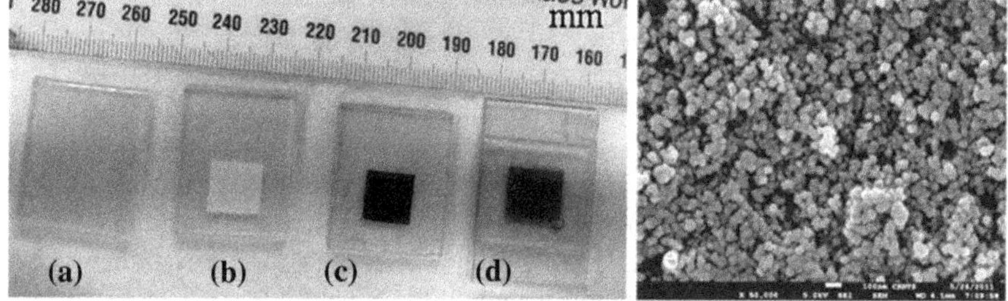

Fig. 2 **a** Bare FTO coated glass, **b** sintered TiO_2 film (white area) on FTO glass, **c** TiO_2 film-stained with N3 dye, **d** the DSSC made with the TiO_2–N3 film and Pt coated FTO, **e** FESEM micrograph of the sintered TiO_2 film (scale bar:100 nm)

1.3 Impact of Milling Technique on Particle size

The primary particle size of the as-received Degussa P25 TiO_2 powder was found to be in between 20 to 30 nm but these were highly agglomerated, as could be seen in the TEM image of the as-received P25 powder in Fig. 3a. Again, the mean particle size of the as-received P25 powder measured by the dynamic light scattering technique was found to be above 1,000 nm, which proved that the TiO_2 nanopowder from Degussa had large agglomerates composed of more than 40 primary particles. These agglomerates were broken down to different sizes when the TiO_2 slurry was milled by different techniques (roller milling and ball milling). The particle size and size distribution in the TiO_2 slurries prepared by roller milling for 24 h (RM24), ball milling for 3 h (BM3) and ball milling for 3 h followed by roller milling for 24 h (BM3RM24), ball milling for 3 h followed by roller milling for 5 days (BM3RM5D) are given in Fig. 3b. The peak of size distribution curve shifted to lower diameter when the slurry was prepared by ball milling for three hours. The mean size of the agglomerates in the TiO_2 slurry prepared by roller milling was determined to be 640 nm while that of the slurries prepared by ball milling for 3 h was found to be about 415 nm (Table 1). It can be inferred from the above results that ball milling was more effective in breaking down the agglomerates than roller milling. Moreover, there was no significant reduction in mean size of the agglomerates when the slurries prepared by ball milling were milled again in the roller mill for 24 h and 5 days. But, the size distribution became narrower on further milling of the slurries obtained after ball milling (Fig. 3b). It is worth noting that ball milled slurries when subjected to roller milling over a longer duration contributed to continued reduction in coarser agglomerates while finer particles formed during ball milling reagglomerated, contributing to narrower size distribution.

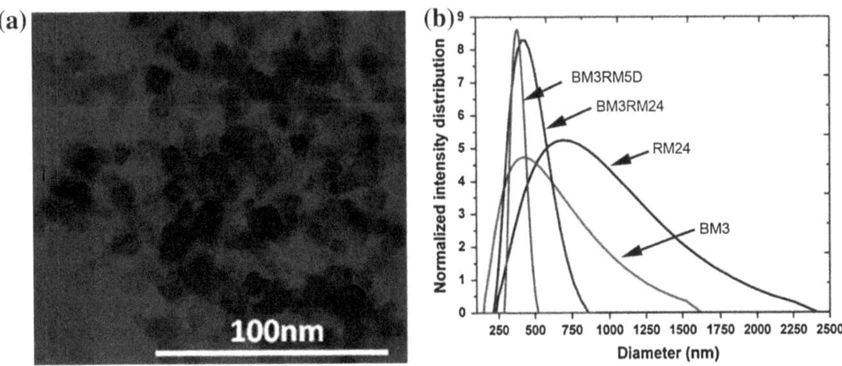

Fig. 3 **a** TEM image of as-received Degussa P25 TiO_2 powder, **b** particle size distribution of P25 slurries prepared by roller milling for 24 h (RM24), ball milling for 3 h (BM3), ball milling for 3 h followed by roller milling for 24 h (BM3RM24), ball milling for 3 h followed by roller milling for 5 days (BM3RM5D)

Table 1 Mean particle size of TiO_2 in the slurries prepared by roller milling and ball milling. Photovoltaic properties of DSSCs made from these slurries

Samples	Mean particle size (nm)	Thickness (μm)	J_{sc} (mA/cm^2)	V_{oc} (mV)	FF (%)	H (%)
RM24	642	8.8	3.1	680.1	58.8	1.2
BM3	415	9.8	4.8	680.1	48.5	1.6
BM3RM24	403	8.1	5.9	680.0	47.1	1.9
BM3RM5D	387	9.7	6.3	680.0	42.7	1.8

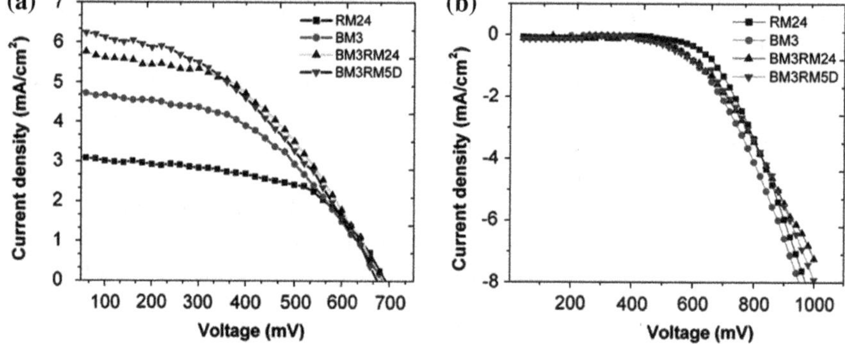

Fig. 4 I-V curves of the DSSCs made from slurries prepared by roller milling and ball milling (roller milling for 24 h; RM24, ball milling for 3 h; BM3, ball milling for 3 h followed by roller milling for 24 h; BM3RM24, ball milling for 3 h followed by roller milling for 5 days; BM3RM5D) **a** under illumination and **b** in dark

1.4 I–V Characteristics of the DSSCs

Thickness of the TiO_2 films made from all the above slurries was found to be very similar and it varied between 8 to 10 μm (Table 1). Figure 4 shows the I–V curves of all the DSSCs fabricated by using the TiO_2 films made from the slurries prepared by different milling techniques. It is very conspicuous from the photovoltaic properties of the above cells that there was no change in the open circuit voltage (V_{oc}) of the cells while a remarkable enhancement in the photocurrent density (J_{sc}) was witnessed for the cells made from the slurries prepared by ball milling. About 100 % increase in J_{sc} (from 3.1 to 6.3 mA/cm^2) was observed for the cell that was made from the slurry prepared by ball milling for 3 h followed by roller milling for 5 days (BM3RM5D). The J_{sc} increased with reduction in the mean agglomerate size and width of particle size distribution of TiO_2 in the slurries.

It was initially believed that there might have been an increase in the surface area of the films on reduction in the size of agglomerates, which was achieved by ball milling the slurry. Surprisingly, the specific surface areas of the TiO_2 powder, as received and those obtained after heat treatment of the slurries prepared by

Table 2 Surface area, dye loading in the films made from slurries prepared by roller milling for 24 h(RM24), ball milling for 3 h (BM3), ball milling for 3 h followed by roller milling for 24 h (BM3RM24), ball milling for 3 h followed by roller milling for 5 days (BM3RM5D) and % TiO_2 loss during sonication of these films

Samples	Specific Surface area(m^2/g)	Thickness (μm)	Dye loading (moles/cm^2)	% loss of TiO_2 after sonication of the films
RM24	43.3	8.8	7.7×10^{-8}	80
BM3	44.1	9.8	6.7×10^{-8}	25
BM3RM24	43.3	8.1	6.0×10^{-8}	27
BM3RM5D	42.8	9.7	6.2×10^{-8}	28

roller milling and ball milling were found to be very similar (Table 1). This indicates that the agglomerates in the P25 TiO_2 powder are porous and the surfaces of the primary particles are exposed in the agglomerates. Therefore reduction in size of the agglomerates does not bring any significant change in the surface area. A similar internal surface area of the TiO_2 films led to adsorption of an equivalent amount of N3 dye in all the films (Table 2). It was thus clear that increase in the current density of the cells on ball milling could not be due to surface area, dye loading.

1.5 Nanoindentation of Sintered Titania Photoanodes

Interestingly, there was a remarkable difference in the interparticle connectivity and adherence of the films to FTO for the films made from slurries prepared by roller milling and ball milling. About 80 % (by weight) of TiO_2 was removed off the substrate by sonication of the film that was made from the slurry prepared by roller milling while there was only about 25 % loss during sonication for all the films made from slurries prepared by ball milling. Hence it is clear that the decrease in size of the agglomerates by ball milling improved the interparticle connectivity and the adherence of the film to FTO. The better interparticle connectivity in the films made from slurries prepared by ball milling was also proved by the force versus displacement curves of the TiO_2 films, obtained by nanoindentation (Fig. 5). It is believed that as agglomerates are broken to finer sizes with ball milling TiO_2 particles are packed better in the film. This was evident from smaller displacement of the nanoindenter at a given applied force on the films made from slurry prepared by ball milling than that of the films made from slurry prepared by roller milling. This result indicated that films made of smaller agglomerates (ball milling) are 'harder' and thus resist penetration of nanoindenter more effectively. In addition, the indents at the three different sites of the film made from ball milled slurry were found to give closely spaced force v/s displacement curves (Fig. 5), indicating a more uniform packing of the particles.

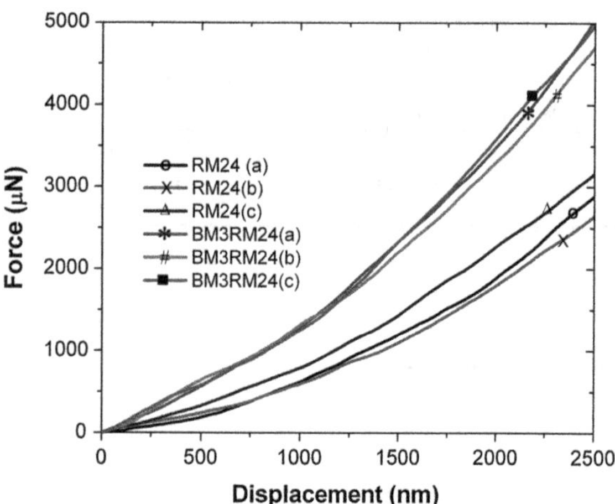

Fig. 5 Force and displacement curves obtained from indenting at three different sites on each of the films made from slurries prepared by roller milling (RM24) and ball milling followed by roller milling (BM3RM24) by a nanoindenter

On the other hand the three indents done at three different sites on the film made from roller milled slurry were not so closely spaced and were found to result in different depth of penetration at the same load. This might be due to more irregular packing of the particles in the films made from slurry prepared by roller milling.

1.6 Impact of Milling Technique on Microstructure of Titania Photoanode

The FESEM micrographs of the surface of the films prepared from slurries—RM24 and BM3RM24, when processed by image analysis software (Fig. 6) show that the film made from the slurries prepared by either roller milling or ball milling are predominantly mesoporous. While both the films showed a range of pore sizes, the film made from slurry prepared by ball milling (BM3RM24) had a narrower pore size distribution and greater uniformity over the film area. It was interesting to note that the cumulative pore volume over the pore size range 1.7–300 nm in films prepared from BM3RM24 slurry was more (0.6 cm^3/g) than that for films prepared with RM24 slurry (0.5 cm^3/g) which could be correlated to the higher area fraction of pores on film surface as determined by image analysis of the FESEM micrographs (Fig. 6). The area fraction of the pores (marked red in Fig. 6) on the surface of the films prepared with BM3RM24 and RM24 slurries were found to be 33.3 and 18 % respectively. Despite the apparently greater volume of pores in films made from slurries prepared by ball milling the small pores might have reduced the electrolyte permeability and hence increased the series resistance of the cell.

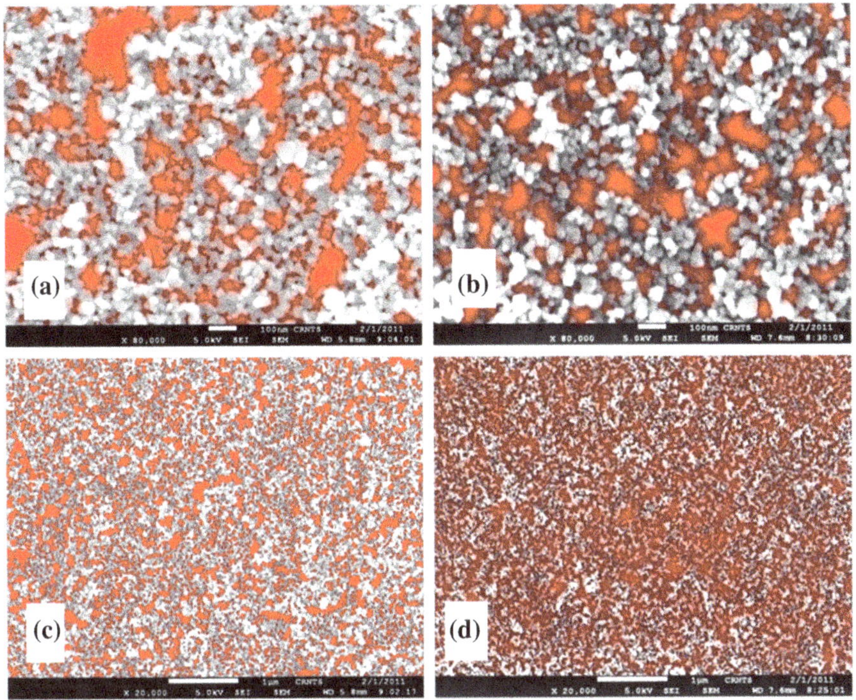

Fig. 6 Images obtained after processing the FESEM micrographs of the surface of the films made from slurry prepared by roller milling (RM24) **a**, **c** and ball milling followed by roller milling (BM3RM24) **b**, **d** through the image analysis software

2 Effect of Sintering Temperature and time

2.1 Fabrication and Characterization of DSSCs with Titania Photoanodes Sintered at Different Temperatures

Required amount of the as-received Degussa P25 TiO$_2$ powder (6 vol. %) was milled on a roller mill for 24 h in ethanol containing 1 ml of PEG 600 per 1 g of TiO$_2$ powder. The doctor bladed TiO$_2$ films were dried overnight at 40 °C before subjecting them to different sintering profiles. For lower temperature-longer time (450 °C, 550 °C, 650 °C for 60 min) sintering, the dried films were heated to the final temperature at a rate of 3 °C per minute and soaked for 1 h followed by furnace cool to room temperature. For all higher temperature-shorter time sintering (700, 800 °C for 10 and 20 min), the films were first heat treated to burn out the PEG (450 °C for 1 h), followed by subjecting the samples to 700 and 800 °C treatment respectively in a preheated furnace with a dwell time of 10 and 20 min at each of the two temperatures. Following the dwell at 700 and 800 °C respectively the samples were removed from the furnace and cooled in air. The sintered films were then dipped in 0.3 mM N3 dye solution for 24 h.

Microstructure of the sintered films was seen with an FESEM. Differences in connectivity between the titania particles with sintering conditions were evaluated through nanoindentation of sintered films at five different sites on each of the sintered films using a Berkovich tip (Hysitron turbo nanoindenter, USA). DSSCs were fabricated as described earlier in Sect. 1.1.

2.2 Nanoindentation of Titania Photoanodes

Microstructural examination could not reveal whether higher temperature shorter time treatments had any significant impact on interparticle connectivity. Also, it was difficult to measure differences in densification in any of the porous titania films which were typically 8–10 μm thick. Nanoindentation was used to characterize the films to highlight the impact of different heat treatments which could then be correlated to measured electrical properties of the dye sensitized solar cells. The results of nanoindentation are shown in Fig. 7. It can be seen that the slope of the force v/s displacement curves for nanoindentation on samples differed, which was attributed to difference in interparticle connectivity. Films with better interparticle connectivity resisted penetration of the indenter more than the loosely interconnected particles and showed smaller depth of penetration at the same force. The TiO_2 films sintered at 650 °C/60 min showed lower depth of penetration than that of the films sintered at 450 °C/60 min for the same force applied. The best interparticle connectivity as seen in terms of lowest depth of penetration of the nanoindentor was seen for films sintered at 700 °C/10 min than films sintered at 450 °C/60 min, 650 °C/60 min, 800 °C/10 min.

Greater depth of penetration of the nanoindenter into TiO_2 film sintered at 800 °C for 10 min than that of TiO_2 film sintered at 700 °C for 10 min (Fig. 7)

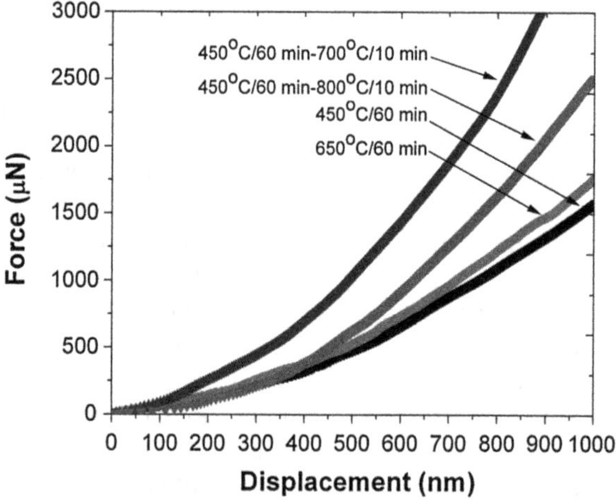

Fig. 7 Load-displacement plot for nanoindentation on TiO_2 films sintered at 450, 650 °C for 60 min and 700, 800 °C for 10 min

was possibly due to intra-agglomerate sintering that caused significant shrinkage leading to creation of separation between agglomerates. The titania powder used here for making the doctor bladed films (Degussa P 25, Crystallite size 25 nm) is known to be produced by flame pyrolysis and thus contains agglomerates that are not broken down easily to individual crystallites or even clusters of a few crystallites.

2.3 Electrical and Electrochemical Characterization of DSSCs

DSSCs prepared with sintered titania films except for the ones sintered at 800 °C for 10 and 20 min showed expected I–V curve at all temperatures. The film sintered at 800 °C experienced intragglomerate shrinkage as suggested by reduced specific surface area (11.9 m^2/g) and nanoindentation. Thus, the dye loading in the 800 °C sintered samples was low leading to significantly low J_{sc}, V_{oc} and fill factor, deviating from the typical behavior of a DSSC (Table 3). The values of short circuit current density (J_{sc}) and open circuit voltage (V_{oc}) varied with sintering temperature and time, as shown in Table 3. J_{sc} increased slightly for films sintered at 550 °C for 60 min (6.5 mA/cm^2) over that of film sintered at 450 °C for 60 min (6.1 mA/cm^2) while J_{sc} decreased for film sintered at 650 °C for 60 min (Table 3). Highest J_{sc} (7.2 mA/cm^2) was observed for cells prepared with films sintered at 700 °C for shorter time (10 min) while sintering at 700 °C for longer time (20 min) showed J_{sc} lower than that of sintering at 450 °C for 60 min (Table 3). Lower J_{sc} for films sintered at 650 °C/60 min, 700 °C/20 min and 800 °C/10 and 20 min was due to lower dye loading which was attributed to reduced TiO_2 surface area. The dye loading was found to be similar (Table 3) for films sintered at 450 and 550 °C for 60 min and 700 °C for 10 min although there was a slight

Table 3 DSSC cell performance characteristics for cells prepared with titania films sintered at different sintering schedules

Sintering schedule (Dwell time in min at peak temp)	J_{sc} (mA cm^{-2})	V_{oc} (mV)	FF (%)	Phtovoltaic conversion efficiency (%)	Dye loading (moles/ cm^{-2})	Pt/ Electrolyte resistance (Ohm cm^2)	TiO_2/ Electrolyte resistance (Ohm cm^2)
450 °C/60 min	6.1	740.2	62.0	2.8	5.9×10^{-8}	7.1	8.0
550 °C/60 min	6.5	751.4	59.4	2.9	5.5×10^{-8}	10.5	9.8
650 °C/60 min	4.8	784.4	63.7	2.4	3.7×10^{-8}	8.9	11.3
700 °C/10 min	7.2	745.6	62.3	3.2	5.9×10^{-8}	13.4	13.2
700 °C/20 min	3.8	684.8	65.3	1.7	3.2×10^{-8}	13.9	13.7
800 °C/10 min	1.1	43.1	–	–	2.9×10^{-8}	15.9	15.0
800 °C/20 min	0.3	4.9	–	–	1.4×10^{-8}	–	–

reduction in surface area, from 44 m^2/g for 450 °C to 36 and 40 m^2/g for 550 °C and 700 °C respectively.

Since cells with films sintered at higher temperatures that showed higher J_{sc} (550 °C/60 min and 700 °C/10 min) had same amount of dye, the increment in J_{sc} was suspected to be due to better light harvest by improved scattering or faster electron transport through the film. However, UV–Vis transmission spectra of the dye-loaded films showed that light harvest remained same for all these films. Therefore, increase in J_{sc} was attributed to faster electron transport achieved by improved interparticle connectivity in the films sintered at higher temperature (550 °C/60 min and 700 °C/10 min).

3 Coprecipitated Nanocrystalline Yttria Stabilized Zirconia Nanopowders

Coprecipitation has been among the most popular wet chemical routes to synthesize nanocrystalline zirconia as it can be scaled up using low cost accessories (Kaliszewski and Heuer 1990; Wang and Zhai 2006; Furlani et al. 2009). But a major disadvantage of the aqueous coprecipitation route is associated with the step of driving away water to dry coprecipitated mass (Lauci 1997; Shi et al. 1991; Srdic and Radonjic 1997). During drying of the precipitate particles are driven to agglomeration and the follow up step of high temperature crystallization treatment (calcination) leads to further enhancement and strengthening of agglomerates (Kaliszewski and Heuer 1990; Readey et al. 1990). The high surface tension of water (73 mN/m) causes zirconium hydroxide nanoparticles to approach closer eventually resulting into hard agglomerates during later stages of processing. One of the ways which has been adopted commonly to minimize agglomeration is to wash the zirconium hydroxide precipitate with ethanol (Shi et al. 1994; Sagel-Ransijn et al. 1996; Wang et al. 2006). Ethanol replaces water and owing to its low surface tension (24 mN/m) minimizes agglomeration of particles. Particles are not forced together during drying and the hydroxyl bridging as seen in water washed powder is avoided and hence lower degree of agglomeration (Kaliszewski and Heuer 1990; Mercera et al. 1992).

3.1 Nano YSZ Synthesis by Coprecipitation and Powder Characterization

Three mole % yttria stabilized zirconia was synthesized by reverse-strike coprecipitation method followed by drying and calcination of the coprecipitated mass at 900 °C. Zirconium oxychloride solution (0.5 M) was prepared in deionized water and yttrium nitrate solution was prepared by adding Y_2O_3 in stirred mixture of deionized water and nitric acid (1:2 ratio). The precursor solution containing zirconium and yttrium was added drop wise into ammonia solution while continuously stirring it ensuring the pH remained above 11 till addition of precursor solution to ammonia was complete. The precipitate was allowed to sediment overnight (~12 h) and then filtered for removal of excess ammonia. The precipitate was washed with distilled water by addition of measured quantity of water and stirring for 30 min. The washing was continued until Cl^- ion was completely removed from the precipitate.

Water washed precipitate was divided into several parts. One part of the water washed precipitate was taken as it is and the crystalline powder obtained on its calcination is referred to as water washed powder. Other six parts of the precipitate were washed with different amounts of ethanol. Quantity of ethanol used for washing the precipitate was expressed as a ratio with respect to the residual water present in the precipitate (93 wt %). Ethanol amount used for washing the precipitate was in following ratios 1:1, 1:2, 1:2.5, 1:3, 1:4 and 1:5 (residual water : ethanol amount). The 1:2.5 (residual water in precipitate to ethanol ratio) was included in the experimental plan after early experiments indicated that the critical water to ethanol ratio lay between 1:2 to 1:3. The precipitate was washed with the respective ethanol amounts over two cycles followed by drying at 100 °C for 12 h. The dried precipitate was calcined at 900 °C for each of the powders washed with different ethanol amounts. Water and ethanol washed powders refer to crystalline YSZ powder obtained by calcination of water and ethanol washed precipitate respectively. The powders were compacted at 200 MPa in a hardened steel die of 6 mm diameter (internal cavity). Stearic acid was used as a die lubricant to minimize powder-die frictional forces during powder compaction. The compacted samples were subjected to nanointendation (Hysitron Inc, USA, TriboIntender TI900) up to peak load of 3000 μN at loading rate of 300 $μN.s^{-1}$ to characterize particle packing in green compacts. For each of the samples, indentation was made at three different regions.

3.2 Influence of Ethanol Amount on Deagglomeration of 3YSZ Powders

From the FESEM of water washed and ethanol washed powders (Fig. 8) differences, if any, in extent of deagglomeration cannot be perceived. All powders appear to be similar in terms of deagglomeration. In order to understand any possible differences between the different powder compacts (water washed, ethanol washed) the corresponding compacts were subjected to nanoindentation. It was expected that penetration of nanoindentor into the compacts for the same applied load may differ owing to differences in particle packing (Raichman et al. 2006). Figure 9 shows the load–displacement curves (three on each sample) for nanoindentation carried out on green compacts produced from water washed and ethanol washed powders. Nanoindentation behavior appeared to be similar for all ethanol washed powder compacts (Fig. 9) while it was significantly different for

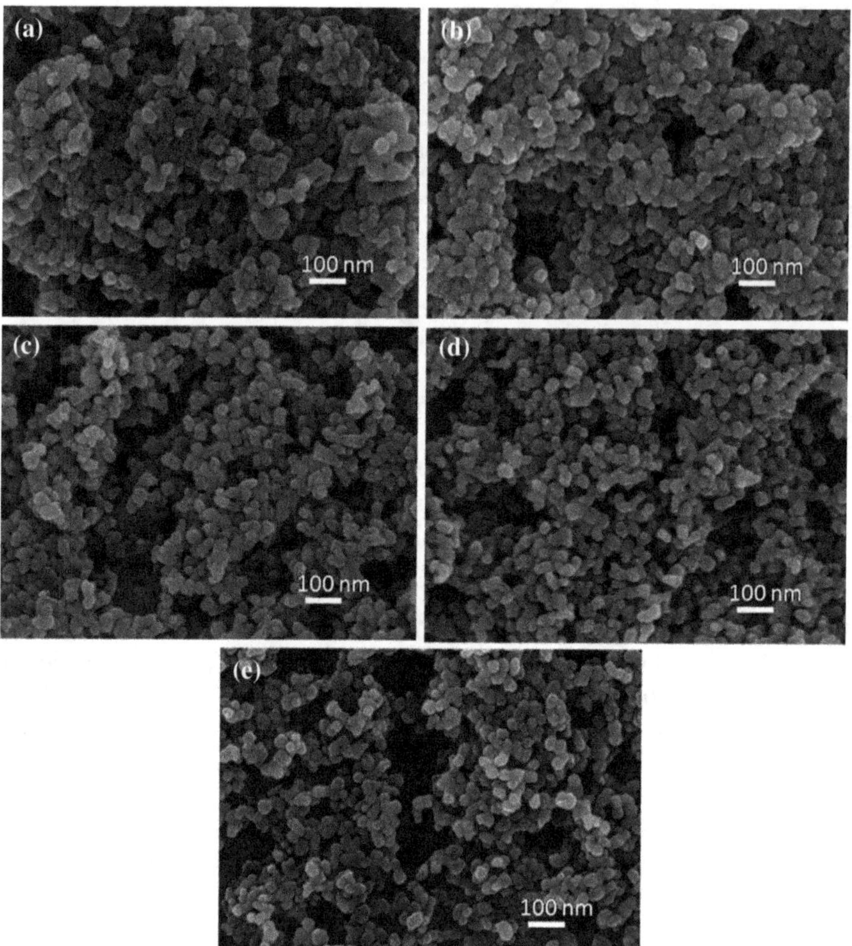

Fig. 8 FE-SEM images of 3YSZ powders obtained by washing the yttrium-zirconium hydroxide precipitate with **a** water and ethanol washing in water to ethanol amount ratio of **b** 1:1 **c** 1:2 **d** 1:2.5 **e** 1:3, calcined at 900 °C

compacts produced from water washed powders. The water washed powder compacts allowed greater penetration of the nanoindentor for the same applied load indicating a low 'apparent' hardness of the green compact. Ethanol washed powders packed better seeming to be 'harder' than the water washed powders which apparently had poorer packing owing to poorer particle deagglomeration or presence of agglomerates. Again, microscopy of green powder compacts (Fig. 10) did not allow distinction to be made between particle packing in compacts prepared from water and ethanol washed powders.

The differences in particle packing as examined by nanoindentation also affected sintered density in expected manner. It was seen that the water washed samples

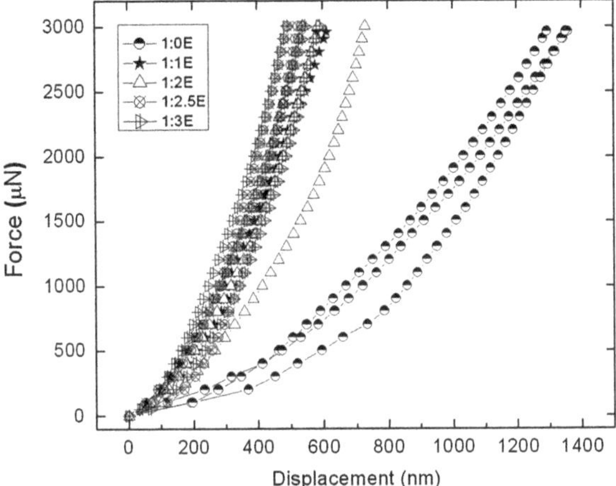

Fig. 9 Load-displacement for nanoindentation on green compacts prepared with powders produced from precipitate washed with water and different amounts of ethanol and calcined at 900 °C. ww (1:0E)water washed powder, water of precipitate to ethanol amount ratio 1:1, 1:2, 1:2.5 and 1:3. Three indentations were carried out at randomly selected region of the samples after Patil et al. (2012)

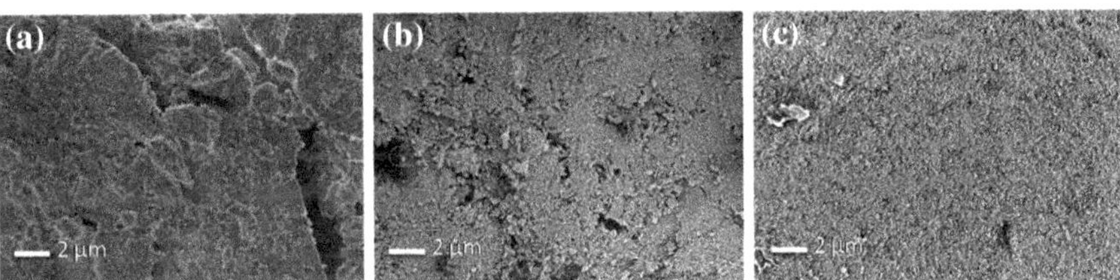

Fig. 10 SEM images of fracture surface of green compacts obtained by compaction (200 MPa) of powders calcined at 900 °C: **a** water washed powder and ethanol washed powder, **b** 1:1 water to ethanol ratio, **c** 1:2.5 water to ethanol ratio

reached a sintered density of 88 % owing to poorer packing and presence of agglomerates while samples made from 1:2.5 ethanol wash reached a sintered density of 96 % of theoretical density.

3.3 *Effect of Calcination Temperature on Deagglomeration for Ethanol Washed Powders*

Synthesis of 3 mol % yttria stabilized zirconia was carried out by reverse-strike coprecipitation method as described in Sect. 2.1 earlier followed by drying and calcination of the coprecipitated mass at various temperatures between 470 °C, the minimum temperature needed for crystallization to 900 °C and soaking time of 120 min was maintained for each of the calcination temperature.

Again as observed earlier direct observations of powders calcined at different temperatures (Fig. 11) did not allow distinction between powders with higher or lower degree of particle agglomeration. To assess differences between the powders, the corresponding green compacts were subjected to nanoindentation. The load–displacement curves (three on each sample) for nanoindentation on green compacts prepared from powders calcined at different temperatures are shown in Fig. 12. Representative images of cavities created by nanoindentation of green compacts are shown in Fig. 13. It can be seen that the nanoindentor penetration in compacts of ethanol washed powders calcined at lower temperature (470 and 600 °C) was limited to lower range (500–600 nm) while the compacts

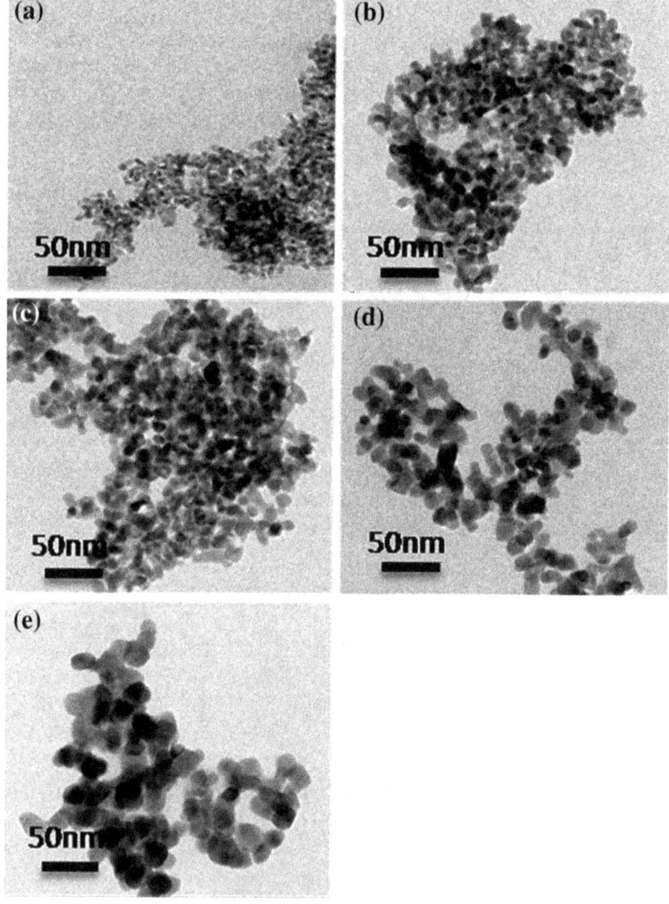

Fig. 11 TEM micrographs of ethanol washed powders calcined at different temperature; 470 °C (**a**); 600 °C (**b**); 700 °C (**c**); 800 °C (**d**) and 900 °C (**e**)

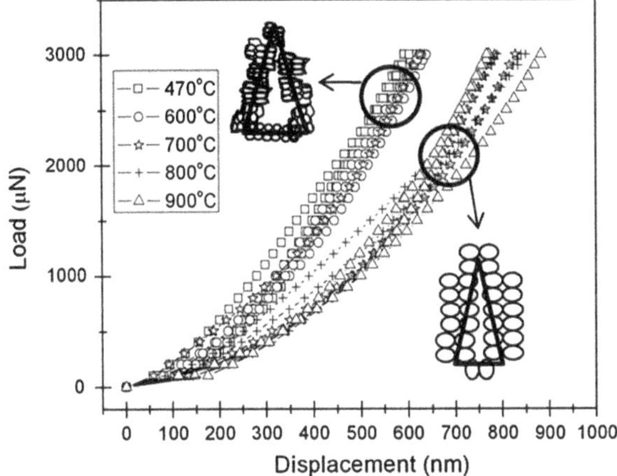

Fig. 12 Load-displacement curves for nanoindentation on green compacts made from powders calcined at different temperatures and compacted uniaxially at 200 MPa. For each sample nanoindentation was carried out at three different widely separated randomly chosen regions after (Patil and Bhargava 2012)

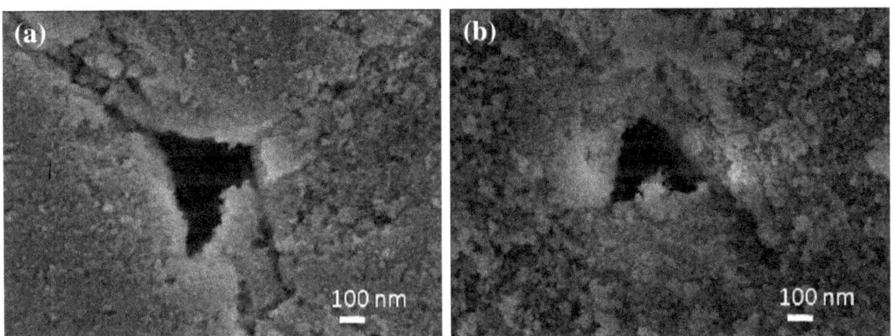

Fig. 13 SEM images of green compacts showing impression of nanoindentation penetration for powders prepared with ethanol (1:2.5 water to ethanol amount ratio) washed Y-Zr precipitate and calcined at **a** 470 °C **b** 600 °C

prepared from powders calcined at 800 and 900 °C allowed deeper penetration (700–900 nm max.) for the same load. Powders calcined at lower temperature have much smaller primary particles (crystallites) which have a greater tendency to agglomerate than the powders calcined at higher temperatures which have coarser primary particles (crystallites). During nanoindentation of compacts produced with powders calcined at low temperature the indenter would have to break or crush the agglomerates in order to penetrate the sample resulting in greater increase in load with penetration. In compacts with powders calcined at higher temperature the nanoindenter would cause separation of weakly agglomerated coarser powder and thus experience lower increase in load in the initial stages of nanoindentor penetration (See inset schematic in Fig. 12).

The above hypothesis was supported by sintering behavior of compacts prepared and sintered under identical conditions. Powders with weakly agglomerated particles (higher calcination temperature) usually pack better during compaction and thus are expected to sinter to higher density. Powders with agglomerated particles (low calcination temperature) do not pack well and densify poorly during sintering. The sintered density of these samples was measured by Archimedes principle. The powders that were calcined at 800 and 900 °C showed highest sintered density of 94 % of theoretical density. The powders calcined at 470 °C showed lowest sintered density of about 88 % TD. For powders calcined at intermediate temperatures the sintered density lay in between that of 470 and 800/900 °C. These observations confirm that obtaining highly densified bodies with fine grained microstructure requires use of powders with weakly agglomerated particles rather than just finer particle size.

4 Concluding Remarks

Nanoindentation served as an effective tool to evaluate particle packing and interparticle connectivity in green and sintered powder compacts. It thus became a means to indirectly evaluate the degree of agglomeration/deagglomeration of powder particles which directly affects particle packing/connectivity. Reduction in size of the TiO_2 agglomerates, achieved by centrifugal ball milling, enhanced the J_{sc} in the DSSC cells because of better interparticle connectivity and better adherence of TiO_2 to the substrate. Nanoindentation curves confirmed existence of improved connectivity between particles in titania films (DSSC photoanodes) produced from slurries with narrower particle size distribution achieved by centrifugal ball milling or a combination of centrifugal ball milling and roll milling. Also, nanoindentation confirmed enhanced connectivity in terms of greater resistance to nanoindentor penetration for films sintered at higher temperatures (650 °C/60 min, 700 °C/10 and 20 min).

The impact of solvent washing and calcination temperature on deagglomeration in nano YSZ powders was clearly distinguishable through nanoindentation of green compacts. Green compacts produced from ethanol washed powders when subjected to nanoindentation were found to be "harder" than compacts produced from water washed powders due to superior particle packing. This was also was also evident from the SEM observations of the green compacts. The compacts from the same powders thus sintered to higher density. Calcination temperature had a significant impact on the state of agglomeration of the 3YSZ nanopowders as produced by coprecipitation. The powders calcined at lower temperature with higher degree of agglomeration offered greater resistance to penetration to the nanoindentor while powders calcined at higher temperature with low agglomeration allowed easier penetration of nanoindentor as seen from gradual rise in load with nanoindentor penetration. Inferences from nanoindentation correlated well with green density and sintered density values. Powders calcined at lower temperatures with higher degree of agglomeration showed lower green and sintered densities.

References

Berginc M, Opara KraÅ¡ovec U, Jankovec M, TopiÄ M (2007) The effect of temperature on the performance of dye-sensitized solar cells based on a propyl-methyl-imidazolium iodide electrolyte. Sol Energy Mater Sol Cells 91:821–828

Chou TP, Zhang Q, Russo B, Fryxell GE, Cao G (2007) Titania particle size effect on the overall performance of dye-sensitized solar cells. J Phys Chem C 6296–6302

Chiba Y, Islam A, Watanabe Y, Komiya R, Koide N, Han L (2006) Dye-sensitized solar cells with conversion efficiency of 11.1%. J App Phys 45 638

Durr M, Schmid A, Obermaier M, Rosselli S, Yasuda A, Nelles G (2005) Low-temperature fabrication of dye-sensitized solar cells by transfer of composite porous layers. Nat Mater 4:607–611

Furlani E, Aneggi E, Maschio S (2009) Effects of milling on co-precipitated 3Y-PSZ powders. J Euro Ceram Soc 29:1641–1645

Ingli Ma TK, Akiyama M, Inoue K, Tsunematsu S, Ken Yao HN, Abe E (2003) Preparation and properties of nanostructured TiO_2 electrode by a polymer organic-medium screen-printing technique. Electrochem Commun 5, 369–372

Ito S, Murakami TN, Comte P, Liska P, Gratzel C, Nazeeruddin MK, Gratzel M (2008) Fabrication of thin film dye sensitized solar cells with solar to electric power conversion efficiency over 10 %. Thin Solid Films 516:4613–4619

Kaliszewski MS, Heuer AH (1990) Alcohol interaction with zirconia powders. J Am Ceram Soc 73(6) 1504–1509

Lauci M (1997) Powders agglomeration grade in the ZrO_2–Y_2O_3 coprecipitation process. Key Eng Mater 132–136:89–92

Mercera PDL, Van Omen JG, Doesburg EBM, Burggraaf AJ, Ross JRH (1992) Influence of ethanol washing of the hydrous precursor on the textural and structural properties of zirconia. J Mater Sci 27:4890–4898

O'Regan B, Gratzel M (1991) A low-cost, high-efficiency solar cell based on dye-sensitized colloidal TiO_2 films. Nature 353:737–740

Patil SB, Bhargava P (2012) Characterization of agglomeration state in 3YSZ nanocrystalline powders through pressure-displacement curves and nanoindentation of green compacts. Powder Technol 228:272–276

Patil SB, Jena A, Bhargava P (2012) Influence of ethanol amount during washing on deagglomeration of coprecipitated calcined nanocrystalline powders. Int J Appl Ceram Technol. doi:10.1111/j.1744-7402.2012.02813.x

Pichot FO, Pitts JR, Gregg BA (2000) Low-temperature sintering of TiO_2 colloids: application to flexible dye-sensitized solar cells. Langmuir 16:5626–5630

Raichman Y, Kazakevich M, Rabkin E, Tsur Y (2006) Inter-nanoparticle bonds in agglomerates studied by nanoindentation. Adv Mater 18:2028–2030

Readey MJ, Lee RR, Halloran JW, Heuer AU (1990) Processing and sintering of ultrafine MgO–ZrO_2 and [MgO–Y_2O_3]-ZrO_2powders. J Am Ceram Soc 73(6):1499–1503

Saito Y, Kambe S, Kitamura T, Wada Y, Yanagida S (2004) Morphology control of mesoporous TiO_2 nanocrystalline films for performance of dye-sensitized solar cells. Sol Energy Mater Sol Cells 83:1–13

Srdic V, Radonjic L (1997) Synthesis and sintering behavior of Nanocrytalline ZrO_2-3 mol% Y_2O_3 powders. Key Eng Mater 132–136:45–48

Shi JL, Gao J-H, Xiang Z, Yen TS (1991) Sintering behavior of fully agglomerated zirconia compacts. J Am Ceram Soc 74(5):994–997

Shi JL, Lin ZX, Qian WJ, Yen TS (1994) Characterization of agglomeration strength of coprecipitated superfine zirconia powders. J Euro Ceram Soc 13:265–271

Sagel-Ransijn CD, Winnubst AJA, Burggraaf AJ, Verwij H (1996) The influence of crystallization and washing medium on the characterization of nanocrytalline Y-TZP. J Euro Ceram Soc 16:759–766

Toyoda T, Sano T, Nakajima J, Doi S, Fukumoto S, Ito A, Tohyama T, Yoshida M, Kanagawa T, Motohiro T, Shiga T, Higuchi K, Tanaka H, Takeda Y, Fukano T, Katoh N, Takeichi A, Takechi K, Shiozawa M (2004) Outdoor performance of large scale DSC modules. J Photochem Photobiol A 164 203–207

Wang S, Zhai Y (2006) Coprecipitation syntheis of MgO-doped ZrO_2 nano powder. J Am Ceram Soc 89(11):3577–3581

Wang S, Li X, Zhai Y, Wang K (2006) Preparation of homodispersed nano zirconia. Powder Technol 168:53–58

Nanoindentation of Micro Weld Formed Through Thin Nanolayered Filler

Julia Khokhlova, Maksym Khokhlov, Alla Tunik
and Anatoliy Ishchenko

Abstract The aim of this work is to find optimal fillers for nano-layered foils of Ti/Al, Ti/Ni, Ni/Al, Al/Cu for diffusion joining of hard-weldable γ-TiAl alloy by means of nanoindentation test. Nano-layered foils have the non-equilibrium state of structure and are prone to the development of self-propagating high-temperature synthesis reaction during heating. Reaction propagates extremely fast and can be characterized by exothermal effect with a transformation of material structure and with a change in micro/nano mechanical properties.

1 Introduction

Titanium aluminide alloys are advanced heat resistant materials and classified as hard-weldable alloys. The interest to these alloys is due to the possibility of their use in the aerospace industry as an alternative to titanium and nickel "super alloys". Objective of this work is to search optimal filler of nano-layered foils of Ti/Al, Ti/Ni, Ni/Al, Al/Cu for diffusion joining of γ-TiAl by investigating the results of nanoindentation test and microstructure analysis by optical and scanning electron microscopy (SEM).

1.1 Titanium Aluminide Alloys

Mono-titanium aluminide γ-TiAl have an ordered face-centered tetragonal lattice, which persists up to 1,440 °C. It displays high resistant to oxidation and did not burn until 900 °C, which distinguishes γ-TiAl from titanium alloys (Khokhlova

J. Khokhlova (✉) · M. Khokhlov · A. Tunik · A. Ishchenko
Paton Electric Welding Institute of National Academy of Sciences of Ukraine,
Kiev-03680, 11, Bozhenko Street, Kyiv 150, Ukraine
e-mail: khokhlova.julia@gmail.com

A. Tiwari (ed.), *Nanomechanical Analysis of High Performance Materials*,
Solid Mechanics and Its Applications 203, DOI: 10.1007/978-94-007-6919-9_13,

and Khokhlov 2008). Microhardness of the alloy is 3,000–4,000 MPa. Young modulus of elasticity of γ-TiAl is 160–175 GPa at room temperature, which is higher than for titanium alloys, but lower than for nickel.

However, commercial use of γ-TiAl is limited due it's brittleness and low plasticity at room temperature ($\delta = 0.2$–0.5 %). Processing of titanium aluminide is complicated due to the high resistance to the material deformation.

After reports about rising of plasticity in two-phase titanium aluminide, these alloys have become the object of research. Maximum low-temperature plasticity of these alloys is a result of formation of fine-grained duplex structure ($\alpha2$-Ti$_3$Al + γ-TiAl) + γ-TiAl, and maximum resistance to creep at elevated temperatures and low temperature toughness provided by the formation of a fully lamella structure (Fig. 1a) of γ-TiAl and $\alpha2$-Ti$_2$Al phases. This is large (60–120 μm) homogeneous slots of defined orientation (Fig. 1b). The lamella structure includes disperse dark phases with high content of niobium, which were uniformly distributed in volume.

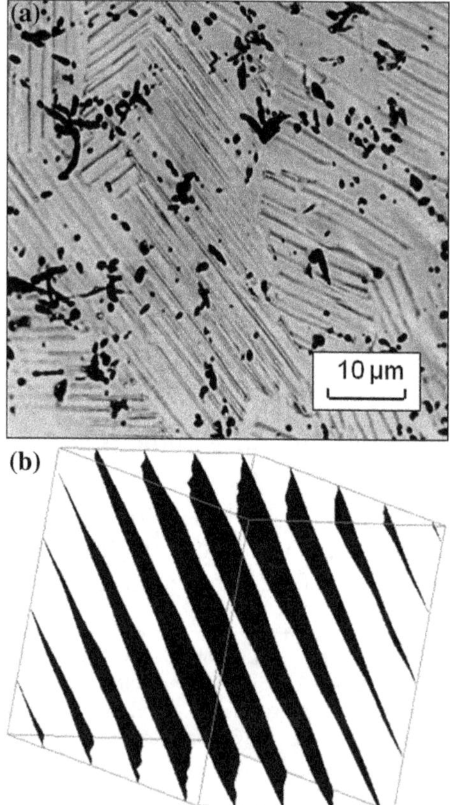

Fig. 1 Microstructure of γ-TiAl alloy (**a**) with lamellar crystals (**b**)

Fig. 2 Nano-layered Ti/Al
foil (**a**) and layers in foils (**b**)
at velocity of consolidation
10 rpm (SEM)

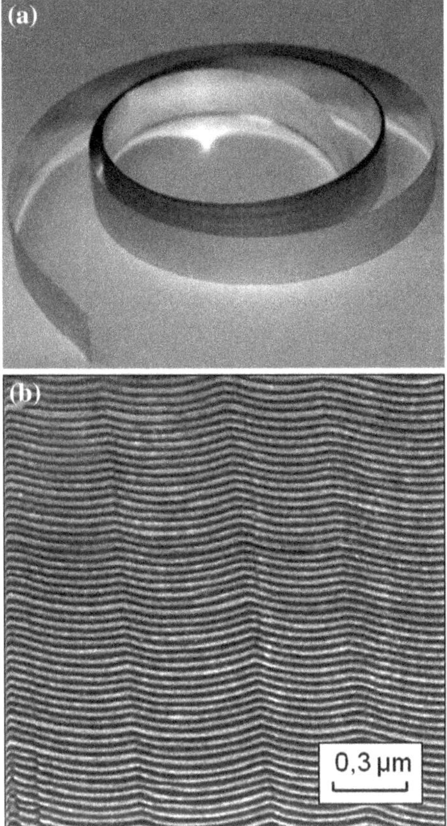

The alloys for practical interest are γ-TiAl alloys with (45–49) at. % Al to which Cr, Nb, Hf, Ta, Zr, W, V, Mn are being added. The increase of the low-temperature plasticity is the results of alloying by Mn, Cr, V, which reduces aluminium content by replacing it's atoms in γ-TiAl lattice. Modern TiAl alloys usually contain one element that increases elasticity, and one which increases heat and oxidation resistances such as Ti - (47–50) at. %, Al - (l–3) at. %, V, Cr, Mn - (2 − 4) at. %, Nb, Ta, Mo, W.

2 Experimental

2.1 Nano-layered Foil as Welding Filler

Thin nanostructured foils (Fig. 2a) from multilayer compositions of various metals (Ti/Al, Ni/Al, Cu/Al, etc.) have been used as welding fillers for solid-phase joining of intermetallic alloys as γ-TiAl.

Their manufacturing is based on the process of layer-by-layer consolidation of elements from the vapour phase using electron beam vacuum technology (Khokhlova and Khokhlov 2009). The deposition technology enables to adjust the process of a layered structure formation in a broad range of thicknesses of individual layers 70–150 nm (Fig. 2b).

Such materials have the non-equilibrium state of structure (Khokhlova et al. 2009a). The nanostructured foils are prone to the development of self-propagating high-temperature synthesis reaction (SPHTS) of it's intermetallic compounds at heating. Reaction is rapid (Fig. 3) and characterized by exothermal effect with transformation of material's structure and micromechanical properties.

The necessary condition for joining of materials due to activation of diffusion processes in solid state is a stable running of SPHTS reaction. This requires application of multilayer foils of more than 30 μm thickness (Khokhlova et al. 2009b).

Figure 4a shows an image of the microstructure of cross-sectional samples of foils directly after deposition. It is seen that in the initial state foils have alternate layers of dark and light contrast. Since etching of samples took place in the electrolyte, which primarily reacts with aluminum, therefore, the light and dark layers are, respectively, the layers of titanium and aluminum. Figure 4b, c shows microstructure of foils after SPHTS reaction. The structure of the reaction product is characterized by heterogeneity, possibly due to the undulating nature of fusion reaction spreading along the foil.

Visualization of SPHTS reaction in volume was modeled by ChemSite 3.01 program (Fig. 5). The selected time of heating was up to 600 °C is 14 ps. The dynamics of Al and Ti atoms in layers, which consist of 3 periods of the crystal lattice, is similar to SPHTS reaction.

It is obvious that the character of structural formation in foils during the SPHTS reaction was not dependent of Ti and Al layers thicknesses. Localization of the reaction and it's undulating passage can cause the additional mass transfer after wave of reaction. Moreover, it contributes to the formation of heterogeneous structure.

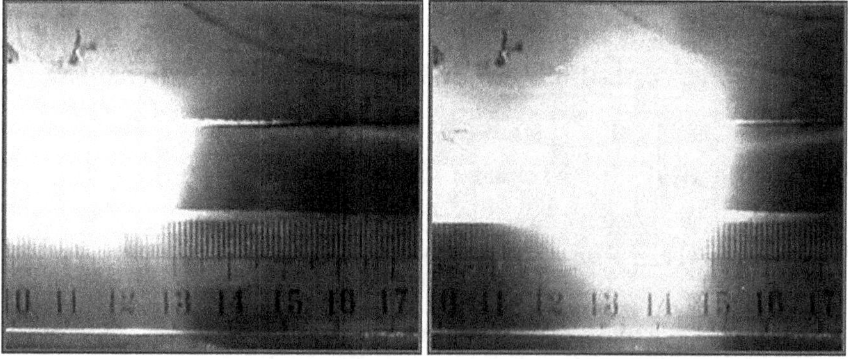

Fig. 3 Velocity of SPHTS reaction in Ti/Al is 5–10 m/sec

Fig. 4 Microstructure of nano-layered Ti–Al foil before (**a**) and after SPHTS reaction (**b**, **c**) (SEM)

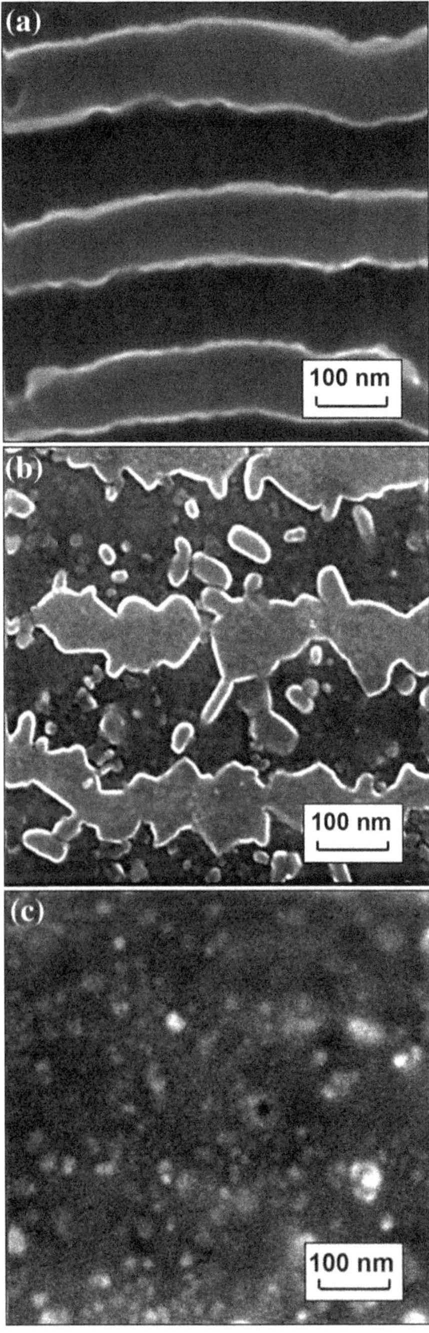

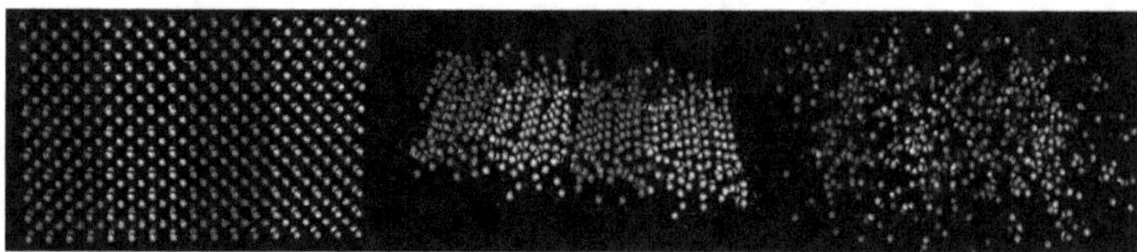

Fig. 5 Visualization (3d) of SPHTS reaction in Ti/Al

2.2 Diffusion Microwelding

Diffusion microwelding of four 6 mm thick plates of Ti-48Al-2Nb-2Mn at. % (γ-TiAl) alloy was accomplished in the free state in a vacuum chamber. Thin foils (15–25 μm) of Ti/Al, Ti/Ni, Ni/Al, Al/Cu with nano-layered structure were obtained by electron-beam technology of consolidation of elements from the vapour phase and were used as fillers. Foils were placed between plates, pressed, heated and withstood at a temperature for a certain time. Electron-beam heater was used as a source of heat. Welding temperature was T = 1,200 °C, welding time was τ = 20 min, pressure was P = 45 MPa and working vacuum in a chamber was B = 1.33 × 10^{-3} MPa.

Samples were grinded in a standard way; polished in acid-chlorine electrolyte of 1,000 ml of glacial acetic acid, 70 ml HClO$_4$, and chemically etched in an aqueous solution of 2 %HNO$_3$ + 2 %HF.

3 Results and Discussion

3.1 Micro/nano Hardness of Welded Joint

Micro/nano hardness tests were performed by using of Vickers and Berkovich indenters (Fig. 6) at 0.002 and 0.0001 N of loading on indenter (Khokhlova et al. 2012; Povarova 2004; Ustinov 2008; Ustinov et al. 2008a, b).

Metallographic investigation of the transition zone of welded joint of Ti-48Al-2Nb-2Mn at. % with nano-layered foil of Ti/Ni showed a continuous layer of heterogeneous structure (20–50 μm of thickness) (Fig. 7). Central region of layer with thickness of 3–4 μm corresponds to intermetallic compound of TiNi, alloyed with manganese. Peripheral part of layer contains triple compound of AlTiNi, alloyed with niobium and manganese. The hardness of the layer was approximately between 9000–10400 MPa.

Welding of Ti-48Al-2Nb-2Mn at. % alloy with micro-layered foil of Ni/Al forms joint (Fig. 8b) with high hardness up to 600 GPa on the border with base material as a result of SPHTS reaction. In the center of intermediate foil, hardness is low (80–90 GPa) (Fig. 9) compared to the base metal (300−400 GPa).

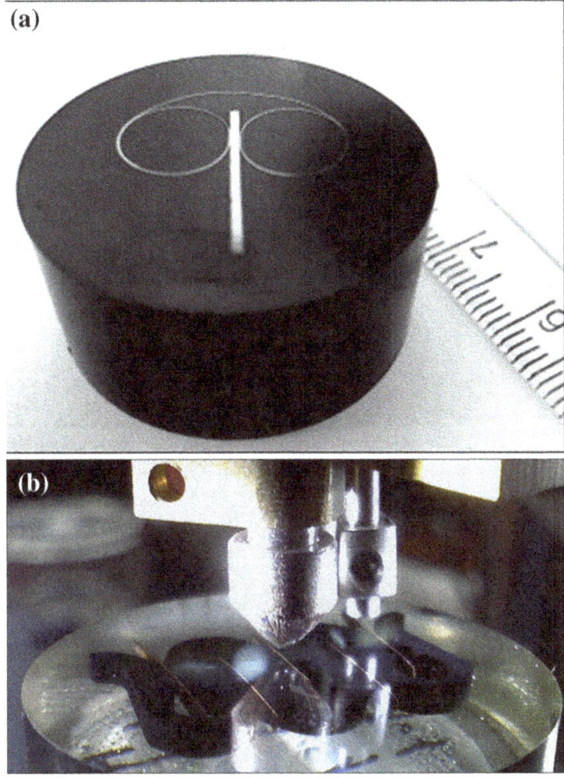

Fig. 6 Nanoindentation of base foil

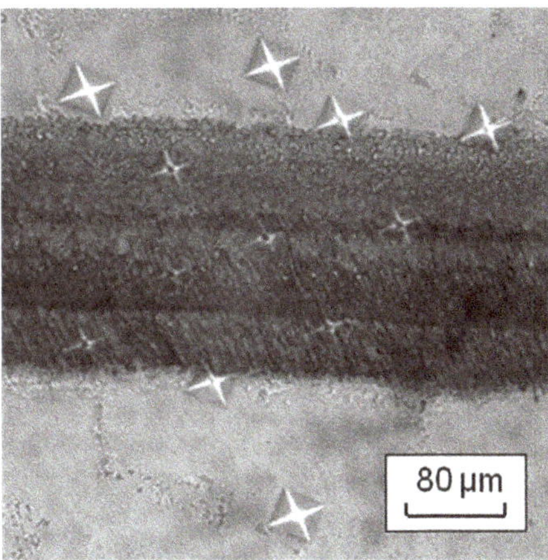

Fig. 7 Indenter's print in diffusion zone of joint through nanolayered Ti/Ni foil

The decrease in strength of welded joint is the cause of gradient in hardness that appears in nanoindentation curve (Fig. 8a).

Nano-layered foil Al/Cu of eutectic composition Al+33 %Cu with a total thickness of 20–25 μm was used for diffusion welding of Ti-48Al-2Nb-2Mn at. %

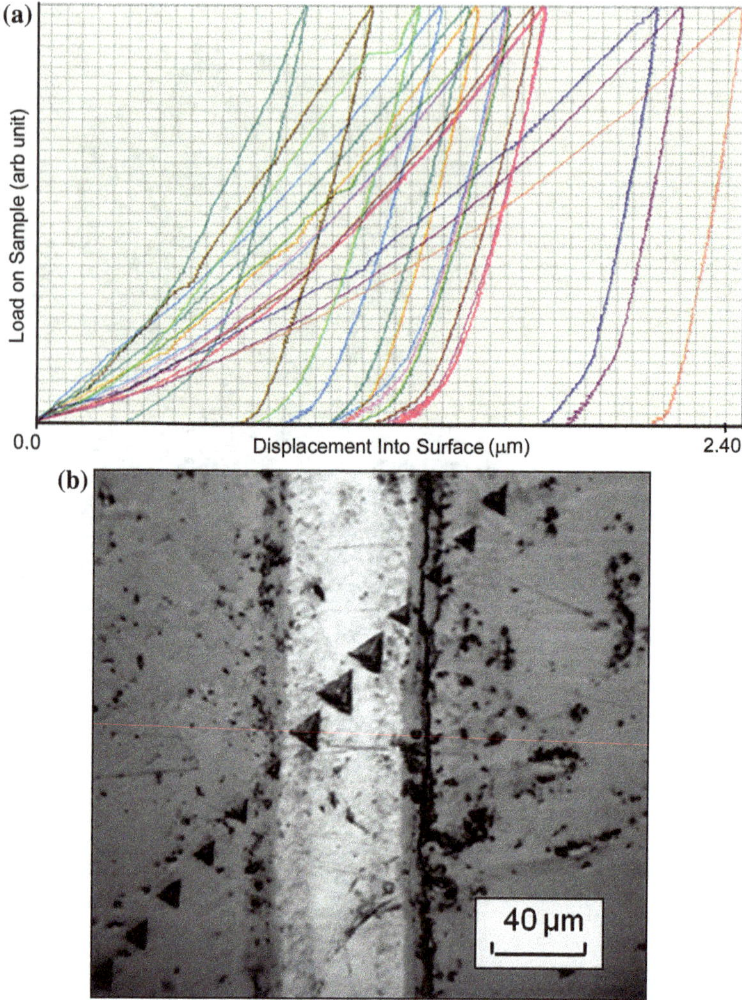

Fig. 8 Diagrams of indentation (**a**) diffusion zone of joint through nano-layered Ni/Al foil (**b**)

alloy. It was anticipated that the application of Al/Cu foil with eutectic composition will result in it's complete dissolution at welding temperature. However, dissolving of the foil in the area of joint does not happen (Fig. 10). Microscopic research showed that the joint zone is a continuous layer with thickness between approximately 15–20 mm. The hardness is between approximately 390–420 GPa and remains practically at the level of base metal. Perhaps, the formation of layers with such composition makes welded joints unsuitable for use in high temperature environments.

Welding of Ti-48Al-2Nb-2Mn at. % alloy with micro-layered foil of Ti/Al (15–20 μm of thickness) forms joint with a solid solution, based on intermetallic

Fig. 9 Indenter's print in diffusion zone of joint through nano-layered Ni/Al foil

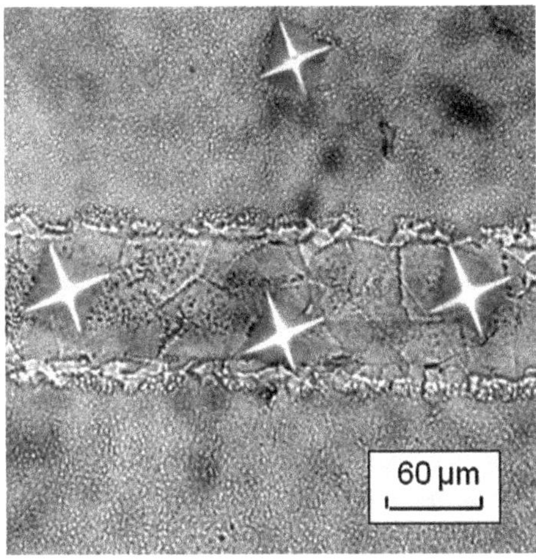

Fig. 10 Dissolving of Al/Cu foil in joint and prints of indenter in microwelding zone

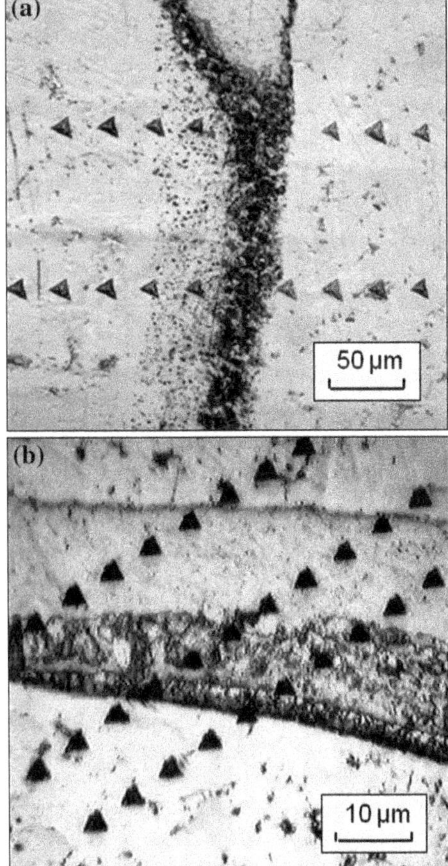

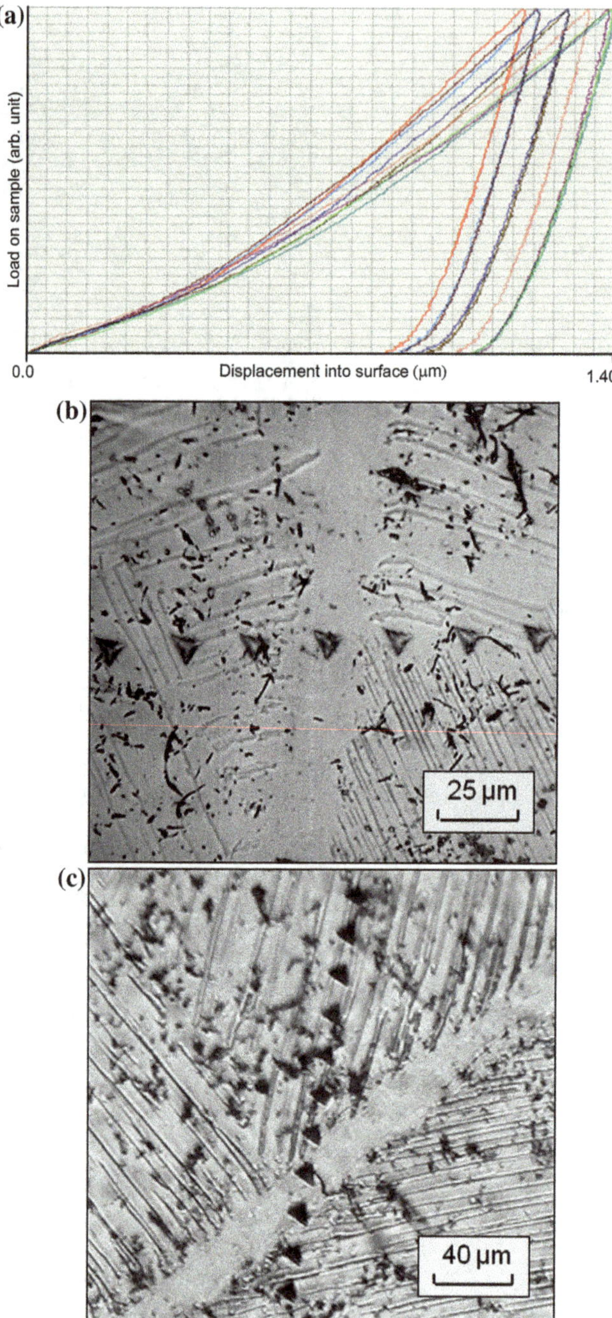

Fig. 11 Diagrams of indentation (**a**) of diffusion zone of joint through nano-layered Ti/Al foil (**b**)

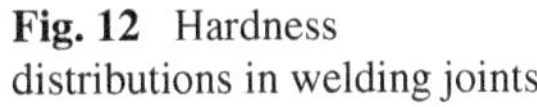

Fig. 12 Hardness distributions in welding joints

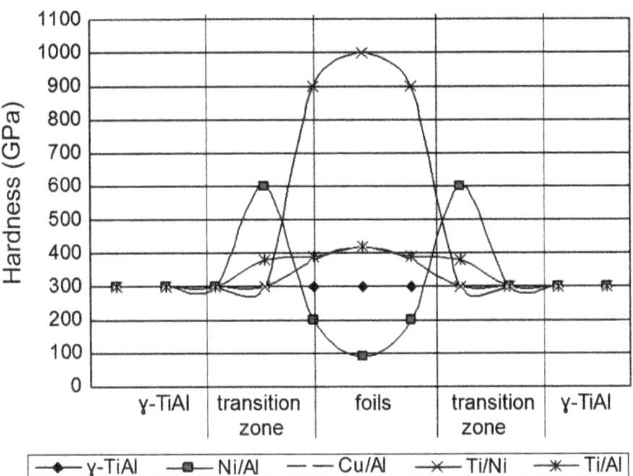

Ti$_3$Al and alloyed with niobium and manganese. The formation of such homogeneous intermetallic layer in welded joint corresponds to the composition of base material of γ-TiAl.

The changes in the structure of the original metal in areas adjacent to the layer were observed, which were related to the γ-TiAl recrystallization processes during diffusion welding. On indentation curves, it was seen that hardness of the joint area remains at the level of base metal (Fig. 11). The hardness of the layer was between approximately 350–420 GPa while hardness of base metal was approximately between 300–400 GPa.

Figure 12 shows the distribution of hardness in welding joints of Ti-48Al-2Nb-2Mn at. % (γ-TiAl) alloy formed by using Ti/Al, Ti/Ni, Ni/Al, Al/Cu foils as fillers. The gradient of hardness contributes to the formation of residual stresses (Khokhlova and Khokhlov 2008, 2009) in welded joints resulting in negative influence on strength. Therefore, the best option is to use Ti/Al as filler due to uniformity in the indentation curves.

4 Concluding Remarks

Several different metallic joints were investigated with nanoindentation technique. Best results were obtained when Ti/Al nano-layered foil was used as filler for microwelding. A uniform intermetallic layer that forms in welded joint was similar in composition to the base material i.e., γ-TiAl. The hardness of the foil-filler after welding was similar to the hardness of the base material.

References

Khokhlova JA, Ishchenko AY, Khokhlov MA (2012) Sub-micromechanical test of grain boundaries after intergranular reactive diffusion. Paper presented at the 9th International Conference of Residual Stresses ICRS, Garmisch-Partenkirchen, Germany, 7–9 Oct 2012

Khokhlova Y, Klochkov I, Grinyuk A, Khokhlov M (2009a) Verification of the values of the Young modulus of elasticity received by nanoindenter Micron-gamma. Non-destr Test Tech Diagn 1:30–32

Khokhlova YA, Khokhlov MA (2008) Method of determining the micromechanical properties of diffusion joint of intermetallic titanium. In: Nikolaev (ed) Ukrainian scientific conference of students and young scientists: welding and allied processes and technology, 3–7 Sep 2008, p 63

Khokhlova YA, Khokhlov MA (2009) Investigation of nanoscale effect at indentation of nano-structured copper condensates. In: Panasyuk VV (ed) Fracture mechanics of materials and structural integrity. Physico-Mechanical Institute of G.V.Karpenko NAS of Ukraine, Lviv, pp 581–585

Khokhlova YA, Khokhlov MA, Fesiun EV (2009b) Assessing the impact of the scale factor on the microhardness at indentation of nanostructured foils. In: V-Ukrainian Scientific conference of young scientists that specialists: welding and relating technology, 2–7 May 2009, p 166

Povarova KB (2004) Powder metallurgy of tungsten alloys (alloying, pretreatment, sintering. NVN, structure, properties). In: Euro PM2004, p 831–837

Ustinov A, Olikhovska L, Melnichenko T, Shyshkin A (2008a) Effect of overall composition on thermally induced solid-state transformations in thick EB PVD Al/Ni multilayers. Surf Coat Technol 202(16):3832–3838

Ustinov AI (2008) Dissipative properties of nanostructured materials. Strength Mater 40(5):571–576

Ustinov AI, Falchenko YV, Ishchenko AY, Kharchenko GK, Melnichenko TV, Muraveynik AN (2008b) Diffusion welding of γ-TiAl based alloys through nano-layered foil of Ti/Al system. Intermetallics 16(8):1043–1045

Effects of Residual Stress on Nano-Mechanical Behavior of Thin Films

M. Sebastiani, E. Bemporad, N. Schwarzer and F. Carassiti

Abstract In this chapter, we present an overview of an optimized method for the determination of surface elastic residual stress in thin ceramic coatings by instrumented sharp indentation. The methodology is based on nanoindentation testing on focused ion beam (FIB) milled micro-pillars. Finite element modeling (FEM) of strain relief after FIB milling of annular trenches demonstrates that full relaxation of pre-existing residual stress state occurs when the depth of the trench approaches the diameter of the remaining pillar. Under this assumption, the average residual stress present in the coating can be calculated by comparing two different sets of load-depth curves: the first one obtained at the center of stress-relieved pillars, the second one on the undisturbed (residually stressed) surface. The influence of substrate's stiffness and pillar's edges on the indentation behavior can be taken into account by means of analytical simulations of the contact stress distributions. Finally, the effect of residual stress on fracture toughness and deformation modes of a TiN PVD coating is analyzed and discussed here.

1 Introduction

Intrinsic (or residual) stresses (Korsunsky 2009; Withers and Bhadeshia 2001), resulting from manufacturing or processing steps, mostly define the performance and limit the lifetime of nanostructured materials (Dye et al. 2001), thin films (Bemporad et al. 2007, 2008; Bull 2005; Bull and Berasetegui 2006; Dye et al. 2001; Espinosa et al. 2003; Fischer et al. 2005; Pauleau 2001; Roy and Lee 2007),

M. Sebastiani (✉) · E. Bemporad · F. Carassiti
Mechanical and Industrial Engineering Department, University of Rome "ROMA TRE",
Via della Vasca Navale 79, 00146 Rome, Italy
e-mail: seba@stm.uniroma3.it

N. Schwarzer
Saxonian Institute of Surface Mechanics, SIO, Tankow 1, 18569 Ummanz, Rügen, Germany

A. Tiwari (ed.), *Nanomechanical Analysis of High Performance Materials*,
Solid Mechanics and Its Applications 203, DOI: 10.1007/978-94-007-6919-9_14,
© Springer Science+Business Media Dordrecht 2014

coatings and MEMS devices (Espinosa et al. 2003) and bulk metallic glasses (Wang et al. 2011). The importance of residual stress for micro-systems is relatively more important than in case of conventional materials and structures, due to the size effects in crystal plasticity, fatigue and fracture behavior that have been recently observed for a wide range of materials.

It has been widely reported in the literature that residual stress can affect significantly the adhesion and fracture toughness of thin films for a variety of industrial applications, ranging from wear resistance coatings to thin films for Solid Oxide Fuel Cells (SOFC) and coatings for biomedical applications.

They can also affect load bearing capacity, elastic strain to failure and ductility of Bulk Metallic Glasses (BMGs), the reliability of micro-welds and other metal interconnects, crack propagation and charge carrier mobility in semiconductor BEoL systems with ultra-low-k (ULK) nano-porous films (Ye et al. 2006), and the resonant frequency and lifetime of micro/nano-electro-mechanical systems (MEMS/NEMS) (Bull 2005).

In the specific case of hard nanostructured coatings (Bemporad et al. 2007; Pauleau 2001; Teixeira 2002), strong compressive in-plane stresses is usually observed (both due to the growth process and from different thermal expansion coefficients between coating and substrate (Bemporad et al. 2007), which often involve buckling of the coating or interfacial failure under in-service load conditions.

Notwithstanding these industrial requirements, the evaluation of the residual stress in (sub)micron layers is still an extremely challenging task from a metrology perspective, especially in the case of nano-crystalline, strongly textured, complex multiphase or amorphous materials and thin films.

For such reasons, the development of site-specific micro-scale evaluation techniques of residual stress still represents a critical issue for design and characterization of small scale systems.

However, the established techniques for micron-scale measurement of residual stress still have strong limitations, e.g. in terms of spatial resolution, lack of depth sensing, their applicability on non-crystalline materials or accessibility to industry.

Strong efforts are also needed in the sense of developing a portable and flexible semi-destructive method that would be applicable down to the microscopic scale, and would allow routine determination of residual stress with high spatial resolution (Bolshakov et al. 1996; Suresh and Giannakopoulos 1998; Tsui et al. 1996).

The aim of this work was to develop an effective and reproducible methodology for the assessment of residual stress by the use of sharp nanoindentation testing on a (sub)micron scale (Oliver and Pharr 1992, 2004).

A brief review of the currently available methods for residual stress analysis by nanoindentation is presented (Suresh and Giannakopoulos 1998), whose principle is depicted in Fig. 1.

It is shown that the available methods are affected by a major limitation, consisting of the difficulty in determining the load–displacement curve for the stress-free condition in case of thin films.

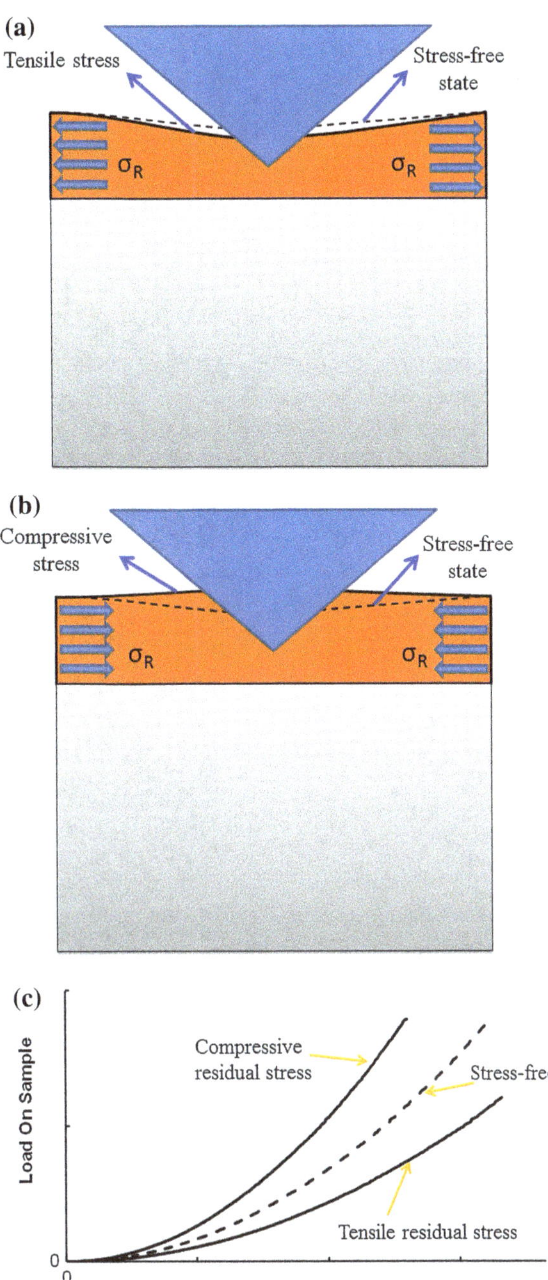

Fig. 1 Concept of residual stress analysis by sharp indentation. **a** a tensile residual stress induces a more pronounced sinking-in with respect to the contact profile corresponding to the stress-free state; **b** a compressive stress induces a piling-up; **c** in both cases, a deviation of the load–displacement curve with respect to the stress-free state is observed, due to the modification of the contact area described by Eqs. (1–2)

To solve this issue, it is proposed to use a focused ion beam (FIB) to produced stress-free micro-pillars on the specimen's surface (Fig. 2).

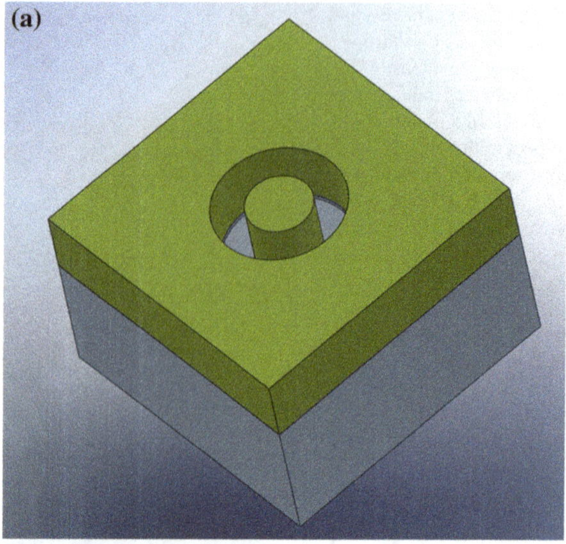

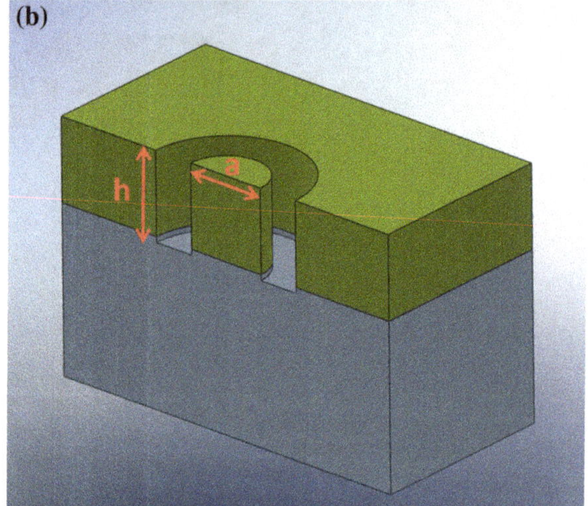

Recent publications by some of the authors (Korsunsky et al. 2009, 2010; Sebastiani et al. 2011; Song et al. 2012) showed that milling of annular trenches on a residually stressed surface gives full stress relaxation at specimen surface of the central stub for h/a > 1 (Fig. 1), thus allowing to obtain, for any kind of material and coating, a local stress-free reference volume (Korsunsky et al. 2010).

This result was obtained by Finite Element Modeling (FEM) of surface relaxation strain distribution after ring-core milling (Korsunsky et al. 2010).

The indentation on the stress relieved pillars gives the reference load–displacement data to be used for the application of the models and the evaluation of the

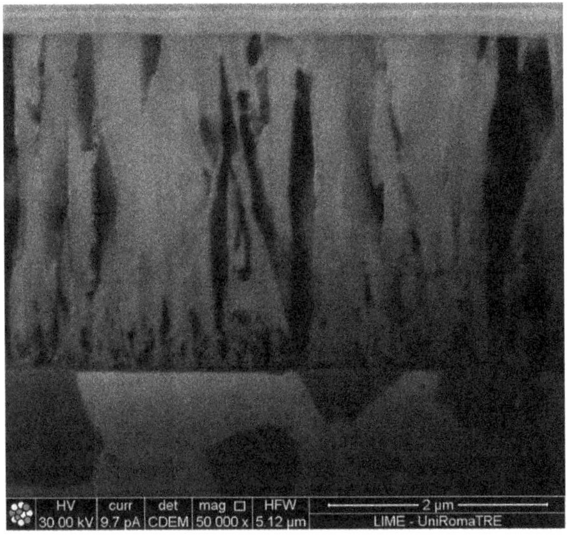

Fig. 3 Microstructure of the PVD TiN coating (columnar) on cemented carbide WC–Co substrate (Dye et al. 2001)

pristine residual stress by comparison with the load–displacement data on the residually stressed material.

The proposed methodology has been applied and validated in this work on a 3.8 µm CAE-PVD TiN coating on WC–Co substrate (see Fig. 3).

Most of the data reported in this book chapter were already presented in international conferences and published in the related conference proceedings (Sebastiani et al. 2010).

2 Models for the Analysis of Residual Stress by Instrumented Indentation and Effects on Fracture Behavior

Literature papers (Bemporad et al. 2008; Suresh and Giannakopoulos 1998; Tsui et al. 1996) have already demonstrated that the presence of a residual stress field at specimen surface can induce measurement inaccuracies in the conventional hardness and elastic modulus analysis procedures by nanoindentation.

Indeed, the influence of applied stress on hardness and apparent modulus was firstly analyzed by Tsui, Bolshakov and Pharr (Bolshakov et al. 1996; Tsui et al. 1996) by nanoindentation testing on 8009 Aluminum alloy and Finite Element Modeling.

These authors concluded that the observed influence of applied stress on measured hardness and modulus is actually due to changes in the real contact areas (measured by optical methods) as a function of the applied stress; in particular, higher compressive stress give higher real contact area, leading to an overestimation of hardness and elastic modulus, as measured by the conventional Oliver-Pharr method(Oliver and Pharr 1992).

When the real contact areas are used for hardness and modulus calculation, they found that both the elastic modulus and hardness were independent of applied stress (Bolshakov et al. 1996; Tsui et al. 1996).

Therefore, sharp indentation testing can be used for the experimental evaluation of the average residual stress state at specimen surface (Suresh and Giannakopoulos 1998).

Several methods still exist in the literature for the evaluation of residual stress and strain by sharp indentation testing, all mainly based on the comparison between the load-depth curve obtained on a residually stressed surface with the corresponding results on the stress-free (or virgin) reference material (Suresh and Giannakopoulos 1998).

Suresh and Giannakopoulos (Suresh and Giannakopoulos 1998) presented a general methodology for the determination of surface (equi-biaxial) elastic residual stress by instrumented sharp indentation testing, based on the original observations made by Pharr's group (Bolshakov et al. 1996; Tsui et al. 1996).

This methodology (Suresh and Giannakopoulos 1998) is based on the main assumption that the average contact pressure due to indentation is unaffected by any preexisting elastic residual stress (invariance of indentation hardness).

A detailed explanation of why the real hardness cannot be affected by any biaxial residual stress is given in detail in the Suresh and Giannakopoulos paper (Suresh and Giannakopoulos 1998).

The concept of this method is depicted in Fig. 1. The principle is that a biaxial residual stress will involve a modification of the contact area during indentation, leading to either a more pronounced sinking-in (Fig. 1a) in case of tensile residual stress or a more a piling-up (Fig. 1b) in case of compressive residual stress, in comparison to the contact area corresponding the stress-free state.

The corresponding modifications of the load displacement curve in comparison to the stress-free state are reported in Fig. 1c.

Basing on this principle, a general relationship between the ratio of contact areas at a fixed depth between the stressed surface and the stress-free (virgin) reference material (or equivalently the ratio of penetration depths at a fixed load) was then given (Fischer et al. 2005):

$$\frac{A}{A_0} = \left(1 + \frac{\sigma_{res}}{H_0}\right)^{-1} \tag{1}$$

$$\frac{h^2}{h_0^2} = \left(1 - \frac{\sigma_{res}}{H_0}\right)^{-1} \tag{2}$$

where σ_{res} is the average biaxial residual stress (negative value if compressive), H_0 is the average contact pressure due to indentation (i.e. the hardness), (A_0, h_0) and

(A, h) are the real contact area and penetration depth for the stress free reference material and the stressed material, respectively.

The proposed methods were successfully applied (Suresh and Giannakopoulos 1998) on artificially strained samples, where the reference curve for the stress-free material is easily measurable and external controlled stress states can be simply applied.

Nevertheless, strong limitations still exist when such methods are applied for the analysis of real residual stress states on polycrystalline materials and thin coatings, essentially due to the fact that a reference load-depth curve for the unstressed material is often not available (or achievable after complex sample preparation procedures), especially in case of nanostructured materials and thin films.

In the methodology proposed here, the existing limits of the available methods for stress calculation from sharp indentation testing are overcome by introducing an original methodology for the local residual stress relieving over a stressed surface.

A complete relief of residual stress is induced by FIB milling of annular trenches at specimen surface, if the depth of the milled trenches are higher than their characteristic diameter (Korsunsky et al. 2010).

Nanoindentation testing over stress relieved stubs then gives the stress-free reference load-depth curve to be used for residual stress calculation by Eqs. (1) and (2) (Suresh and Giannakopoulos 1998).

The additional boundary conditions given by the presence of the edges of pillar and by coating/substrate interface can also be considered during the stress calculation by analytical modeling of contact stresses, Film Doctor® software, as described in the next chapter.

In addition to this, nanoindentation testing over the stress relieved pillars also allows (1) a more proper evaluation of hardness and modulus of the coating by using the conventional Oliver-Pharr method and (2) to analyze the effects of residual stress on fracture toughness of the coating.

To evaluate this latter effect (Anstis et al. 1981; Toonder et al. 2002), nanoindentation under load-controlled conditions is performed on pillar structures, up to a maximum load where a controlled fracture of the pillar can be induced.

Fracture toughness can be then evaluated by the measurement of the dissipated energy by brittle fracture through the coating during nanoindentation testing (Chen and Bull 2009; Toonder et al. 2002).

According to the model proposed by Toonder et al. (2002), a through thickness cracking during sharp instrumented indentation will cause a drop in the measured displacement, in case of load-controlled nanoindentation.

(1) Lower and (2) upper bounds for the dissipated energy U_{fr} can be obtained by direct measurement on the load-depth curve in correspondence of the failure, by assuming that material's behavior is either (1) fully elastic or (2) fully plastic before and after cracking (Chen and Bull 2009).

Fracture toughness can be therefore evaluated by the following equation:

$$K_C = \left[\frac{E_f \cdot U_{fr}}{\left(1 - v^2\right) \cdot A_{crack}} \right]^{\frac{1}{2}} \tag{3}$$

where A_{crack} is the actual surface of crack in the coating (Toonder et al. 2002), which is often measured directly by optical or SEM microscopy observations of the cracks.

Obviously, indentation on pillars gives the fracture toughness of the stress-free material.

Consequently, the actual toughness of the residually stressed coating can be evaluated by the following equation, which takes into account the influence of residual stress on fracture toughness:

$$K_C = K_C^0 + Z \cdot \sigma_R \cdot c^{1/2} \qquad (4)$$

where σ_R is the average surface residual stress, c is the radial dimension of the crack and Z is a crack-geometry parameter, which is equal to 1.26 when the radial dimension of the crack is equal to its depth.

3 Software Assisted Correction for Substrate/Edge Effects and Residual Stress Calculation by the Use of Analytical Modeling (Film Doctor®)

The models for residual stress analysis by indentation that were presented in the previous chapter are based on the geometrical boundary conditions of a homogeneous half-space being indented by an axisymmetric indenter.

Conversely, the experimental methodology that is being presented here consists of the indentation on micro-pillars realized on a coating-substrate system.

In order to assess the validity of those models in the present case, it is necessary to take into account (1) the layered character of the system under examination and (2) the effect of edges (Jakes et al. 2009) in case of indentation on pillar structure; this means two additional boundary conditions that could affect the results of residual stress calculation.

A commercial software package FilmDoctor® (FilmDoctor® 2013; Schwarzer et al. 2001) was adopted for the analytical modeling of the experimentally measured unloading curves for the cases of (1) residually stressed coating and (2) stress relieved coating.

The adopted software evaluates the complete elastic stress field in the moment of beginning unloading by the use of an Extended Hertzian Theory (FilmDoctor® 2013; Schwarzer et al. 2001), and allows to consider the additional boundary conditions given by (1) substrate and edges and (2) to include the presence of residual stress.

The final procedure for residual stress analysis can be summarized as follows:

- **Step-1:** Determination of the correct coating Elastic modulus by analyzing the experimental unloading curves obtained on stress relieved pillar structures.
- **Step-2:** Determination of the true coating yield strength by analyzing the experimental unloading curves obtained on stress relieved pillar structures, after

evaluation of the von Mises stress in the moment of beginning unloading from the unloading curve of the pillars.

- **Step-3:** Determination of the apparent coating yield strength by analyzing the unloading curves obtained on the stressed (undisturbed) coating, after evaluation of the von Mises stress in the moment of beginning unloading from the unloading curve of the half space (no pillar). This gives apparent higher coating yield strength (in the case of compressive residual stress) because the high intrinsic stresses also explain higher hardness and thus, lower indentation depth at same load in comparison with pillar structure.
- **Step-4:** Determination of the intrinsic stress by fitting the biaxial stress to the true yield strength for same loading situation as given in step 3; evaluation of the von Mises stress in the moment of beginning unloading from the unloading curve of the half space but now with biaxial intrinsic stress taken into account. Fitting the biaxial stress to the true yield strength from step 2. In order to obtain a quantitative estimation of the effect of adopted corrections for edge and substrate effects, residual stresses were also calculated by Eqs. (1–2) at several penetration depths, performing also a quantitative evaluation of the critical penetration depth, below which the edge effects are not relevant anymore.

4 Experimental Details

The experimental activities were focused on a 3.8 μm TiN coating deposited on a WC–Co substrate by CAE-PVD using the following deposition parameters (also reported in (Bemporad et al. 2007)): pressure 1.5 Pa, deposition temperature 450 °C, applied bias voltage 150 V, current 70 A. The coating and substrate microstructures are illustrated in Fig. 3.

Annular FIB milling was performed using FEI Helios Nanolab 600 DualBeam FIB/SEM equipment, using current of 48 pA at 30 kV and adopting outer-to-inner path of the ion beam to reduce re-deposition over the island surface. A series of regular cross sections was also realized before ring milling in proximity of the pillar volume in order to minimize re-deposition over the pillar surface. The beam drift was also monitored during milling and correction applied as necessary.

Based on the FEM results reported in a previous work by some of the authors (Korsunsky et al. 2010), the maximum milling depth and the outer diameter of the island were fixed at 3.8 μm (equal to the coating thickness) to ensure that complete strain relief was achieved.

A complete dimensional characterization of each produced pillar (surface diameter, overall depth, lateral slope) was also performed by high resolution in situ SEM imaging.

Fifteen pillars were realized over a surface area of about 0.2×0.2 mm^2.

Indentation testing was then performed on both pillar structures and undisturbed (residually stressed) surface by a Nano Indenter G200 (Agilent technologies), with a Berkovich indenter calibrated on certified fused silica reference sample.

Indentations on pillar structures (ten tests) were performed in a continuous stiffness measurement (CSM) mode, under a constant strain rate of 0.05 s^{-1}. A maximum penetration depth of 200 nm was set for all tests, in order to avoid cracking of the stress relieved pillars during indentation and to minimize the edge and substrate effects.

The actual shapes of the residual indents were in all cases analyzed by SEM-FEG analysis. The real contact area in both cases of stress relieved pillars and undisturbed surface was also numerically estimated by image analysis on high resolution SEM micrographs; hardness and elastic modulus were consequently re-calculated.

Deformation mechanisms of the TiN coating under normal indentation were investigated as a function of the applied load by FIB-TEM techniques.

A lamella for TEM observation was also prepared by FIB technologies in correspondence of a high load (5 N) Vickers indentation: microstructural observations were performed both by in situ SEM after FIB sectioning and by TEM-SAED analysis. In this latter case, the influence of substrate plastic deformation on the deformation mechanisms of the coating was also investigated.

As reported in the following of this paper, a strong tendency to brittle failure was observed in the stress relieved pillars, while no cracking was observed in the residually stressed coating even for relatively high applied load (up to 500 mN). This suggested that residual stress can play a crucial role in determining the cracking behavior of the TiN coating under investigation.

Therefore, five more tests were performed on pillar structures under load controlled conditions. A maximum applied load of 30 mN was set, in order to have radial cracking of the TiN coating without any further damaging of the Pillar.

Fracture toughness was analyzed by the Toonder bound model (Toonder et al. 2002), which was briefly described in the previous chapter of this paper.

The average residual stress inside the coating was finally independently measured by a Dmax-RAPID Rigaku micro-diffractometer equipped with a cylindrical image plate (IP) detector and a collimator diameter of 300 μm, using Cu-Kα radiation. Average residual stresses were calculated by the analysis of a single Debye ring via the conventional d versus $\sin^2\psi$ plot (Gelfi et al. 2004). The depth of the X-ray gauge volume used in this case corresponded to the coating thickness.

5 Results

The summary of obtained results is reported in the Table 1. An SEM micrograph of some of the FIB-produced pillars is presented in Fig. 4a–b: it is worth noting that no surface modification are induced on the pillar surface by the FIB milling and that the actual shape of the pillar (dimensions, slope of the lateral walls) is essentially identical to the ideal one used for modeling and stress calculations (lateral slope lower than 2°, Fig. 3c).

Figure 3d shows pillars with Berkovich indentation at their center, as indicated by arrows; the average off-center of indentation was of the order of 0.4 μm (which is in any case lower than the nominal positional accuracy of the nanoindenter, which is 1 μm).

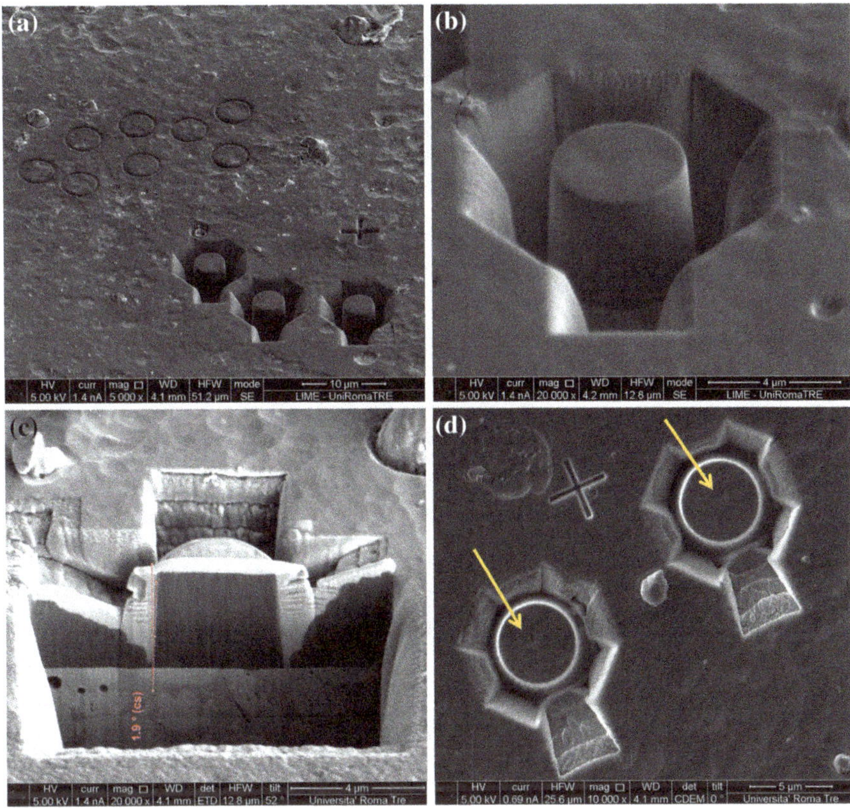

Fig. 4 **a** Example of some of the realized pillar structures before indentation and **b** detail of one of the pillars; **c** FIB/SEM cross section observation of one of the pillar; **d** pillars after indentation (Berkovich indentation can be seen at pillar's centers)

Table 1 *Left column*: apparent properties of the TiN coating under investigation as measured by conventional nanoindentation on the undisturbed surface and *Right column*: real properties (including residual stress and fracture toughness) after indentation testing on stress relieved pillars and the application of the new proposed methodology

	Apparent properties of the TiN coating as measured by conventional nanoindentation	Real properties as evaluated after nanoindentation testing on stress relieved pillars and modeling
E (GPa)	543.00 ± 23.35	500.17 ± 21.34
H (GPa)	31.88 ± 2.80	27.05 ± 2.80
σ_{res} (GPa)	–	-5.63 ± 0.85
K_{IC} (mPa·m$^{0.5}$)	–	$6.09 \leq K_C \leq 7.46$
K_{IC} including σ_{res} (mPa·m$^{0.5}$)	–	$19.54 \leq K_{C(\text{with stress})} \leq 20.91$

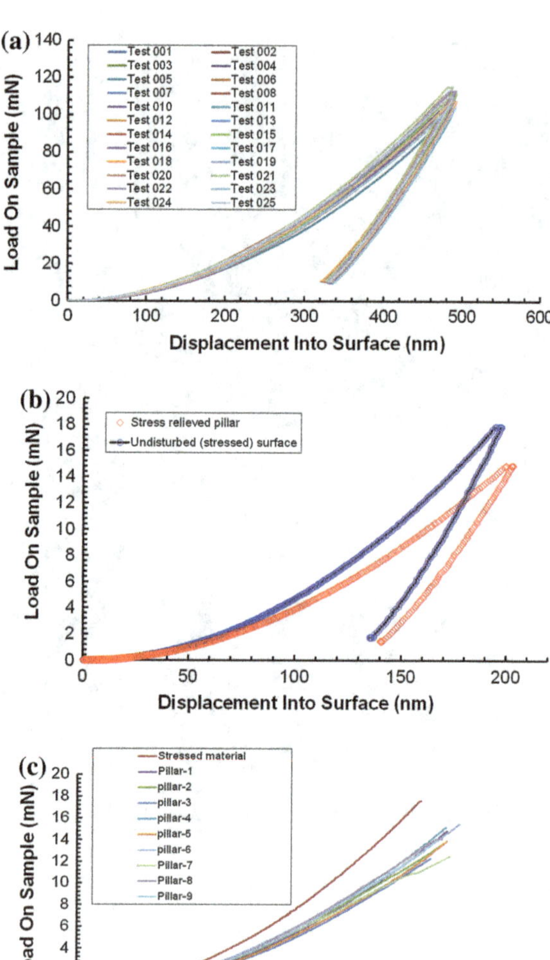

Fig. 5 **a** set of load displacement cures on the residually stressed TiN coating; **b** comparison between the load–displacement curves obtained on the stressed surface and the stress relieved pillars; **c** full data set for indentation on pillars

The results for nanoindentation on the undisturbed (residually stressed) coating are presented in Fig. 5a, while a comparison between load-depth curves obtained on the stressed surface and on stress relieved pillars is then reported in Fig. 5b: this results clearly show that strong compressive residual stresses are present in the TiN coating under examination. Figure 5c also shows a good repeatability over nine different measurements on the pillar structures.

On the basis of nine repeated tests carried out in this study, the average value of residual stress in the coating, obtained with the stress calculation procedure reported in the previous chapter (substrate and edge effect on indentation behavior taken into account), was found to be equal to -5.63 ± 0.85 GPa, as also reported

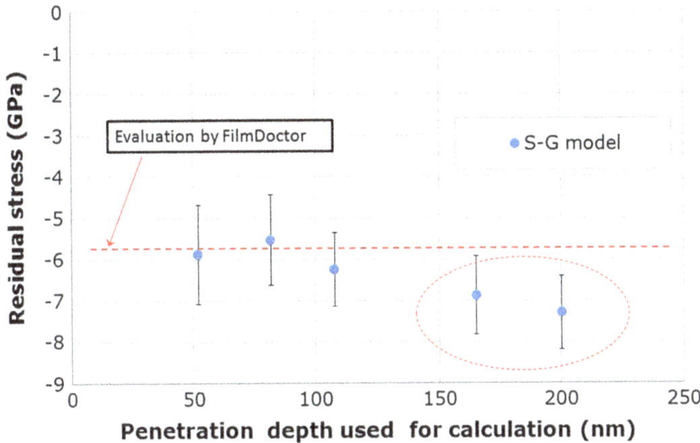

Fig. 6 Stress calculation by the Suresh-Giannakopoulos model (Suresh and Giannakopoulos 1998) at different penetration depths (Eq. 2) and comparison (*dashed line*) with results obtained by FilmDoctor® (Schwarzer et al. 2001)

in Table 1 and highlighted by a dashed line in Fig. 6. The results for contact stress distribution obtained by film doctor are reported in the Fig. 7.

The obtained residual stress value is in good agreement with the estimate obtained by XRD analysis of -5.84 GPa (Korsunsky et al. 2010), although XRD data analysis contained a greater uncertainty due to the strong texture of the TiN coating.

Figure 6 also reports results of stress calculations by the conventional Suresh-Giannakopoulos (S-G) model (Suresh and Giannakopoulos 1998) in comparison with the stress value obtained by FilmDoctor® analysis: it is evident that close agreement is found between the two methods when the stress calculation with the S-G model is performed for penetration depths lower than 120 nm, while for higher penetration depths some discrepancies between the two methods are found (at maximum penetration depth a value of -7.4 GPa is calculated by the S-G model).

This is most probably due to the influence of both the edges and the substrate (not considered in the S-G model) which involve an overestimation of the compressive residual stress for a relative penetration depth higher than 1/30 of the pillar diameter, that can be assumed as the critical penetration depth (for this particular coating/substrate combination), below which the effects of edges and substrate are not relevant anymore.

Figure 8(a–d) show the FEG-SEM micrographs of the indentation marks performed on (a–b) undisturbed (residually stressed) coating and (c–d) stress relieved pillar: a strong difference between the actual indent morphology is clearly visible.

In particular, the occurrence of piling-up due to compressive stress (as observed in previous works (Bolshakov et al. 1996; Tsui et al. 1996) is evident for the residually stressed surface, as clearly visible in Fig. 8a and in the FIB cross-section of an indentation mark reported in Fig. 8e.

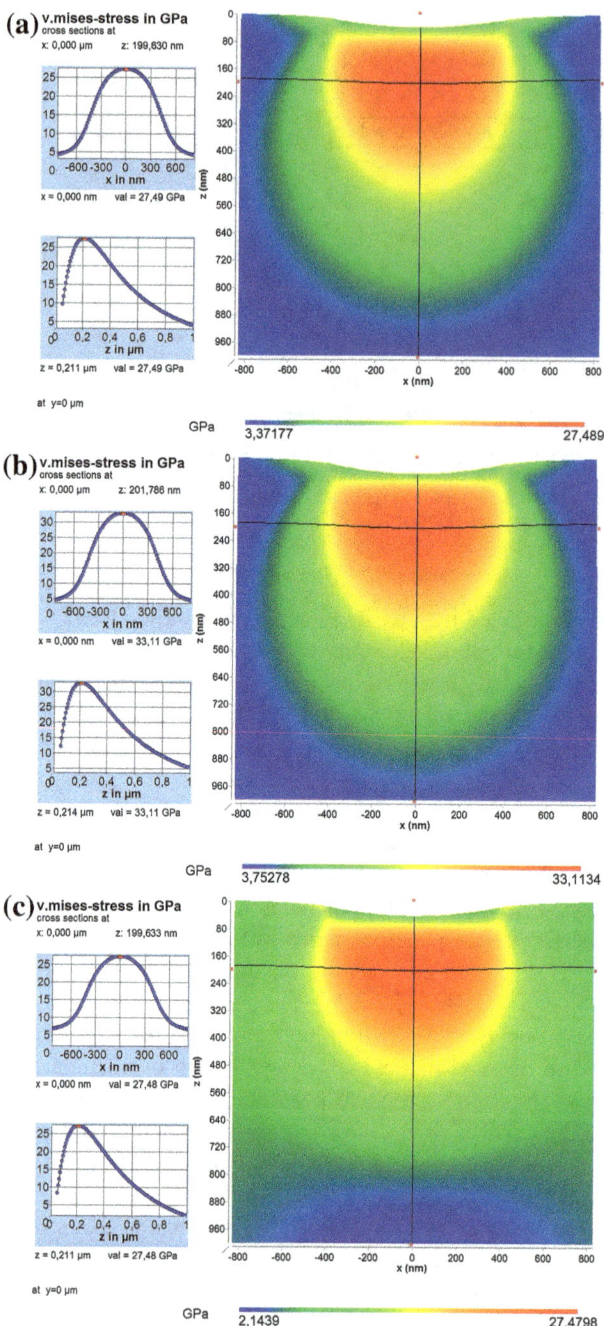

Fig. 7 **a** Example of modeling by Filmdoctor® (Schwarzer et al. 2001): True Yield strength of coating material with $Y = 27.461$ GPa, **b** apparent higher yield strength of stressed coating with 33.111 GPa and **c** true yield strength of stressed coating by taking the biaxial intrinsic stress of -5.6 GPa into account

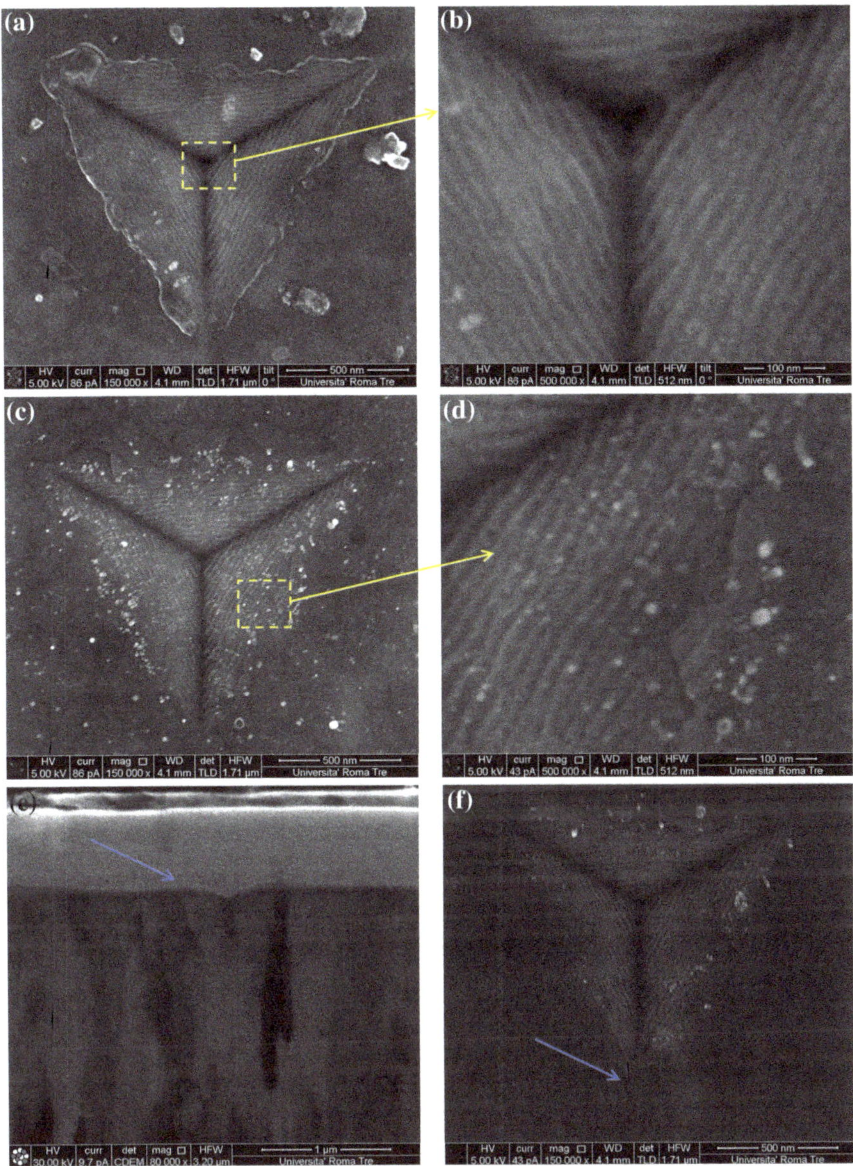

Fig. 8 SEM-FEG observation of produced indentation marks on (**a–b**) undisturbed (stressed) surface and (**c–d**) pillar structure; (**e**) FIB cross-section in correspondence of one indentation and (**f**) evidence of cracks during indentation on the stress-relieved pillars

SEM-FEG analyses also allowed to do a quantitative evaluation of contact area and perform a re-evaluation of coating's hardness and elastic modulus, which were significantly over-estimated (due to piling-up) when calculated by the conventional Oliver-Pharr method after nanoindentation on the undisturbed surface (see Table 1).

For the indentations reported in Fig. 8a–d, the measured contact areas were of $8.98 \cdot 10^3$ nm^3 for indentation on the virgin material (Fig. 8a) and $6.55 \cdot 10^3$ nm^3 for the stress relieved pillar (Fig. 8d). When the actual contact area is known, residual stress can also be calculated from the Eq. (2). In this case, the value of -7.30 GPa was calculated, which is close to the average value of -7.4 GPa calculated by the Eq. (1) at maximum penetration depth (Fig. 6).

It is worth noting that after re-evaluation of contact areas by SEM observation, the hardness and elastic modulus measured on the undisturbed (stressed) surface were very similar to the ones obtained on the stress relieved pillars, thus confirming that hardness and elastic modulus are essentially independent of surface elastic residual stress (Suresh and Giannakopoulos 1998).

The FEG-SEM micrographs of the indentations performed on both pillar structure and stressed surface (details showed in Figs. 8b, d) clearly show that plastic deformation at the nano-scale essentially occurs by formation of nano-shear bands (average size 15 nm) inside the columnar grains with no reciprocal sliding of the grain boundaries, independently of the presence of residual stress.

On the other hand, a complete modification of the deformation mechanisms has been observed in case high load Vickers indentation testing: observing the FIB/SEM cross section analysis reported in Fig. 9.

In particular, TEM-SAED analysis confirmed that deformation mechanisms essentially consist of (1) grain boundary cracking and reciprocal sliding when deforming over the Cobalt matrix (Fig. 9b) and (2) complete grain deformation, recrystallization (see SAED pattern in Fig. 9c) and interface delamination over the (harder) WC-grains.

It is therefore clear that deformation mechanisms at high load are essentially driven by the mechanical properties of the substrate (Fig. 9a).

Conversely, strong influence of residual stress on the fracture behavior of the coating was observed.

Figure 10 shows one the indentations load-depth curves realized under load-controlled conditions for fracture toughness analysis (Anstis et al. 1981; Chen and Bull 2009), and the observation of a broken pillar after testing.

The horizontal drop of the displacement during indentation indicates the complete brittle failure of the coating.

Figure 10b also show that failure of the pillar essentially occurs by brittle failure into three identical segments, following the three-side geometry of the Berkovich indenter: this observation allows the estimation of the actual area of fracture to be used in Eq. (3) for fracture toughness calculation, which is (under the assumption of smooth surface, see Fig. 2b for symbols):

$$A_{fr} = k \cdot 3 \cdot \left(\frac{a}{2} \cdot h \right) \tag{5}$$

where k is a correcting factor (always ≥ 1) which takes into account that the fracture surfaces are not smooth at all.

In this case, fracture surfaces appear to be very irregular, due to the fact that it propagates along the grain boundaries of the coating: it is therefore likely that the actual area of fracture is higher than the ideal one ($k > 1$ in Eq. (5)): an estimated

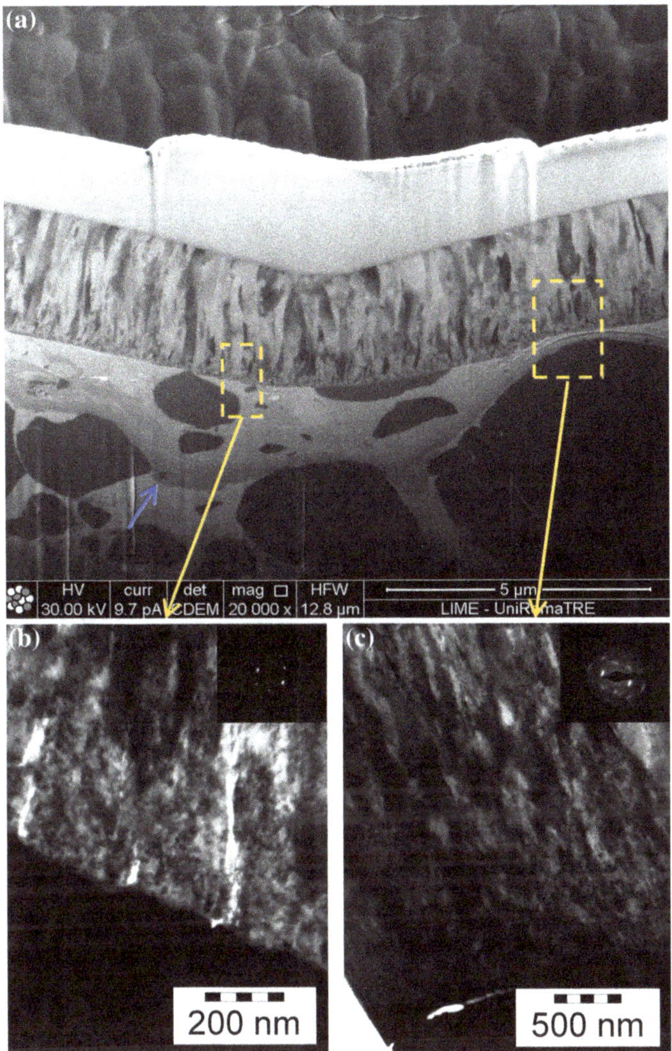

Fig. 9 **a** FIB/SEM observation of the cross section of a Vickers indentation mark (5 N) showing severe plastic deformation of the Cobalt matrix in the substrate, **b–c** TEM observation of deformation mechanisms in different areas of the coating. In this case, deformation mechanisms essentially depend on the mechanical behavior of the substrate

value of $k = 2$ (which is likely in case of fracture propagation along a (100) direction) was therefore taken for toughness calculation.

Fracture toughness of the (stress relieved) TiN coating was then calculated by the use of the Toonder et al. bound model (Eq. (3), (Toonder et al. 2002)) as:

$$6.09 \leq K_C \leq 7.46 \left(\mathrm{MPa} \cdot \mathrm{m}^{0.5} \right) \tag{5}$$

Fig. 10 Measurement of fracture toughness by load-controlled nanoindentation on pillars. **a** example of a load-depth curve showing the brittle failure of the pillar and **b** SEM observation of one of the broken pillars, revealing fracture morphology to be used for toughness calculation [34]

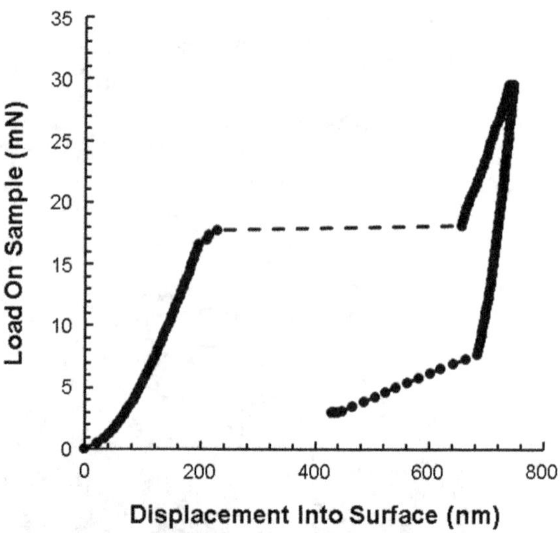

In addition, the influence of residual stress on fracture toughness was evaluated by the use of Eq. (4) (Anstis et al. 1981): in this case, the presence of a compressive stress of −5.63 GPa leads to a fracture toughness (stress dependent) of the coating of:

$$19.54 \leq K_{C(with\,stress)} \leq 20.91 \left(\mathrm{MPa} \cdot \mathrm{m}^{0.5}\right) \tag{6}$$

This explains why no cracking under sharp indentation is usually observed on highly stressed hard PVD coatings, even under very high applied normal load.

A summary of mechanical characterization activities is reported in Table 1, where the effect of residual stress on the nano-mechanical characterization of the coating can be evaluated by the comparison between different results obtained on stress relieved pillars and undisturbed surface.

6 Discussion

In this work, an optimized procedure for the analysis of residual stress effects on the nano-mechanical behavior of thin coatings is presented, based on the integrated adoption of FIB milling and nanoindentation testing.

The methodology allows for the quantitative evaluation of the surface elastic residual stress on a micron scale, by using well-established experimental techniques that can be easily reproduced on FIB equipments (Bei et al. 2007; Uchic et al. 2004).

The application on a PVD TiN coating showed a good agreement between the new method and the estimation obtained by the conventional XRD $\sin^2\psi$ method.

It is important to remind that only the in-plane average stress can be evaluated by this procedure: nonetheless, residual stresses are usually assumed as equi-biaxial in case of thin coatings (Chason et al. 2002; Davis 1993; Detor et al. 2009; Marks et al. 1996; Pao et al. 2007; Uchic et al. 2004; Windischmann 1987), so the proposed method can be a reliable way for stress analysis in this specific case.

At this point, some more considerations are necessary on the adopted assumption during stress calculation. The adoption of Eqs. (1–2) is based on the assumptions that (1) the elastic behavior of the analyzed coating is homogeneous and isotropic and (2) that continuum mechanics is still valid at the considered scale.

An SEM micrograph of the TiN coating is reported in Fig. 3, where a strong columnar microstructure is evident: this indicates that some inaccuracies in stress calculation could arise as a consequence of a not proper estimation of the actual elastic anisotropy of the coating.

However, the observed deformation mechanisms at the nanoscale (Fig. 8) show that grain boundaries are not involved in the deformation mechanisms for penetration depths below 200 nm.

Therefore, the assumption of isotropic elastic behavior of the coating could be a reasonably choice, at least at the considered scale, where grain boundaries are not involved in the deformation mechanisms.

Anyway, the use of similar elastic constant and a similar probing volume guarantee the consistency between residual stress values measured by the XRD-sin2ψ and nanoindentation method.

It is also important to comment on standard deviation of the calculated residual stress, that was about 0.85 GPa: such variation of experimentally measured values is likely due also to a real variation of residual stress from point to point, probably due to the non-homogeneity of the WC–Co substrate (see Fig. 3). This effect is surely superimposed to the uncertainty of the method itself and further work is now ongoing on simpler systems in order to clarify this point.

Table 1 reports a comparison between the apparent properties as measured by conventional analysis of nanoindentation data obtained on the stressed surface and the actual properties (including residual stress and toughness) as measured by indentation on stress relieved pillars.

The obtained results on fracture toughness show that residual stress plays a significant role in determining the "in-service" toughness of the PVD coating (including stress) applied to a specific substrate. As a consequence, toughness data without knowing the real residual stress of the specific sample cannot give affordable values to be used in different context, nor this information can be used in different systems (i.e. different substrate, even if with a coating obtained with the same deposition process and parameters).

On the other hand, indentation testing on stress relieved pillars also gives the "stress-free fracture toughness" of the coating (i.e. without the effect of residual stress) giving an unbiased value to better predict the in-service fracture behavior of the coatings when external forces (i.e. contact load) will be applied.

It is also worth noting that the application of the Toonder et al. (Toonder et al. 2002) bound model revealed to be particularly effective when applied to the well confined geometry which characterizes the stress relieved pillars.

In fact, in case of indentation of pillars the geometry of the cracks is very well defined and can be evaluated by SEM observation after testing; a more accurate evaluation of fracture toughness can be therefore performed, in comparison with the conventional procedure (which requires radial chipping of the coating).

In addition, it is also important to remind that in case of strong compressive stress (e.g. the TiN coating under investigation), the measurement of fracture toughness by sharp indentation on the undisturbed surface is often not possible at all, simply because residual stresses inhibit any radial cracking, even at very high applied load.

The proposed methodology gives then a more complete framework of the effect of residual stress on the nano-mechanical and fracture behavior of thin coatings and can be a robust support for a more proper prediction of in-service mechanical behavior and failure modes.

7 Concluding Remarks

In this work, a new methodology for residual stress measurement in thin films and small scale systems is presented, based on nanoindentation testing on FIB-produced pillars.

First results of stress measurement on a TiN hard coating reported in this paper are promising, and a good agreement was found with the estimation obtained by XRD $\sin^2\psi$ analysis.

The new method gives a statistically reliable, robust and very site-specific evaluation of the surface residual stress field, giving also further information on the effect of residual stress on nano-mechanical behavior of hard coatings.

In fact, the proposed procedure also allows a more correct evaluation of other key-properties, such as hardness and modulus (which are usually affected by significant measurement errors due to the presence of residual stress) and fracture toughness as a function of the residual stress state.

Acknowledgments Authors would like to acknowledge Daniele De Felicis for technical assistance during FIB analyses, performed at the interdepartmental laboratory of electron microscopy of university of Roma Tre, Rome Italy (http://www.lime.uniroma3.it), and prof. Laura Depero (University of Brescia) for XRD residual stress measurements.

References

Anstis GR, Chantikul P, Lawn BR, Marshall DB (1981) A critical evaluation of indentation techniques for measuring fracture toughness: i, direct crack measurements. J Am Ceram Soc 64(9):533–538. doi:10.1111/j.1151-2916.1981.tb10320.x

Bei H, Shim S, Miller MK, Pharr GM, George EP (2007) Effects of focused ion beam milling on the nanomechanical behavior of a molybdenum-alloy single crystal. Appl Phys Lett 91(11):111915

Bemporad E, Sebastiani M, Casadei F, Carassiti F (2007) Modelling, production and characterisation of duplex coatings (HVOF and PVD) on Ti–6Al–4 V substrate for specific mechanical applications. Surf Coat Technol 201(18):7652–7662. doi:http://dx.doi.org/10.1016/j.surfcoat.2007.02.041

Bemporad E, Sebastiani M, Staia MH, Puchi Cabrera E (2008) Tribological studies on PVD/HVOF duplex coatings on Ti6Al4 V substrate. Surf Coat Technol 203(5–7):566–571. doi:http://dx.doi.org/10.1016/j.surfcoat.2008.06.055

Bolshakov A, Oliver WC, Pharr GM (1996) Influences of stress on the measurement of mechanical properties using nanoindentation: Part II. Finite element simulations. J Mater Res 11(03):760–768. doi:10.1557/JMR.1996.0092

Bull SJ (2005) Nanoindentation of coatings. J Phys D Appl Phys 38(24):R393

Bull SJ, Berasetegui EG (2006) An overview of the potential of quantitative coating adhesion measurement by scratch testing. Tribol Int 39(2):99–114. doi:http://dx.doi.org/10.1016/j.triboint.2005.04.013

Chason E, Sheldon BW, Freund LB, Floro JA, Hearne SJ (2002) Origin of compressive residual stress in polycrystalline thin films. Phys Rev Lett 88(15):156103

Chen J, Bull SJ (2009) Modelling the limits of coating toughness in brittle coated systems. Thin Solid Films 517(9):2945–2952. doi:http://dx.doi.org/10.1016/j.tsf.2008.12.054

Davis CA (1993) A simple model for the formation of compressive stress in thin films by ion bombardment. Thin Solid Films 226 (1):30-34. doi:http://dx.doi.org/10.1016/0040-6090(93)90201-Y

Detor AJ, Hodge AM, Chason E, Wang Y, Xu H, Conyers M, Nikroo A, Hamza A (2009) Stress and microstructure evolution in thick sputtered films. Acta Materialia 57(7):2055–2065. doi:http://dx.doi.org/10.1016/j.actamat.2008.12.042

Dye D, Stone HJ, Reed RC (2001) Intergranular and interphase microstresses. Curr Opin Solid State Mater Sci 5(1):31–37. doi:http://dx.doi.org/10.1016/S1359-0286(00)00019-X

Espinosa HD, Prorok BC, Fischer M (2003) A methodology for determining mechanical properties of freestanding thin films and MEMS materials. J Mech Phys Solids 51(1):47–67

FilmDoctor® (2013) software for the evaluation of mechanical contact for homogeneous and layered materials. www.siomec.de/FilmDoctor. 2013

Fischer W, Malzbender J, Blass G, Steinbrech RW (2005) Residual stresses in planar solid oxide fuel cells. J Power Sources 150:73–77. doi:http://dx.doi.org/10.1016/j.jpowsour.2005.02.014

Gelfi M, Bontempi E, Roberti R, Depero LE (2004) X-ray diffraction Debye Ring Analysis for STress measurement (DRAST): a new method to evaluate residual stresses. Acta Materialia 52(3):583–589. doi:http://dx.doi.org/10.1016/j.actamat.2003.09.041

Jakes JE, Frihart CR, Beecher JF, Moon RJ, Resto PJ, Melgarejo ZH, Suárez OM, Baumgart H, Elmustafa AA, Stone DS (2009) Nanoindentation near the edge. J Mater Res 24(03):1016–1031. doi:10.1557/jmr.2009.0076

Korsunsky AM (2009) Eigenstrain analysis of residual strains and stresses. J Strain Anal Eng Des 44(1):29–43. doi:10.1243/03093247jsa423

Korsunsky AM, Sebastiani M, Bemporad E (2009) Focused ion beam ring drilling for residual stress evaluation. Mater Lett 63(22):1961–1963. doi:http://dx.doi.org/10.1016/j.matlet.2009.06.020

Korsunsky AM, Sebastiani M, Bemporad E (2010) Residual stress evaluation at the micrometer scale: Analysis of thin coatings by FIB milling and digital image correlation. Surf Coat Technol 205(7):2393-2403. doi:http://dx.doi.org/10.1016/j.surfcoat.2010.09.033

Marks NA, McKenzie DR, Pailthorpe BA (1996) Molecular-dynamics study of compressive stress generation. Phys Rev B 53(7):4117–4124

Oliver WC, Pharr GM (1992) An improved technique for determining hardness and elastic modulus using load and displacement sensing indentation experiments. J Mater Res 7(06):1564–1583. doi:10.1557/JMR.1992.1564

Oliver WC, Pharr GM (2004) Measurement of hardness and elastic modulus by instrumented indentation: advances in understanding and refinements to methodology. J Mater Res 19(01):3–20. doi:10.1557/jmr.2004.19.1.3

Pao C-W, Foiles SM, Webb EB III, Srolovitz DJ, Floro JA (2007) Thin film compressive stresses due to adatom insertion into grain boundaries. Phys Rev Lett 99(3):036102

Pauleau Y (2001) Generation and evolution of residual stresses in physical vapour-deposited thin films. Vacuum 61(2–4):175–181. doi:http://dx.doi.org/10.1016/S0042-207X(00)00475-9

Roy RK, Lee K-R (2007) Biomedical applications of diamond-like carbon coatings: A review. J Biomed Mater Res B Appl Biomater 83B(1):72–84. doi:10.1002/jbm.b.30768

Schwarzer N, Hermann I, Chudoba T, Richter F (2001) Contact modelling in the vicinity of an edge. Surf Coat Technol 146–147:371–377. doi:http://dx.doi.org/10.1016/S0257-8972(01)01418-9

Sebastiani M, Bemporad E, Carassiti F, Schwarzer N (2010) Residual stress measurement at the micrometer scale: focused ion beam (FIB) milling and nanoindentation testing. Phil Mag 91(7–9):1121–1136. doi:10.1080/14786431003800883

Sebastiani M, Eberl C, Bemporad E, Pharr GM (2011) Depth-resolved residual stress analysis of thin coatings by a new FIB–DIC method. Mater Sci Eng: A 528(27):7901–7908. doi:http://dx.doi.org/10.1016/j.msea.2011.07.001

Song X, Yeap KB, Zhu J, Belnoue J, Sebastiani M, Bemporad E, Zeng K, Korsunsky AM (2012) Residual stress measurement in thin films at sub-micron scale using Focused Ion Beam milling and imaging. Thin Solid Films 520(6):2073–2076. doi:http://dx.doi.org/10.1016/j.tsf.2011.10.211

Suresh S, Giannakopoulos AE (1998) A new method for estimating residual stresses by instrumented sharp indentation. Acta Materialia 46(16):5755–5767. doi:http://dx.doi.org/10.1016/S1359-6454(98)00226-2

Teixeira V (2002) Residual stress and cracking in thin PVD coatings. Vacuum 64(3–4):393–399. doi:http://dx.doi.org/10.1016/S0042-207X(01)00327-X

Toonder JD, Malzbender J, With GD, Balkenende R (2002) Fracture Toughness and Adhesion Energy of Sol-gel Coatings on Glass. J Mater Res 17(01):224–233. doi:10.1557/JMR.2002.0032

Tsui TY, Oliver WC, Pharr GM (1996) Influences of stress on the measurement of mechanical properties using nanoindentation: part I. experimental studies in an aluminum alloy. J Mater Res 11(03):752–759. doi:10.1557/JMR.1996.0091

Uchic MD, Dimiduk DM, Florando JN, Nix WD (2004) Sample dimensions influence strength and crystal plasticity. Science 305(5686):986–989. doi:10.1126/science.1098993

Wang L, Bei H, Gao YF, Lu ZP, Nieh TG (2011) Effect of residual stresses on the hardness of bulk metallic glasses. Acta Mater 59(7):2858–2864. doi:http://dx.doi.org/10.1016/j.actamat.2011.01.025

Windischmann H (1987) An intrinsic stress scaling law for polycrystalline thin films prepared by ion beam sputtering. J Appl Phys 62(5):1800–1807

Withers PJ, Bhadeshia HKDH (2001) Residual stress. Part 1—Measurement techniques. Mater Sci Technol 17(4):355–365

Ye J, Shimizu S, Sato S, Kojima N, Noro J (2006) Bidirectional thermal expansion measurement for evaluating Poisson's ratio of thin films. Appl Phys Lett 89(3):031913

Multiscale Modeling of Nanoindentation: From Atomistic to Continuum Models

P. S. Engels, C. Begau, S. Gupta, B. Schmaling, A. Ma and A. Hartmaier

Abstract Nanoindentation revealed a number of effects, like pop-in behavior or indentation size effects, that are very different from the classical mechanical behavior of bulk materials and that have therefore sparked a lot of research activities. In this contribution a multiscale approach is followed to understand the mechanisms behind this peculiar material behavior during nanoindentation. Atomistic simulations reveal the mechanisms of dislocation nucleation and multiplication during the very start of plastic deformation. From mesoscale dislocation density based models we gain advanced insight into how plastic zones develop and spread through materials with heterogeneous dislocation microstructures. Crystal plasticity models on the macroscale, finally, are able to reproduce load-indentation curves and remaining imprint topologies in a way that is directly comparable to experimental results and, thus, allows for the determination of true material properties by inverse methods. The complex interplay of the deformation mechanisms occurring on different length scales is described and the necessity to introduce the knowledge about fundamental deformation mechanisms into models on higher length scales is highlighted.

1 Introduction

Nanoindentation has been one of the first methods to characterize material properties by nanomechanical testing. However, despite many years of experience it is still a challenge to correlate results of nanoindentation experiments to true material behavior. The reasons for this lie on the one hand side in the complicated triaxial and locally concentrated mechanical stress state underneath a nanoindenter, which

P. S. Engels · C. Begau · S. Gupta · B. Schmaling · A. Ma · A. Hartmaier (✉)
Interdisciplinary Centre for Advanced Materials Simulation (ICAMS),
Ruhr-Universität Bochum, 44780 Bochum, Germany
e-mail: Alexander.Hartmaier@icams.rub.de

A. Tiwari (ed.), *Nanomechanical Analysis of High Performance Materials*,
Solid Mechanics and Its Applications 203, DOI: 10.1007/978-94-007-6919-9_15,
© Springer Science+Business Media Dordrecht 2014

makes it difficult to compare the results for mechanical properties to uniaxial testing methods. On the other hand side, the arising difficulties are also inherent to nanomechanical testing itself. Since only very small volumes are tested the heterogeneous microstructure of materials causes a large scatter in the results, depending for example on microstructural details like whether pre-existing dislocations are present in the tested volume or not, see for example recent work by Morris et al. (2011) or Lodes et al. (2011). This has been attributed as one kind of indentation size effect (ISE) by Shim et al. (2008). A more classical ISE is caused by the strain gradients arising from the inhomogeneous deformation during nanoindentation (Nix 1989). However, even this ISE reveals itself in different forms.

For sharp indenters, like Berkovich, cube corner or conical ones, ISE means that indentation hardness of metallic materials increases with decreasing depth of indentation; see for example (Nix 1989; Oliver and Pharr 1992; De Guzman et al. 1993; Stelmashenko et al. 1993; Ma and Clarke 1995; Atkinson 1995; Poole et al. 1996; McElhaney et al. 1998; Suresh et al. 1999, Zagrebelny et al. 1999). In contrast, for blunt indenters, like spherical indenters, at a fixed ratio of contact radius versus indenter radius the indentation hardness increases with decreasing indenter radius, see for example (Tymiak et al. 2001; Swadener et al. 2002). Of course, size effects are not only experienced in indentation experiments, but also observed in micro-torsion of thin copper wires (Fleck et al. 1994), micro-bending of micron thick nickel foils (Stölken and Evans 1998; Shrotriya et al. 2003), and in aluminum, reinforced by micron-size silicon carbide particles (Lloyd 1994). Since the pioneering work by Uchic et al. (2004), research in nanomechanics has become more divers and mechanical testing of nanopillars in compression, nanowires in tension, beams in bending, etc. has attracted a lot of attention. Since, however, the main focus of this contribution is on nanoindentation, only a rather recent overview article by Kraft et al. (2010) shall be referenced here to cover this topic.

The main aim of nanomechanical testing is to gain information on local material properties, as well as on fundamental deformation mechanisms. From the diversity of size effects described in the literature it is clear that many different mechanisms must occur on different length scales and that hence a proper understanding of nanoindentation and ISEs must integrate experimentation and modeling over those length scales as well. This multiscale nature of nanoindentation is highlighted in the present contribution where modeling methods from atomistics, over dislocation density based mesomodels, to continuum mechanical approaches are applied to unveil the hitherto unknown interplay of mechanical loading and material response over the length scales, see Fig. 1. Throughout this work results of numerical modeling are compared to experimental data from the literature on different scales, to highlight the possibilities of realistic material descriptions with advanced simulation methods, but also to point out their current limitations and shortcomings. It will be demonstrated that only by a proper combination of modeling and experiment a profound physical understanding of the deformation mechanisms occurring during nanoindentation can be achieved. Furthermore, it is shown that this combination of experiment and simulation allows us to derive true material properties from nanoindentation.

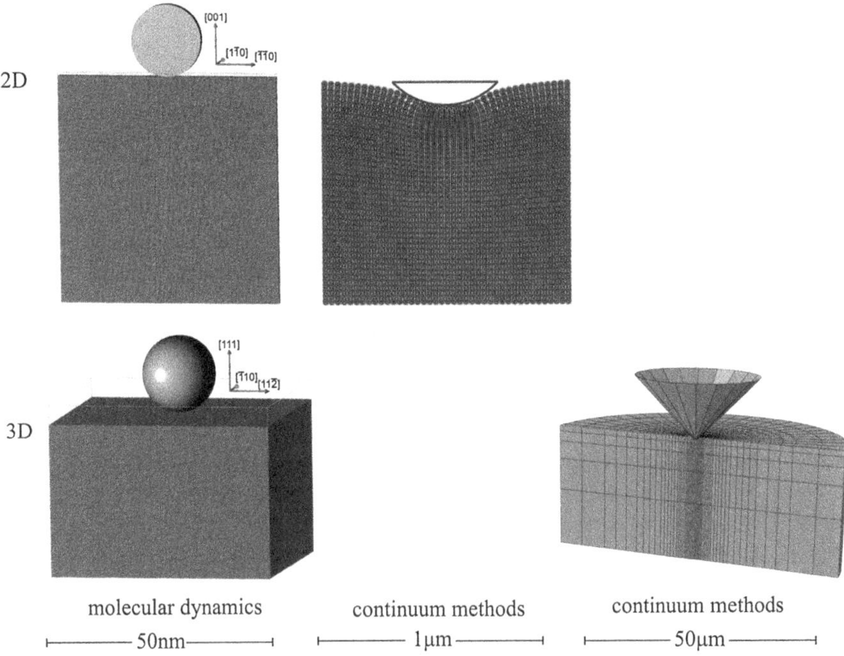

Fig. 1 Methods and length-scales used in this work. Atomistic simulation methods cover length scales of up to 50 nm, whereas dislocation density-based mesomodels can describe volumes on the micron scale (yet, in this work only a 2D model is presented on this scale). Continuum models, finally, cover much larger volumes and can be used to simulate nano and micro indentations with very high accuracy

2 Atomistic Aspects of Nanoindentation

Simulation methods on the atomic scale such as molecular dynamics (MD) are well-established tools to investigate the deformation processes in crystalline materials. In these simulations, an interatomic potential defines the interaction between atoms and thus the material properties. As a consequence, MD simulations do not require any assumptions to be made on dislocation nucleation or dislocation glide mechanisms, rendering them a powerful tool to study the fundamentals of plastic deformation. However, classical MD simulations are limited with respect to the length- and the timescales that can be accessed with today's computers. While billion-atom simulations have become feasible using large computing clusters, the physical time scale that can be simulated remains limited to a few nanoseconds, because the time-step of a simulation must be on the order of the atomic lattice vibrations, and thus is commonly 1fs. Another limitation of atomistic methods arises from the large amount of raw data that is produced during such simulations. This raw data comprises energies, position- and velocity-vectors per atom and per time-step, however no higher order information concerning mechanical or physical material behavior is available immediately. Thermodynamic quantities, like temperatures or stresses, can be assessed as ensemble averages given within the framework of statistical mechanics. Advanced data analysis methods need to be applied in order to identify further physical or mechanical quantities. To correlate microstructure and mechanical behavior, it is for example necessary to determine the positions of dislocations and to compute the Burgers vector of each dislocation segment. Once this kind of data is made available, it is becoming possible to obtain information of plastic deformation that can serve as the basis of theory building or provide useful input for simulation on larger scales, as it is described in this work.

The MD results on nanoindentation in single crystals, described in the following section, have been obtained by a combination of the Nye-tensor method, a method to numerically approximate Burgers vectors (Begau et al. 2012), and the method of dislocation skeletonization (Begau et al. 2011). The latter algorithm simplifies the set of atoms identified as part of a dislocation core during the Nye-tensor method into a set of connected geometrical lines, representing the dislocation network. Each curve is associated with a Burgers vector that is derived from the set of atoms in the dislocation core. In total this method enables a very detailed description of dislocation networks to be derived directly from atomistic configurations. Further analysis are performed with a modified version of the Bond Angle Analysis (Begau et al. 2011), originally proposed by Ackland and Jones (2006).

In the following sections, it will be demonstrated how atomistic simulations of nanoindentation into copper single crystals can serve as a rich basis for the understanding of the initial stages of plastic deformation in very small volumes. To accomplish this, the analysis of two simulations of nanoindentation is described in detail. Simulation 1 is designed to study the processes of dislocation nucleation in detail, see also (Begau et al. 2011), while simulation 2 demonstrates how the evolution of dislocation densities is determined quantitatively and how the results can be compared with the model of Nix and Gao (1998), for further details see (Begau et al. 2012).

All simulations are performed using the IMD code (Stadler et al. 1997) and an EAM (Embedded Atom Method) style interatomic potentials for copper developed by Mishin et al. (2001). A constant temperature of 30 K is maintained during the simulations by a Nosé-Hoover thermostat. The crystal is indented into the free (111) surface, the degrees of freedom of the opposite surface are blocked to avoid rigid body motion, see also the lower left image in Fig. 1. Periodic boundary conditions are applied on all lateral faces of the simulation domain. Further details concerning the simulation setup are given in Table 1 and in (Begau et al. 2011).

2.1 Mechanisms of Dislocation Nucleation

The initial deformation mechanism of plastic deformation and especially the initial homogeneous nucleation of dislocations at sites of highest shear stress underneath the indenter tip have been studied in detail with MD simulations (Kelchner et al. 1998; Vliet et al. 2003; Liang et al. 2003; Zhu et al. 2004; Zimmerman et al.

Table 1 Parameters for nanoindentation simulations in copper

	Setup 1	Setup 2
Box size	$44.9 \times 44.6 \times 40$ nm^3	$60.1 \times 60.2 \times 40$ nm^3
Atoms	6,825,984	12,324,864
Indenter size	12 nm, spherical	10 nm, spherical
Indenter velocity	30 m/s	23 m/s
Maximum indentation depth	3 nm	5.98 nm
Duration of loading	0.1 ns	0.26 ns
Duration of unloading	–	0.08 ns

2001). During these typically displacement-controlled simulations a sharp drop in the measured hardness versus indentation depth curves is observed immediately after the nucleation events (see Fig. 3 (left)). A similar behavior is observed in recent experimental studies (Morris et al. 2011; Bei et al. 2005, 2008) on oxide-free surfaces. Thus it is now widely accepted, that the nucleation of dislocations is responsible for the pop-in effect of well-prepared, oxide-free surfaces.

Up to now, the deformation mechanisms following this initial onset of plasticity are studied in far less detail, primarily due to a lack of proper analysis methods that characterize the dislocation network. By a combination of the methods mentioned above, this limitation has now been overcome and the mechanisms of dislocation multiplication and heterogeneous nucleation events after the onset of plasticity have been analyzed in some detail. Furthermore, local stress tensors could be evaluated as indicators of nucleation events. These stress values are derived from spatial and temporal ensemble averages of the virial stress tensors of the atom configuration, and thus can be related to the Cauchy stress as discussed by Zimmermann et al. (2004) and Subramaniyan and Sun (2008). For details on the implementation of all methods the reader is kindly referred to Begau et al. (2011).

By the virtue of these analysis methods, several heterogeneous nucleation events have been observed at surface steps placed at the contact zone between indenter and sample. The surface steps are produced by screw dislocations ending on the free surface, while gliding on the $\{111\}$ planes in $\langle 1\bar{1}0 \rangle$ directions away from the nanoindenter. During glide, they leave behind a perfect lattice inside the crystal, but create slip steps on the surface. Heterogeneous nucleation events at such slip steps are observed by the formation of new dislocation segments in the derived dislocation network and are accompanied by high local von Mises stresses.

An example of such a nucleation process is given in Fig. 2. The surface step in this example is visible as a straight line along the $[\bar{1}01]$ direction as a row of defect atoms in the atomistic data (second row, first image). The approximated Burgers vectors for these atoms derived from the Nye-tensor method do not yield significant values, in contrast to the nearby atoms at dislocation cores, thus they are not yet to be treated as dislocation at this stage. Immediately before the nucleation of a new dislocation, the local von Mises stress in the vicinity of the slip step exceeds 20 GPa. This value is significantly larger than the stresses reached in the surrounding region that amount to about 8–12 GPa. The critical resolved shear stress,

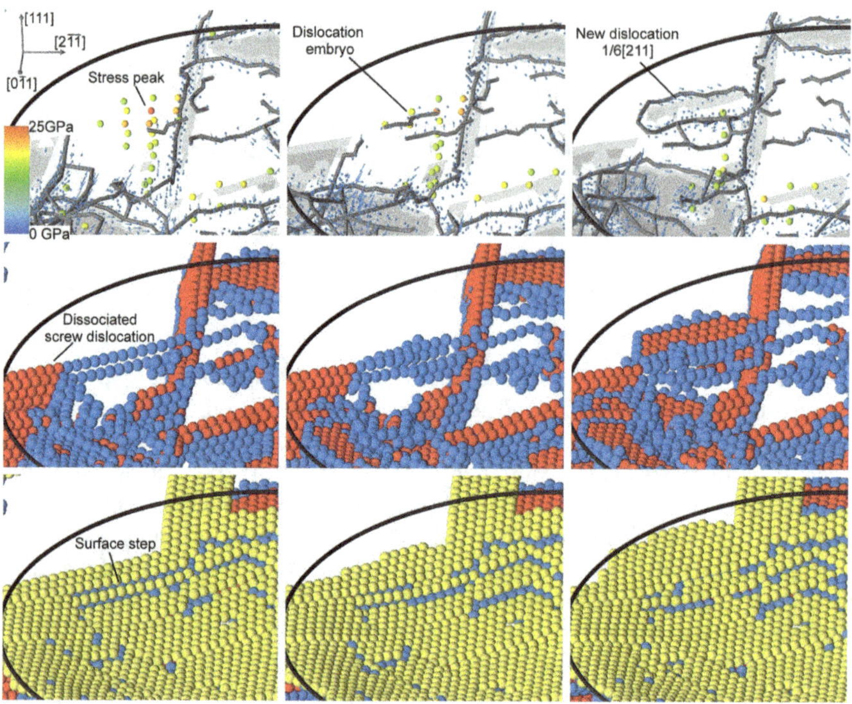

Fig. 2 Nucleation of a dislocation at a surface step. Figures are taken at indentation depths of 1.527, 1.542 and 1.56 nm (corresponding to simulation times of 58.5, 59.0 and 59.6 ps). The first row displays the dislocation network (*lines*), resultant Burgers vectors (*arrows*) and stacking faults (*gray layers*). Small spheres indicate sites of local von Mises stresses exceeding 15 GPa. In the second row defect atoms and stacking faults are shown. The third row includes surface atoms close to defects as well. The black ring indicates the current contact area of the indenter tip (Begau et al. 2011)

which is derived from projecting the stress tensor at the nucleation site onto the glide plane and into the glide direction, corresponds to a value of approximately 2.9 GPa. Noteworthy, this value is somewhat smaller, but on a similar order as the shear stresses required for the two initial homogeneous nucleations. These homogeneous dislocation nucleation events take place at stress levels of 3.8 and 4.7 GPa, which is close to or even exceeds the theoretical strength of the material that for the interatomic potential used here is determined to be 4.0 GPa by homogeneous shearing simulations. After the nucleation event, the von Mises stress values decrease rapidly as the dislocation starts to grow. These results of dislocation nucleation at surface steps are in agreement with observations that have been made during nanoindentation close to artificially created surface steps by Zimmermann et al. (2001). Dislocations nucleate heterogeneously at these steps, rather than homogeneously in the region of highest shear stresses underneath the indenter.

In total, at least six dislocation nucleation events at surface steps similar to the one described above have been identified in the simulation. The knowledge of nucleation

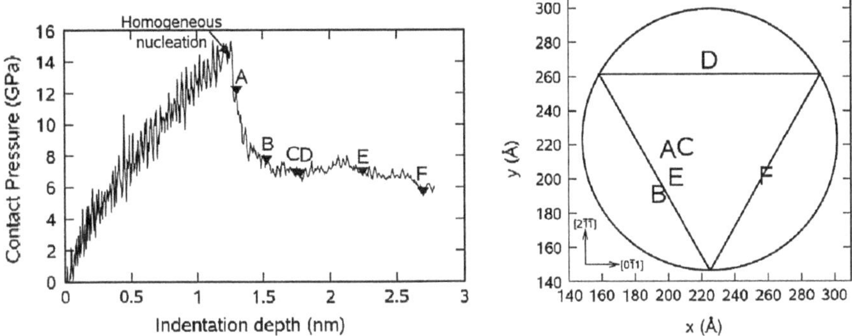

Fig. 3 Measured contact pressure versus displacement (*left*) and sites of heterogeneous dislocation nucleations at surface steps labeled A–F (*right*). The triangle is drawn along the $\langle 1\bar{1}0 \rangle$ directions inside the maximum indentation contact area (Begau et al. 2011)

sites and the corresponding time steps is taken into account to establish a link to the contact pressure versus displacement curve, shown in Fig. 3 (left). The first significant drop in the contact pressure takes place shortly after the initial homogeneous nucleation event. Thereafter, two additional pressure drops coincide with the first two surface nucleation events, although no drop can be identified for the following events. Most likely, the effect of later surface nucleations on the contact pressure is overshaded by dislocations reactions within the bulk whose frequency is increasing with time. However, it can be seen in the curve that the hardening by higher dislocation densities indeed seems to balance the competing mechanism of softening caused by dislocation multiplication during plastic yielding.

Furthermore, the distribution of nucleation sites in space has been examined. The plot in Fig. 3 (right) indicates that nucleation sites are not arbitrarily distributed, but concentrate at specific sites. Inside the triangle with edges along the $\langle 1\bar{1}0 \rangle$ directions on the (111) plane, nucleation sites are located close to the median lines of the triangle. This is geometrically motivated, since these sites are displaced most by the spherical indenter along the $\langle 1\bar{1}0 \rangle$ directions.

2.2 Incipient Plastic Deformation and Distribution of Dislocations

Deriving the dislocation networks including Burgers vectors with the methods mentioned above, enables one to compute the dislocation density tensor α inside an arbitrarily shaped region v as Eq. 1

$$\alpha = \frac{1}{v} \oint_{\perp \text{in } v} d\mathbf{l} \otimes \mathbf{b}, \tag{1}$$

where $\mathbf{l}$ describes the dislocation line direction and $\mathbf{b}$ the Burgers vector. The tensor α describes the deformation caused by a superposition of all dislocations inside the volume and thus is directly related to the density of geometrically necessary dislocations (GND). Thus, the prediction of the model by Nix and Gao (1998) can be tested using MD simulations in simulation setup 2 by measuring GND densities and lattice rotation patterns caused by plastic deformation. Similar experiments have been performed by Demir et al. (2009) using nanoindentation in single crystal copper and measuring the GND density from lattice rotation patterns via three-dimensional electron backscatter diffraction method (EBSD) with serial sectioning. Due to limitations of the current computer performance, the MD sample is by a factor of 100 smaller compared to this experiment.

In order to estimate a scalar density correctly, the tensor α needs to be decomposed into a set of dislocations on the slip system as described by Arsenlis and Parks (1999). However, such a transformation depends on the selected set of possible dislocation for a crystal structure. In the atomistic simulations, the tested volumes are rather small, thus comparable GND densities can be approximated by Eq. 2

$$\rho_{\mathrm{GND}} = |\mathbf{b}|^{-1} \sum_{i=1}^{3} \sum_{j=1}^{3} \left|\alpha_{ij}\right| \tag{2}$$

In addition to the GND densities, it is straightforward to measure the dislocation density ρ_{total} in line length per unit volume. Note that while the total dislocation density measures the length of all dislocations in a certain volumes, by definition in Eq. (2) a GND density requires a net Burgers vector to be present in a certain volume. If the sum of all Burgers vectors of all dislocation segments cancel out in a given volume, only so-called statistically stored dislocations (SSD) are present.

In the simulation, a three-fold symmetry along the $\{111\}\langle 110\rangle$ slip system is observed in the dislocation network under peak load (Fig. 4). By the interaction of dislocations and cross-slipping events, prismatic dislocation loops are frequently created in the strongly plastically deformed zone close to the indenter. The glide of these loops in $[\bar{1}1\bar{1}]$, $[\bar{1}\bar{1}1]$ or $[1\bar{1}\bar{1}]$ direction, causes the distribution of dislocations to be larger than the plastic zone. However, inside the plastic zone underneath the indenter tip, dislocations form a complex network during this advanced stage of plastic deformation. The high density of GNDs is required to compensate the large deformation gradients in this region.

Figure 5 displays the distribution of GNDs and in-plane lattice rotations at the central section in the $(11\bar{2})$-plane under peak load. GNDs are derived from the identified dislocation network and then reduced to scalar values by Eq. (2). Each visible cell represents a volume of $2 \times 2 \times 2$ nm^3. Compared to the experimentally obtained results on GND distributions by Demir et al. (2009), many similarities are found, although the absolute values of the dislocation densities differ by about two orders of magnitude owing to the steeper strain gradients since the length scale of the simulations is two orders of magnitudes smaller than the experimental scale.

In the central section, dislocation densities are concentrated at three sites in both experimental and simulation results in a central peak under the indenter tip and two

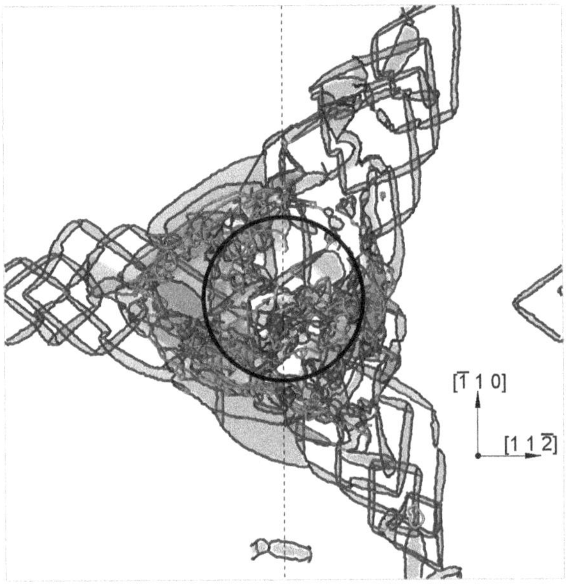

Fig. 4 A top-down view onto the extracted network of dislocations and stacking faults under maximum load. The three-fold symmetry in the $\langle 1\bar{1}0\rangle(111)$ slip system is clearly visible. The circle indicates the indenter contact area (Begau et al. 2011). The cross-section shown in Fig. 5 is taken along the *dashed line*

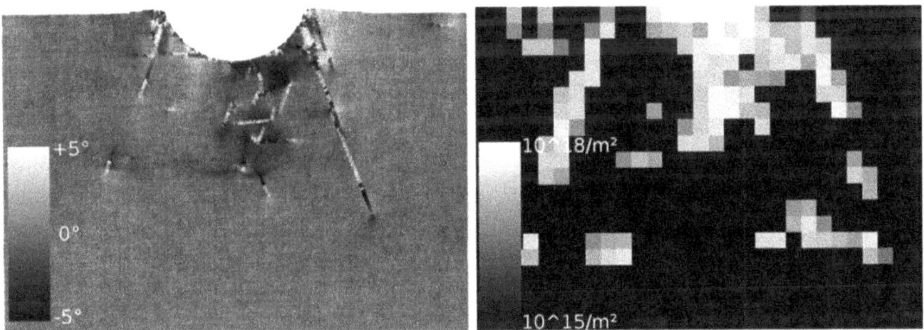

Fig. 5 Lattice rotation patterns about axis $[11\bar{2}]$ (*left*) and the total dislocation density tensors (*right*) at the central cross-section under peak load (Begau et al. 2011)

domains to the left and to the right hand side, separated by regions with significantly lower GND densities. The central peak originates from the complex dislocation network found in the plastic zone underneath the nanoindenter, whereas the outer domains are associated with the spread of prismatic loops. The partial dislocations are bounding extended stacking faults, which approximately represent the boundaries of the plastic zone. In contrast to the prismatic dislocation loops outside the plastic zone, where glide is limited to certain slip systems, inside the plastic zone no preferred slip system could be identified. Due to the extremely high dislocation density, dislocation interaction and multiplication are frequent and result in a highly complex dislocation network. Based on strain-gradient theory (Gao et al. 1999), it is expected that significant lattice rotations are present at domains with high GND densities. Indeed this can be observed in this simulation; however the inverse conclusion is not valid. Very steep rotation gradients are found at thin deformation twins that are produced by the glide of multiple partial dislocations on adjacent atomic planes, while GND densities are low in these regions.

The dislocation densities ρ_{GND} and ρ_{total} have been computed every 5 ps both for the total simulation box and for the plastic zone. The model of Nix and Gao (1982) assumed GNDs to be restricted inside a hemispherical plastic zone underneath the contact zone. However, the simulation results indicate a plastic zone, which is larger than the contact area (see Fig. 4). This is in agreement with observations made by Durst et al. (2005) that propose a larger plastic zone based on a comparison between numerical models and experimental work. In the simplified schematics of Nix and Gao, edge dislocations orthogonal to the indentation direction are compensating the deformation—a mechanism that is not possible in copper due to the $\{111\}\langle110\rangle$ slip system and the chosen (111) indentation plane. Dislocations nucleated at the surface, as described in Sect. 2.1, repel each other and spread outside the contact area as they glide away from the indenter tip. Therefore and in accordance to (Hua and Hartmaier 2010) a factor of $a_{pz} = 1.9$ for the size of the plastic zone has been applied here. Pile-up or sink-in effects are not considered.

These results presented in Fig. 6 reveal that the total dislocation density in the entire simulation domain increases linearly during loading and decreases during unloading, with the exception of a burst of dislocation density during the initial nucleation avalanche occurring at an indentation depth of about 1.1 nm. Inside the plastic zone the total dislocation density increases as well, while the GND density remains fairly constant. This result is in agreement with theoretical estimations of the GND densities using spherical indenters by Swadener et al. (2002), which extends the Nix and Gao model towards spherical indenter tips and predicts a GND density inside the plastic zone that depends on the indenter radius, but not

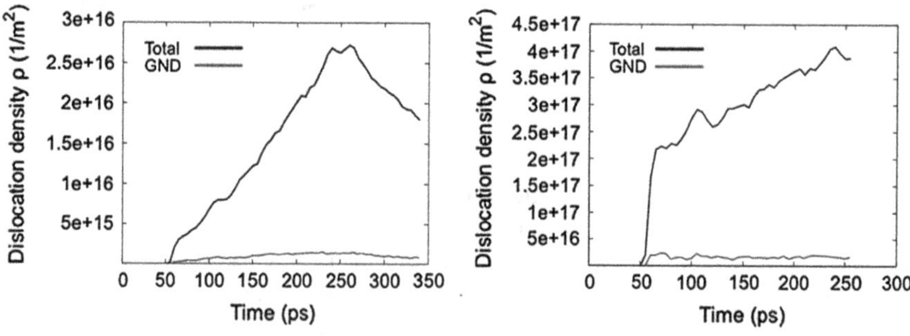

Fig. 6 Total dislocation density and GND density in the entire volume (*left*) and inside the plastic zone (*right*) (Begau et al. 2011)

on the indentation depth. As the total density of dislocations is increasing as well, the SSD density must be increasing inside the plastic zone. The definition of the plastic zone coupled with the indenter contact radius becomes inconsistent during unloading, where the indenter gradually losses contact. Therefore, only values for the total volume are calculated during unloading. Here, the decrease in both GND and total dislocation density is noticeable. Anelastic relaxation takes place, since a certain number of dislocations are disappearing at the contact area.

In conclusion, interlinking dislocation networks, GND density distributions and lattice rotation patterns obtained from atomistic simulations gives a detailed insight into the processes of plastic deformation during nanoindentation. The patterns of dislocation density tensor distribution observed in the simulation have been linked to the geometry of the active slip system of copper as well. GNDs are concentrated in certain domains due to the three-fold symmetry of the $\{111\}\langle 110\rangle$ slip system. A comparison of the dislocation density tensor patterns from simulation with experimental results shows remarkable similarities, although the length scales differ by more than two orders of magnitude. This indicates that in the MD model the same mechanisms are active as during deformation experiments.

2.3 Highlights

Since plastic deformation mechanisms only depend on the interatomic potential and the crystal structure, atomistic simulations do not need to make any artificial assumptions on these mechanisms and hence are a valuable tool to investigate the early stages of plastic deformation during nanoindentation. However, it is necessary to process and analyze the atomistic raw data, consisting only of atomic positions and energies, with suitable methods to derive a quantitative description of the dislocation network developing during plastic deformation. Based on this microstructural information more sophisticated analyses of the mechanisms during incipient plastic deformation are possible, yielding a profound understanding of the underlying physics of initial dislocation nucleation and multiplications, and also the formation of typical dislocation structures, like for example prismatic loops. The initial structure of atomistic simulations is usually a perfect single crystal, which for metallic metals is not very realistic. However, since nanoindentation probes only very small volumes of a material, experimental studies have shown that it is possible to find a defect-free structure in the neighborhood of the nanoindenter (Morris et al. 2011; Lodes et al. 2011; Bei et al. 2008). From the comparison between experiment and atomistic modeling, the conclusion can be drawn that homogeneous dislocation nucleation in perfect crystals occurs at stress levels close to the theoretical strength. Once the first dislocations are created, these dislocations move and multiply at much lower levels of stress inside the bulk of the material. This dislocation motion produces plastic straining of the crystal and hence reduces the average stress level, such that further homogeneous nucleation is suppressed. Consequently, further heterogeneous nucleation of dislocations occurs preferentially at the corner of the contact zone between indenter and free surface, where the stress levels at surface steps or similar defects are still high. This pronounced difference in the very high stress required for homogeneous dislocation nucleation to the much lower stress level necessary to produce a burst of plastic strain by motion and multiplication of the nucleated dislocations, causes the pop-in event frequently observed experimentally on well-prepared oxide-free surfaces. In displacement controlled simulations, this plastic strain burst causes a load-drop in the load-penetration curve.

With the very detailed information concerning dislocations networks from atomistic simulations, it is possible to assess not only individual dislocations, but also to calculate dislocation densities. Furthermore, advanced analysis tools allow us to discriminate between statistically stored dislocation densities and geometrically necessary dislocation densities, the latter occurring in regions of lattice rotations to accommodate strain gradients. Simulations clearly show that even on the atomistic scale the estimate of the GND density underneath the indenter from the model by Nix and Gao (1998) is valid, although the size of the plastic zone and the distribution of GND's are somewhat different. Thus it is concluded that atomistic simulations yield reliable information on the deformation mechanisms occurring during nanoindentation and that the obtained fundamental understanding of plastic deformation can be used to enrich continuum models with appropriate physical descriptions of the mechanisms of plastic deformation.

3 Dislocation Density Based Continuum Model

From atomistic modeling we learned that plastic deformation during nanoindentation of well-prepared, i.e. effectively dislocation-free specimens starts with the homogeneous nucleation of dislocations at very high stress levels, exceeding by far the typical yield strength of a material. This has been demonstrated in detail in Sect. 2.1 as well as in the literature (Begau et al. 2011; Kelchner et al. 1998; Vliet et al. 2003; Miller and Rodney 2008). Translating this understanding into continuum mechanical descriptions of plastic deformation poses a severe challenge, because classical continuum models inherently assume that there is a finite dislocation density at every material point and that plastic deformation occurs by motion of these dislocations at a stress level corresponding to the yield strength of the material. While this assumption is usually fulfilled during the deformation of macroscopic volumes of metallic materials, it fails when directly applied to deformation of nanoscale volumes if the length scale of the tested volume is smaller than the length scale of the heterogeneity of the dislocation microstructure. Deformation of such small volumes can occur during nanoindentation or other nanomechanical testing, but also during fracture mechanical testing of brittle materials.

Apart from the necessity to deal with heterogeneous dislocation densities in continuum models and to take into account homogeneous dislocation nucleation during deformation of dislocation-free volumes, a further challenge for classical, i.e. scale independent continuum models is the description of size effects. As shown in Sect. 2 the indentation size effect can be traced back to the high density of geometrically necessary dislocations (GNDs) in the plastic zone below the indenter. The discrimination into statistically necessary dislocation (SSD) densities and GND densities arises naturally from the atomistic results by evaluating the total Burgers vector content in any volume element of the atomistic domain. If the sum of all Burgers vectors of all dislocation segments of unit length in a certain volume is zero, this volume element only contains SSDs. If, in contrast, this sum yields a certain net Burgers vector within a volume element there must be GNDs present. The sum of SSD density and GND density is always equal to the total dislocation density, i.e. the length of all dislocation segments divided by the volume of the element. In continuum theories, GND densities can be calculated from lattice lattice rotations, based on the Nye theory (Nye 1953) as described in the following Sect. 3.1. In so-called strain gradient or non-local plasticity models GND densities lead to additional work hardening and are the origin of size-dependent plastic material behavior.

In the following we device a rather simple two-dimensional continuum mechanical model, which is applicable to describe the deformation of small volumes that can be either completely or partially dislocation-free and hence to mimic the natural heterogeneity of dislocation microstructure. This mesomodel is developed with special emphasis on a physically sound description of the incipient phases of plastic deformation during nanoindentation. A more general non-local crystal plasticity model will be brought forward in Sect. 4. The model is a natural extension of atomistic models, because it completely takes into account the atomistic findings on deformation mechanisms, but it drastically extends the volumes and also the time scales it can be applied to. To accomplish this, we apply a suited physically-based plasticity model that considers homogeneous dislocation nucleation as well as the creation and fluxes of GND densities. For a more detailed description of the applied methods we refer the reader to the publication by Engels et al. (2012). This model represents of first step towards a completely continuum mechanical description of nanoindentation and other nanomechanical tests, which has two benefits: On the one hand side, modeling of nanoindentation will help us to understand the competition between external loading and plastic stress relief as response of the material on realistic length and time scales. On the other hand side, and more importantly, nanoindentation techniques enable us to parametrize sophisticated crystal plasticity models of poly-crystalline alloys by inverse procedures, see Sect. 5.

3.1 Micromechanical Model of Incipient Plastic Deformation

The simple physically-based crystal plasticity (CP) model of lower-order type (see Sect. 4.1) Eq. 12 is based on the total strain rate tensor, $\dot{\varepsilon}$, for which an additive decomposition can be applied in a small-strain framework (Eq. 3)

$$\dot{\varepsilon} = \dot{\varepsilon}^{e} + \dot{\varepsilon}^{p}(\varepsilon^{e}), \tag{3}$$

where $\dot{\varepsilon}^{e}$ is the elastic strain rate tensor and $\dot{\varepsilon}^{p}$ is the sum of all plastic shear-contributions $\dot{\gamma}^{\alpha}$ on the N_S different glide systems α multiplied by the symmetric part of the Schmidt matrix $\mathbf{P}^{\alpha}$. This yields Eq. 4:

$$\dot{\varepsilon}^{\mathrm{p}} = \sum_{\alpha=1}^{N_S} \dot{\gamma}^{\alpha} \mathbf{P}^{\alpha} = \sum_{\alpha=1}^{N_S} \rho^{\alpha} b v_0 \tau^{\alpha} (\tau_c^{\alpha})^{-1} \mathbf{P}^{\alpha}. \tag{4}$$

Herein the slip rate $\dot{\gamma}^{\alpha}$ is defined according to Orowan's law as the product of Burgers vector norm b, mobile dislocation density ρ^{α} and dislocation velocity $v = v_0 \tau^{\alpha} / \tau_{c^{\alpha}}$. In the latter expression the velocity depends proportionally on the ratio of resolved shear stress τ^{α} and the critical resolved shear stress $\tau_{c^{\alpha}}$ (Eq. 5)

$$\tau_{c^{\alpha}} = c_3 G b \sqrt{\rho^{\alpha}} \tag{5}$$

is formulated based on the isotropic Taylor hardening mechanism with c_3 as a fitting parameter. Such a consideration of dislocation densities is the key ingredient of physically-based crystal plasticity models. The SSD density and the GND density both contribute unweighted and uncoupled to the total dislocation density (Nix and Gao 1998) (Eq. 6)

$$\rho^{\alpha} = \rho^{\alpha}_{\mathrm{SSD}} + \rho^{\alpha}_{\mathrm{GND}}. \tag{6}$$

The evolution of the SSDs is governed by the Kocks-Mecking approach (Mecking and Kocks 1981) (Eq. 7)

$$\dot{\rho}^{\alpha}_{\mathrm{SSD}} = \left(c_1 \sqrt{\rho^{\alpha}_{\mathrm{SSD}} + \rho^{\alpha}_{\mathrm{GND}}} - c_2 \rho^{\alpha}_{\mathrm{SSD}} \right) \dot{\gamma}^{\alpha}, \tag{7}$$

where c_1 is a multiplication and c_2 an annihilation parameter (Roters et al. 2000). Note, only SSD densities are described by this evolution law and that only SSDs take part in annihilation, since the evolution of GND densities is completely described by strain gradients as given by the Nye-tensor. However, both SSDs and GNDs can multiply and thus contribute to the increase in SSD density. In the small-strain framework, the Nye-tensor $\dot{\Lambda}$ is approximated as Eq. 8

$$\dot{\Lambda} = \left(-e_{jkl} \dot{\varepsilon}^{\mathrm{p}}_{il,k} \right)^{\mathrm{T}} \mathbf{e}_i \otimes \mathbf{e}_j \tag{8}$$

where e_{jkl} is the antisymmetric permutation tensor and $\mathbf{e}_i$ ($i \in \{1, 2, 3\}$) are unit vectors spanning the coordinate system. Finally, the density of GNDs is calculated according to Eq. 9

$$\dot{\rho}^{\alpha}_{\mathrm{GND}} = \dot{\rho}^{\alpha}_{\mathrm{GND(edge)}} + \dot{\rho}^{\alpha}_{\mathrm{GND(screw)}}$$
$$= \frac{1}{b} \left(|\mathbf{d}^{\alpha} \dot{\Lambda} \mathbf{l}^{\alpha}| + |\mathbf{d}^{\alpha} \dot{\Lambda} \mathbf{d}^{\alpha}| \right) \tag{9}$$

using the dislocation line and direction vectors $\mathbf{l}^{\alpha}$ and $\mathbf{d}^{\alpha}$ respectively.

In order to activate plasticity in a dislocation-free crystal ($\rho^{\alpha}_{\text{init}} = 0$ everywhere), a nucleation mechanism is formulated with the help of Heaviside's function $\Theta(x)$ (Eqs. 10 and 11)

$$\rho^{\alpha}_{\text{nuc}} = c_0 |\tau^{\alpha}| \Theta(|\tau^{\alpha}| - \tau_{\text{c,nuc}}), \tag{10}$$

where

$$\tau_{\text{c,nuc}} = \frac{Gb}{\pi e^3 r_0} \frac{2 - \nu}{1 - \nu}. \tag{11}$$

This necessary resolved shear stress for the nucleation of a dislocation loop can be derived from energetic considerations (Hirth and Lothe 1982). Herein, r_0 is the radius of the dislocation core and e is Euler's number. This nucleation only creates SSDs and from Eq. (7) it is clear the dislocation multiplication is a purely local process. Hence, the only way how a plastic zone can appear or grow in a dislocation-free material is either through homogeneous dislocation nucleation or through the non-local nature of the GND evolution by strain gradients, as described by Eq. (9). Such strain gradients can occur between purely elastic regions and regions that have already deformed plastically and thus cause fluxes of GNDs to spread throughout the material.

Considering the above model, two-dimensional (plane strain) nanoindentation of a rectangular aluminum single-crystal is simulated. We solve the partial differential equations of mechanical equilibrium with the help of a finite-difference approach, which enables a numerically straightforward incorporation of the gradient terms in Eq. (8). We apply vanishing displacements at the left, lower and right boundary. The load of a circular-shaped tip is mimicked by displacement conditions ($\dot{u} = const$) at the upper free surface. Two active glide systems with an inclination of $\pm\pi/6$ to the vertical median are considered. The full set of geometrical and material parameters can be found in Table 2.

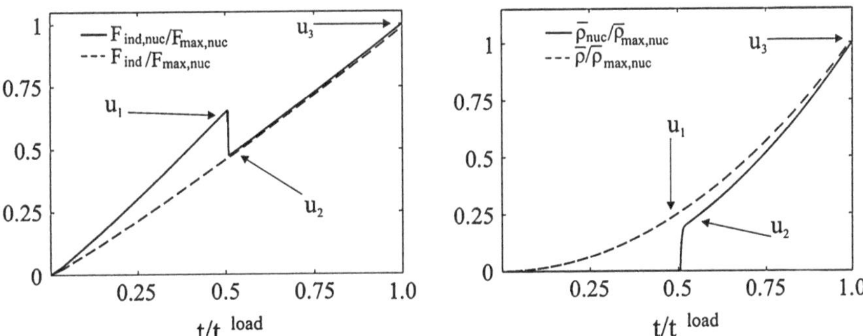

Fig. 7 Comparison of load–displacement curves (*left*) and averaged total dislocation densities (*right*) for an initially dislocation-free crystal (*solid line*) and a crystal with preexisting dislocations (*dashed line*) (Engels et al. 2012)

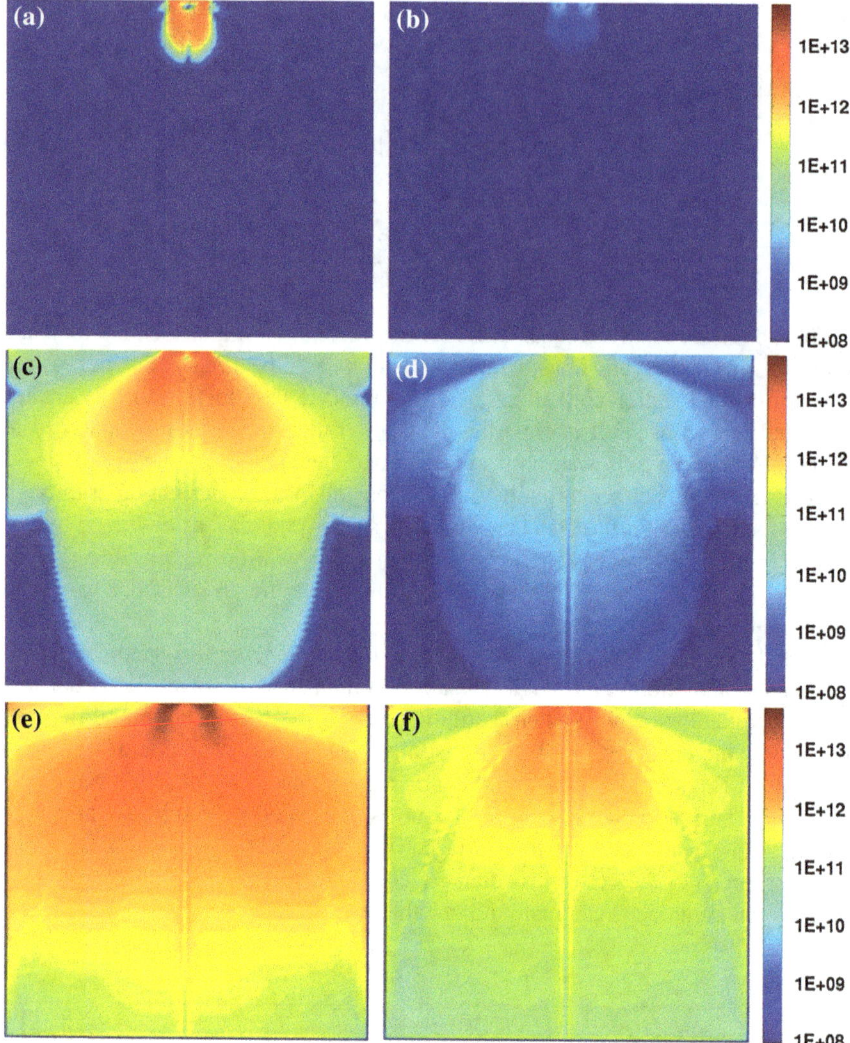

Fig. 8 Comparison of total dislocation densities $\rho(\text{m}^{-2})$ (*left column*) and GND densities $\rho_g(\text{m}^{-2})$ (*right column*) at different indentation depths for $\rho_{\text{init}}^{\alpha}(x,y) = 0$: **a/b** $u_1 = 0.51u_{max}$, **b/c** $u_2 = 0.53u_{max}$, **d/e** $u_3 = u_{max}$ (Engels et al. 2012)

3.2 Nanoindentation into Material with Heterogeneous Dislocation Density

The model derived above is here applied to nanoindentation simulations into a dislocation-free region of a material, a material with homogeneous dislocation density and a material with heterogeneous dislocation density. The results of inden-

Table 2 Summary of the applied model parameters (Sect. 3)

Parameter	Symbol	Value
Domain size	L_x, L_y	$10\,\mu\mathrm{m}$
Discretization spacing	$\Delta x, \Delta y$	$0.1\,\mu\mathrm{m}$
Indenter tip radius		$2\,\mu\mathrm{m}$
Max. indentation depth	u_{max}	$0.07\,\mu\mathrm{m}$
Duration of indentation		$1.5\,\mathrm{s}$
Young's modulus	E	$71{,}000\,\mathrm{N/mm}^2$
Poisson's ratio	ν	0.3
Burgers vector length	b	$2.86 \times 10^{-7}\,\mathrm{mm}$
Ref. dislocation velocity	V_0	$5\,\mathrm{mm/s}$
Critical nucleation stress	$\tau_{c,nuc}$	$G/60\,\mathrm{N/mm}^2$
Glide system angle	φ	$\pm\pi/6$
Nucleation parameter	c_0	$1\,\mathrm{N}^{-1}$
SSD creation parameter	c_1	$0.1/b$
SSD annihilation parameter	c_2	10
Hardening parameter	c_3	0.5

tation simulations of a dislocation-free crystal (represented by symbols with index n, solid lines in figures) and a crystal with a homogeneous initial dislocation density of $\rho^\alpha_{\mathrm{init}}(x, y) = 3.2 \times 10^8\,\mathrm{m}^{-2}$ (symbols without index, dashed lines in figures) are presented in Fig. 7. The plots show the corresponding load-indentation curves and an average dislocation density $\bar{\rho} = (N_x N_y)^{-1} \sum_{x,y} \sum_\alpha (\rho^a(x, y))$ respectively. With the help of our continuum model we are additionally able to resolve the local dislocation density field underneath the indenter (Fig. 8).

During the initial phase of indentation into a dislocation-free region of a material, the mechanical response is completely elastic. Since no dislocations are present, plastic deformation cannot occur until dislocations are nucleated homogeneously. This occurs when the resolved shear stress τ^α reaches its critical value $\tau_{c,nuc}$ in the region immediately below the indenter. At this stage a SSD density ρ^α_{nuc} is generated by homogeneous nucleation on slip system α. As can be seen from Fig. 8a/b, at $u_1 = 0.51 u_{max}$ gradients of plastic strains occur and hence lead to the creation of GNDs as a consequence of the localized plastic deformation due to the nucleation. When the avalanche of dislocations in the crystal has triggered a burst of plastic deformation in a certain area, the macroscopic load-indentation curve exhibits a force-drop of about $15\,\%$ due to the release of elastic energy. Here, the rate of total dislocation generation reaches its peak. At $u_2 = 0.53 u_{max}$, steady state conditions are achieved. As can be seen from the comparison of the microstructure (Fig. 8c/e) almost the total area is plastically deformed and the further increase in the dislocation density is mainly due to SSD multiplication. High densities of GNDs are located in a spherical regime underneath the indenter. In comparison, the load–displacement curve for preexisting dislocation condition appears shallower; stresses are limited by dislocation movement and since the nucleation criterion is not reached, no load-drop can be observed. Note that the final microstructures (Fig. 9) are comparable with respect to the total dislocation density

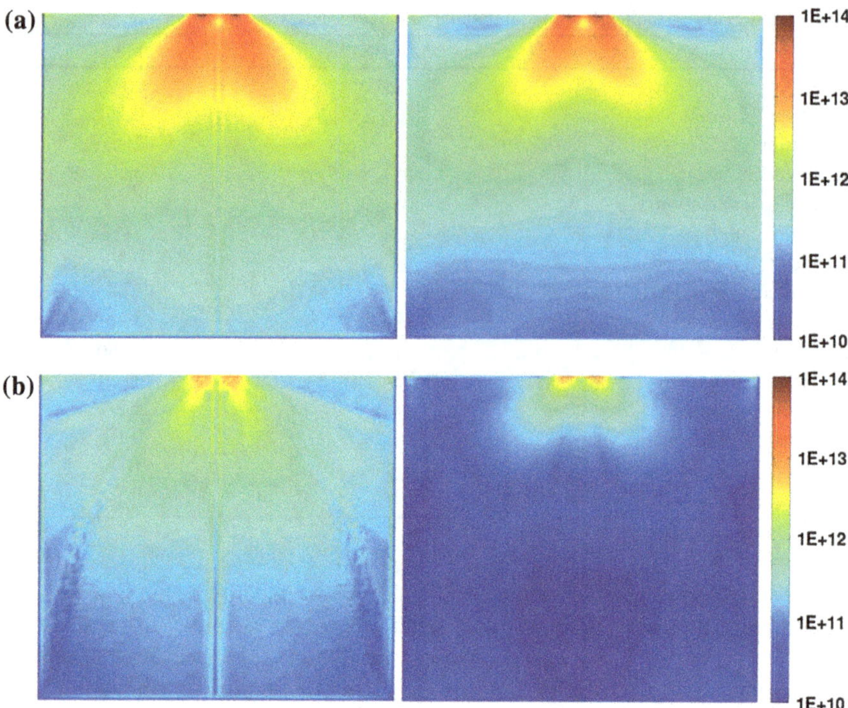

Fig. 9 Comparison of (**a**) total dislocation densities $\rho\,(\mathrm{m}^{-2})$ and (**b**) geometrically necessary dislocation densities $\rho_{\mathrm{GND}}\,(\mathrm{m}^{-2})$ for different initial dislocation distributions at maximum indentation depth. $\rho_{init}^{\alpha}(x,y) = 0$ (*left column*) and $\rho_{\mathrm{init}}^{\alpha}(x,y) = 3.2 \times 10^8\,\mathrm{m}^{-2}$ (*right column*) (Engels et al. 2012)

$\rho(x, y)$, whereas the GND distribution is less localized in the case of $\rho_{\mathrm{init}}^{\alpha}(x, y) = 0$. This behavior is explained by reasoning that for finite initial dislocation densities the more homogeneous work hardening prevents strain gradients from spreading throughout the material, whereas in the initially dislocation-free material the strain gradients can spread freely over large distances.

In addition to homogeneous dislocation distributions we investigate the influence of a *heterogeneous* initial dislocation distribution on the pop-in behavior. In this case most of the domain is dislocation-free, except for a small region in the remote field of the indenter. In this region 5×5 material points in a distance of $0.8L_y$ below the indenter are set to an initial dislocation density of $\rho_{\mathrm{nuc}} = 3.2 \times 10^8\,\mathrm{m}^{-2}$; in Fig. 10 this region is visible due to its high dislocation density. This mimics the case that the highly stressed area below the indenter is smaller than the average spacing between dislocations. In this case there is some initial elastic loading, but before the critical stress for homogeneous dislocation nucleation is reached in the region of highest shear stresses immediately below the indenter, the preexisting dislocations remote from the indenter start to move. This occurs at a local stress level corresponding to the yield strength of the material, which by definition is the stress to move preexisting dislocations. Due to the heterogeneous stress field of the indenter that decays

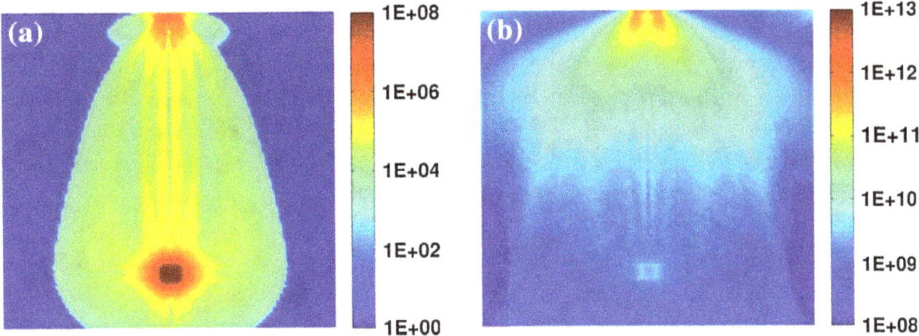

Fig. 10 Total dislocation density $\rho\,(\mathrm{m}^{-2})$ distribution for two indentation depths and *heterogeneous* initial dislocation distribution. **a** Plastic deformations in the highly stressed area below the indenter tip are activated by the GND fluxes. **b** The dislocation distribution after the load drop is comparable to that in Fig. 3a (Engels et al. 2012)

with the distance to the contact area, the stress close to the indenter exceeds the yield strength of the material, but due to the vanishing dislocation density in this region plastic deformation cannot occur. As soon as the local stress in the region with finite initial dislocation density reaches the yield strength, the dislocations in this area start to move and multiply in the usual manner. This yields a successive plastically deformation of the material starting in the region of finite initial dislocation density and spreading towards the highly stressed region below the indenter, as seen in Fig. 10a. A closer analysis reveals that fluxes of GNDs, which start to develop due to strain gradients in the region of finite initial dislocation density, are responsible for this triggering of plastic deformation. The load-indentation curve for this case reveals, that the load-drop is actually shifted towards lower indentation depths and its magnitude is severely reduced (cf. (Engels et al. 2012), Fig. 5).

The comparison with MD simulations performed in Sect. 2 reveals a good qualitative agreement in terms of slope and shape of the load drop (Begau et al. 2011; Zimmerman et al. 2001; Ziegenhain et al. 2010) as well as with displacement-controlled experimental data of Minor et al. (2006). However, we performed additional quasi-two-dimensional MD simulations of aluminum, see Fig. 1, top left. Material parameters, geometry and lattice orientation are chosen such that the setup is comparable with the continuum simulation. The resulting load-indentation curve (Fig. 11) shows the characteristics as the curve from the continuum model for initially dislocation-free material (Fig. 7 (left)), i.e. initial elastic loading followed by a sharp load drop after which the curve continues with a reduced slope.

3.3 Highlights

Based on the results presented above we conclude that the introduction of a criterion for homogeneous dislocation nucleation and an appropriate evolution of GND densities enables us to simulate nanoindentation into materials with heterogeneous

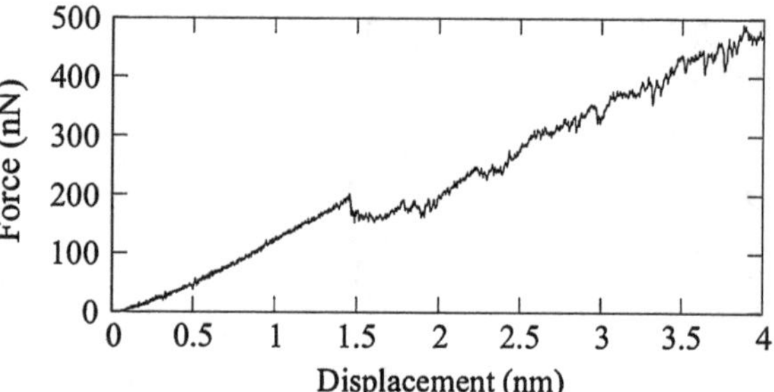

Fig. 11 Force displacement curve in aluminum obtained by MD simulation. The length axis of the cylindrical indenter $r = 12\,\text{nm}$ is parallel to the $[1\bar{1}0]$ axis using periodic boundary conditions in this direction to ensure a setup that is comparable with the CP-simulation

dislocation distributions on all scales, i.e. ranging from completely dislocation-free material, over mostly dislocation-free material exhibiting only local spots of finite dislocation densities, to material with an initially homogeneous dislocation density. Here, the term dislocation-free is not necessarily referring to the entire material, but rather to local regions between areas with finite dislocation densities. The simple continuum mechanical model developed here is capable of reproducing pop-in behavior observed in experiment and can even explain why the pop-in load and pop-in strength is reduced if dislocation motion in areas remote to the indentation occurs before the stress level right below the indenter reaches the critical stress for homogeneous dislocation nucleation. In cases where such homogeneous dislocation nucleation occurs the load at which the pop-in takes place and also the strength of the pop-in reach a maximum value that only depends on the theoretical strength of the material and on the loading situation, i.e. the shape of the indenter and the applied force. In our simulations we can observe a slight delay between the first dislocation nucleation event and the drop of the indentation force, which is consistent with the results of atomistic simulations and supports the conclusions of Knap and Ortiz (2003) who noted that the pop-in is not a reliable indicator for first dislocation activities. Comparable distributions of GND densities can be found in the work of Demir et al. (2009), Huang et al. (2006) and Han et al. (2007) with the assumption of preexisting dislocations. However we found that for an initially dislocation-free crystal the distribution of GNDs is not well-described by the model of Nix and Gao (1998), who assume a very narrow spherical GND field underneath the indenter, which is more appropriate in cases of a material with a finite dislocation density. Hence, for a completely dislocation-free crystal such an assumption might not be a good approximation of the real dislocation structure.

4 Crystal Plasticity Model: Indentation Size Effect and Imprint Topology

From the previous sections it became clear that a correct treatment of statistically stored dislocation (SSD) densities and geometrically necessary dislocation (GND) densities is necessary to describe the strain gradients occurring during nanoindentation correctly. With the help of such non-local plasticity models the so-called indentation size effect (ISE), described in the introduction (Sect. 1), is treated consistently. However, the detailed treatment of the evolution of SSD and GND densities as described in the previous section turns out to cause a prohibitive numerical effort, such that fully three-dimensional models with realistic indenter geometries cannot be simulated with such a demanding material model. In the following we will describe a non-local crystal plasticity model that is also based on dislocation densities, but introduces some simplifications that render it numerically more efficient. Furthermore, this crystal plasticity model is embedded in a large strain formulation such that simulations of deep nanoindentations with realistic indenters geometries, like spherical, sphero-conical or Berkovich indenter tips, become possible. As discussed in the previous section, the inherent assumption of standard plasticity models that there exists a homogeneous dislocation density everywhere and the plastic yielding occurs as soon as the yield strength of the material is reached locally has an influence on the evolution of strain gradients and GND densities. However, it could also be shown that this difference is only pronounced for shallow indents, whereas for deeper indentations the solutions obtained for initially dislocation-free material and material with a homogeneous initial dislocation density approach each other. Since we will be dealing with rather deep indentation in the following, we will henceforth use standard plasticity formulations and thus assume a homogeneous initial dislocation density, i.e. we will be treating the case where the depth of the indentation is large compared with the heterogeneity of the dislocation distribution. The accuracy achieved with such models enables us to compare numerical results directly with experimental data, such that the application of inverse methods to quantify model parameters becomes possible (see Sect. 5).

4.1 Non-Local Crystal Plasticity Model

A fully non-local crystal plasticity finite element model for finite deformations is developed here. Compared to simple plasticity models, like J2 or von Mises plasticity, crystal plasticity models allow for the consideration of physics-based elastic and plastic material models that take into account anisotropic material behavior or even dislocation densities on different slip systems. The explicit treatment of material anisotropy is particularly important for nanoindentation simulations, because in many cases effectively only one grain is indented such that the anisotropic deformation of single crystals needs to be described correctly. In this work only slip based deformation mechanisms are modeled. Elastic and plastic parts of large deformations can be separated by multiplicative decomposition (Lee 1969) of the total deformation gradient tensor $\mathbf{F}$ (Eq. 12)

$$\mathbf{F} = \frac{\partial \mathbf{x}}{\partial \mathbf{X}} = \frac{\partial \mathbf{x}}{\partial \tilde{\mathbf{x}}} \frac{\partial \tilde{\mathbf{x}}}{\partial \mathbf{X}} = \mathbf{F}^{\mathrm{e}} \mathbf{F}^{\mathrm{p}}, \tag{12}$$

where $d\mathbf{X}$, $d\tilde{\mathbf{x}}$ and $d\mathbf{x}$ are line elements in the reference configuration, intermediate (unloaded) configuration and current configuration, respectively. In Eq. (12), $\mathbf{F}^e$ and $\mathbf{F}^p$ are the elastic and plastic deformation gradient, respectively. The rate of elastic $\dot{\mathbf{F}}^e$ and plastic deformation gradient $\dot{\mathbf{F}}^p$ can be defined in terms of the velocity gradient $\mathbf{L}$ as follows (Eqs. 13 and 14)

$$\dot{\mathbf{F}}^e = \mathbf{L}\mathbf{F}^e - \mathbf{F}^e\mathbf{L}^p \tag{13}$$

$$\dot{\mathbf{F}}^p = \mathbf{L}^p\mathbf{F}^p \tag{14}$$

where $\mathbf{L} = \dot{\mathbf{F}}\mathbf{F}^{-1}$ and $\mathbf{L}^p = \dot{\mathbf{F}}^p\mathbf{F}^{p-1}$ are the total and the plastic velocity gradients defined in the current and the intermediate configuration respectively. Elastic and plastic deformation cannot be considered independent because plastic deformation is just the dissipation of the energy stored by elastic deformation. Elastic deformation can be defined with the help of the constant stiffness tensor $\tilde{\mathbb{C}}$ in the intermediate configuration because plastic deformation leaves the crystal lattice orientation unchanged. Using the second Piola–Kirchhoff stress tensor $\tilde{\mathbf{S}}$ and its work conjugate elastic Green strain tensor $\tilde{\mathbf{E}} = \frac{1}{2}\left(\mathbf{F}^{eT}\mathbf{F}^e - \mathbf{I}\right)$ in the intermediate configuration, elastic law can be defined as follows Eq. 15

$$\tilde{\mathbf{S}} = \tilde{\mathbb{C}}_0\tilde{\mathbf{E}} \tag{15}$$

We can always convert the second Piola–Kirchhoff stress $\tilde{\mathbf{S}}$ into current configuration Cauchy stress σ by push forward method (Eq. 16)

$$\vec{\sigma} = \mathbf{F}^e\,\tilde{\mathbf{S}}\,\mathbf{F}^{eT}\,\frac{1}{\det(\mathbf{F}^e)} \tag{16}$$

Non-local behavior of the model appears during the plastic deformation of the material. Plastic deformation occurs through slipping of dislocations on certain planes along certain directions (slip system). The driving force for the plastic deformation can be a function of externally applied loads, isotropic resistance of dislocation interaction and long range back stress of dislocation pile-ups. Commonly used assumption for plastic deformation can be defined in terms of plastic part of velocity gradient $\mathbf{L}^p$ and shear rate at particular slip system $\dot{\gamma}^\alpha$ (Eq. 17)

$$\mathbf{L}^p = \sum_{\alpha=1}^{N_S} \dot{\gamma}^\alpha\tilde{\mathbf{M}}^\alpha \tag{17}$$

where $\tilde{\mathbf{M}}^\alpha$ is the Schmidt tensor for the slip system α and N_s is the number of slip system.

Based on the classical GND density tensor (Nye 1953), a number of non-local constitutive models have been proposed in the literature in order to model the material deformation in presence of gradients of plastic deformation $\mathbf{F}^p$. As examples of the lower order strain gradient constitutive model category, in which no additional governing differential equations and no extra boundary conditions are introduced, Nix and Gao (1998) added the GND densities of plastic strain gradients to the Taylor hardening mechanism; Ohashi (2005) assumed that the mean-free path of mobile dislocations of one slip system depends on the forest GND density of this slip system; and the model proposed by Ma et al. (2006) considered the influence of the GND density on the average jump distance, the activation volume, and the passing stress of a mobile dislocation. Several experiments (Weertman 2002; Mughrabi 2007; Suzuki et al. 2009; Hayashi et al. 2011) clearly support the forest dislocation hardening mechanism of GND densities. The isotropic hardening caused by homogeneously distributed GNDs can be formulated in form of a Taylor law as Eqs. 18 and 19

$$\tilde{\tau}_{\mathrm{GNDp}^\alpha} = c_1 Gb\sqrt{\rho_{\mathrm{GND}}^\alpha},\tag{18}$$

where

$$\rho_{\mathrm{GNDp}}^\alpha = \sum_{\beta=1}^{N_S} \frac{\chi_{\alpha\beta}}{b}\left[\left|\delta_{jkl}\tilde{d}_i^\alpha \tilde{d}_j^\alpha \sin\left(\tilde{\mathbf{n}}^\alpha,\tilde{\mathbf{l}}^\beta\right) F^{\mathrm{p}_{i,k,l}}\right| + \left|\delta_{jkl}\tilde{d}_i^\alpha \tilde{l}_j^\alpha \sin\left(\tilde{\mathbf{n}}^\alpha,\tilde{\mathbf{d}}^\beta\right) F^{\mathrm{p}_{i,k,l}}\right|\right].\tag{19}$$

Here, N_S is the total number of slip systems and $\chi_{\alpha\beta}$ the interaction coefficients between different slip systems. c_1 is the Taylor hardening coefficient (Evans and Hutchinson 2009) or the geometrical factor (Gottstein 2004) and G is the shear modulus. The slip direction, $\tilde{\mathbf{d}}^\alpha$, the slip plane normal direction, $\tilde{\mathbf{n}}^\alpha$ and edge dislocation line direction, $\tilde{\mathbf{l}}^\alpha$, are defined with respect to the undistorted configuration. Note that in this formulation the work hardening by SSDs and GNDs is added up linearly, which is a simplification of the more realistic formulation, where the SSD and GND densities are added to yield the total dislocation density from which the total work hardening is then calculated by the Taylor formula. However, the formulation chosen here allows for some necessary simplification that renders the numerical treatment much more efficient.

In recent micro bending experiments (Suzuki et al. 2009; Hayashi et al. 2011), size dependent kinematic hardening and size dependent isotropic hardening are observed in mechanical tests with multi-planar and multi-slip deformation modes. At the same time, size independent isotropic hardening and only weakly size dependent kinematic hardening are observed in the tests with coplanar double slip deformation mode. These experimental results can only be described consistently by a model taking into account isotropic and kinematic hardening of a GND density properly. To our knowledge such a model has not been formulated yet.

Non-local constitutive models of higher order often consider additional degrees of freedom (DOF) with respect to strain gradients. Their governing partial differential equations are of higher order than those of classical crystal plasticity finite element methods. The Cosserat models (Forest et al. 2000) considered the micro-rotation vector connecting to the elastic spin tensor as additional DOF, the plastic strain gradient models (Gurtin 2002) included the nine components of the plastic deformation gradient as additional DOF. Furthermore, the recently developed micromorphic non-local models (Cordero et al. 2010) adopted a non-symmetric plastic micro-deformation tensor, which can be different from the plastic deformation gradient, as additional DOF. For this type of constitutive model, the governing equation with respect to the additional DOF is the balance of moment of momentum or a similar kind. Currently, the biggest challenge for higher order non-local constitutive model is how to treat the necessary boundary conditions for the additional DOF across grain boundaries and phase boundaries. The internal stress or back stress caused by second order strain gradient is responsible for the kinematic hardening. As an example, through considering the standard stress fields of a dislocation segment (Hirth and Lothe 1982; Wit 1967; Devincre 1995; Geers et al. 2007) in an isotropic elastic material, the internal stress caused by linear GND pile-ups inside a $L_1L_2L_3$ domain has been derived recently as follows Eq. 20

$$S_{ij}^{\text{GND}} = \mathbb{K}_{ijabcd} F_{ab,cd}^{\text{p}} \tag{20}$$

$$\mathbb{K}_{ijabcd} = -(L_1L_2L_3)\delta_{mbc} \sum_{I,J=1}^{3} C_{ija}^{IJ} \bar{t}_m^I \bar{g}_d^J. \tag{21}$$

where $\mathbf{C}^{IJ}$ is the module to calculate the stress at [0, 0, 0] caused by a dislocation segment $[-L_I \bar{\mathbf{t}}^I/2, +L_I \bar{\mathbf{t}}^I/2]$ locating at position $L_J \bar{\mathbf{g}}^J$.

The evolution of the passing stress due to GNDs is governed by lattice curvature at current time. The second non-local term is the internal or back stress $\mathbf{S}_{ij}^{\text{GND}}$ caused by the GND density gradients which have been calculated by considering standard stress fields of a dislocation segment, second order plastic strain gradient around one material point and average dislocation pile-up size. This internal stress provides kinematic hardening effects during complicated loading histories like Indentation, bending and torsion.

In this contribution a very simple phenomenological viscoplastic flow rule is used. As mentioned above some modifications in the flow rule have been made to consider the influence of the GND density on plastic deformation. The resolved shear stress τ is calculated by the projection of macroscopic stress $\tilde{\mathbf{S}}$ and microscopic internal stress $\tilde{\mathbf{S}}_{\text{GND}}$ onto the slip system as shown below (Eq. 22)

$$\tau_\alpha = (\tilde{\mathbf{S}} + \tilde{\mathbf{S}}_{\text{GND}}) \cdot \tilde{\mathbf{M}}^\alpha. \tag{22}$$

Thus, a modified flow rule dealing with non-local effects supplied by GNDs can be described as follows (Eq. 23)

$$\dot{\gamma}^\alpha = \gamma_0 \left| \frac{\tau^\alpha}{\hat{\tau}^\alpha + \tau_{GNDp}^\alpha} \right|^{p_1} \text{sign}(\tau^\alpha), \tag{23}$$

where p_1 is the inverse value of strain rate sensitivity and $\hat{\tau}^\alpha$ the resistance coming from all of the SSD densities at α slip system. Temperature effects have not been included in the flow rule by considering the isothermal test condition. We have two kinds of evolution law for the $\hat{\tau}^\alpha$ (hardening by SSDs). The first one is based on the Kocks-Mecking approach Mecking and Kocks (1981), which has already been described in the previous Sect. 3.1. The second one is a phenomenological evolution law inspired from the work of Kalidindi et al. (1992) (Eq. 24)

$$\dot{\hat{\tau}}^\alpha = \sum_{\beta=1}^{N_S} h_0 \chi^{\alpha\beta} \left(1 - \frac{\hat{\tau}^\alpha}{\hat{\tau}^s}\right)^{p_2} \left|\dot{\gamma}^\beta\right|, \tag{24}$$

where h_0 is the initial hardening rate, $\hat{\tau}^s$ is the saturation slip resistance and p_2 is a fitting parameter. We compared the hardening behavior of a material using two different approaches (dislocation density based work hardening rule and phenomenological work hardening rule) and found the phenomenological approach to be currently more suited for the further calculations presented in this contribution.

4.2 Finite Element Model

Due to the non-local formulation of the model, an additional degree of freedom (DOF) $\mathbf{F}^{p'}$ has been considered. In this model a weak coupling between the conventional DOF $\mathbf{u}$ and the additional DOF $\mathbf{F}^{p'}$ has been established. A user material subroutine UMAT has been developed using the commercial Abaqus software,[1] which can calculate the stress, stiffness and other internal variables at each time increment, taking non-local terms into account.

As described in the introduction (Sect. 1) different ISE occur for different kinds of indenters. In the following deep nanoindentation simulations for spherical have been performed using the model described above. A fully three-dimensional simulation of nanoindentation into an aluminum single crystal is performed. Since the ISE in case of spherical indenter is related to different radii we investigated radii ranging from 200 to 70 μm. A finite sliding, hard contact problem is modeled, where the indenter is assumed to be an analytical rigid body. Displacement boundary conditions are applied at the indenter tip. The bottom of the cylindrical model is constrained in all three translational DOF. A crystal orientation of [100] has been considered for all simulations. Finite element simulations are performed using linear as well as quadratic elements to compare the results with and without internal stress (which can only be considered by using quadratic elements). The crystal plasticity parameters mimicking the behavior of an Al single crystal adopted from Roters et al. 2004) are listed in Table 3.

---

[1] The Abaqus Software is a product of Dassault Systemes Simulia Corp., Providence, RI, USA.

Table 3 Summary of the crystal plasticity parameters mimicking the mechanical properties of an Al single crystal (Roters et al. 2004)

C_{11} (GPa)	C_{12} (GPa)	C_{44} (GPa)	p_1	p_2	c_1	h_0 (MPa)	$\dot{\gamma}_0$ (s^{-1})	$\hat{\tau}^\alpha$ (MPa)	$\hat{\tau}^s$ (MPa)	$\chi^{\alpha\beta}$ Copl.	$\chi^{\alpha\beta}$ Non-copl.
106.75	60.41	28.34	0.02	2.25	0.1	60	0.001	12.5	75	1.0	1.4

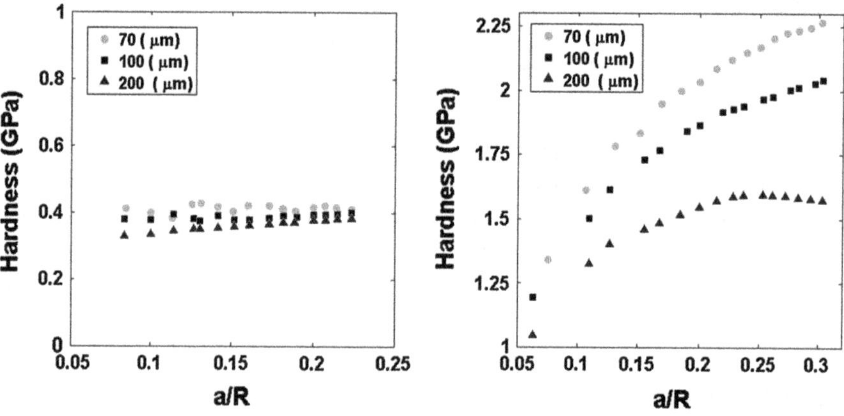

Fig. 12 Indentation hardness H versus a/R ratio; in the legends the indenter radii are given. *Left* simulation performed with local model; *Right* simulation performed with non-local crystal plasticity model

4.3 Indentation Size Effect and Imprint Topologies

The model described above has been applied to describe the ISE and its influence on the resulting imprint topology of the indented surface. Furthermore, the effect of higher order stresses on the pile-up behavior of the material was analyzed. We find that lower order models where only isotropic hardening by GNDs is considered are sufficient for capturing the ISE, but including higher order stresses into the model enables us to predict more realistic indentation topologies. In Fig. 12 we present simulation results capturing the ISE for spherical indenters in form of indentation hardness H at a fixed ratio of a/R. The indentation hardness H is defined as the indentation force divided by the projected contact area between surface and indenter, i.e. the fraction of the contact area that has a normal direction parallel to the indentation direction. Furthermore, a is the contact radius and R is radius of spherical indenter. It is seen that the indentation hardness H is increasing monotonously with the ratio a/R, whereas it decreases with increasing indenter radius. This result is in good agreement with experimental observations shown in (Swadener et al. 2002), cf. Fig. 3, and also with the theoretical expectation. However, if the simulations are performed with local crystal plasticity model the obtained hardness is almost same for all indenters and a/R ratios as shown in Fig. 12 (left). This clearly demonstrates the need to use a non-local plasticity model for nanoindentation simulations if an ISE is expected.

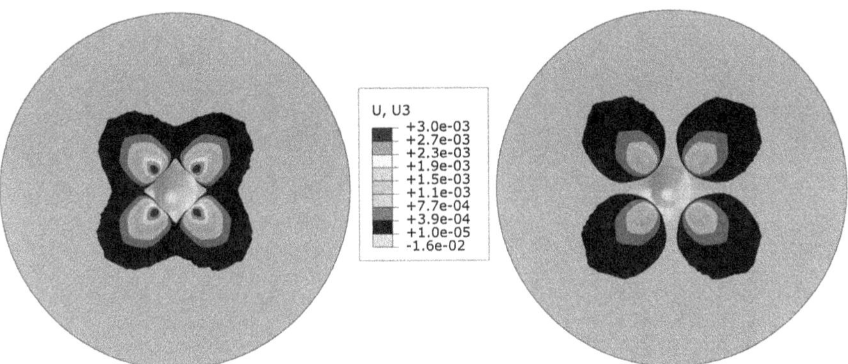

Fig. 13 Surface topology for local and non-local crystal plasticity model; the legends give the displacements with respect to the nominal surface. *Left* significant pile-up for local crystal plasticity model; *Right* decreased pile-up size for non-local crystal plasticity model

Note however, that for very deep indentations experiment shows that the ISE saturates and the indentation hardness converges to a constant bulk value.

The imprint topology on the surface remaining after indentation, i.e. the so-called pile-up or sink-in, is analyzed here as a function of the work hardening behavior of the material. The results of the finite element simulations reveal first of all a very pronounced plastic anisotropy that reflects the four-fold symmetry in the orientation of the slip systems in the material, see also (Eidel and Gruttmann 2007). Furthermore, it is seen that the non-local crystal plasticity model produces a less pronounced plastic pile-up compared to the local crystal plasticity model (see Fig. 13). This is explained by the extra work hardening provided by GNDs and this finding is in line with the observation reported in the literature that a more pronounced work hardening produces smaller pile-up structures, see for example (Alcalá et al. 2000). In the extreme case of very strong work hardening a sink-in is produced by the indentation rather than a pile-up, i.e. the material around the indentation is displaced into the surface instead out of the surface.

Further results from the non-local model show a trend of decreasing pile-up size with an increasing coefficient of GND hardening (Taylor's hardening coefficient c_1) which is consistent with findings reported in the literature (McElhaney et al. 1998; Cheng and Cheng 1998). Figure 14 demonstrates that the pile-up is significantly higher for the value of $c_1 = 0.01$ (left), decreases for $c_1 = 0.05$ (middle) and transforms into a sink-in for $c_1 = 0.1$ (right) for otherwise the same finite element simulation.

4.4 Highlights

The macroscopic simulation of nanoindentation with spherical shape reveals results that are comparable to experiments. Our studies with different non-local models show that the ISE is captured already reliably by lower-order formulations.

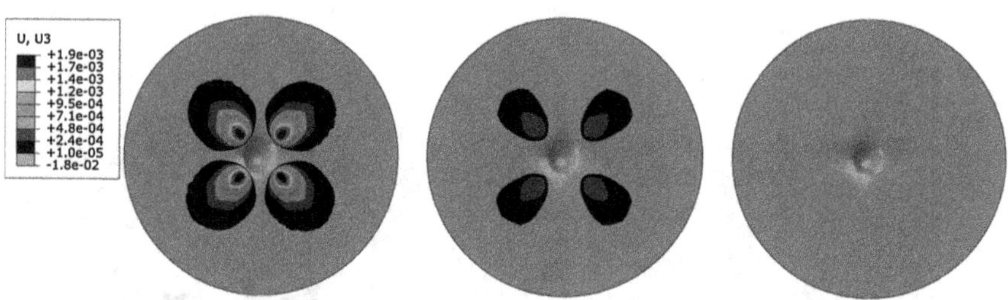

Fig. 14 Pile-up imprint for different values of Taylor's hardening coefficient for GNDs c_1; the legends give the displacements with respect to the nominal surface. *Left* significant pile-up for $c_1 = 0.01$; *Middle* decreased pile-up size for $c_1 = 0.05$; *Right* sink-in behavior for $c_1 = 0.1$

However, higher-order models have an influence on the imprint topology. Consistent with the literature, we find that the imprint profile reflects the crystallographic slip symmetry of the indented crystal. The imprint profile is furthermore very sensitive to the work hardening behavior of the material and thus also to the work hardening contribution of GNDs, that are produced by the strain gradients occurring during the indentation. Hence, we conclude that a comparison of the imprint profiles obtained from experiment and numerical modeling provide a critical test for such models and a possibility to assess model parameters.

5 Crystal Plasticity Parameter Identification by Inverse Method

With the non-local crystal plasticity model a powerful constitutive model for the mechanical behavior of single crystals has been introduced in Sect. 4.1. This model takes into account dislocation slip on individual glide planes, thus describing the plastic anisotropy of the single crystal in a physical way. Since in this section the finite element simulation and inverse analysis is only applied to rather deep microindentations, where the indentation size effect (ISE) is no longer significant, we use a simplified slip rule for a local crystal plasticity model without explicitly taking into account dislocation densities. The main purpose of this simplification was to gain numerical efficiency and thus to simplify the optimization during the inverse analysis of the results. If this procedure is applied to more shallow indents, where the ISE plays a more prominent role, it is essential to apply non-local formulations of higher order, as demonstrated in the previous section.

The inverse analysis itself has the purpose to identify material parameters for the crystal plasticity model in a unique and robust way. To accomplish this, finite element (FE) model of a microindentation is implemented that obeys the same boundary conditions, i.e. indenter geometry, crystal orientation and maximum indentation force, as the experiment that is conducted in parallel. Furthermore, the FE simulation is initially performed with guessed values for the material parameters. From the experiment as well as from the FE simulation the remaining imprint topology after the microindentation is quantified and both results are compared. Then the material parameters are varied by an optimization procure until an optimal agreement between numerical and experimental results is established. These optimal parameters are then considered to be the true material parameters for the crystal plasticity model. In previous work (Schmaling and Hartmaier 2012) the uniqueness and robustness of this method for a simple two-parameter plasticity model has been demonstrated in the case of conventional Rockwell hardness imprints. In this contribution we demonstrate its application to parameterization of a crystal plasticity model for α-iron. Recently, Zambaldi et al. (2012) published the application of a similar method for a crystal plasticity model of α-titanium.

5.1 Inverse Analysis Method

The inverse analysis is expressed as the minimization problem for a discrepancy norm quantifying the deviation between experimental and simulated quantities describing an indentation topology. Minimization is performed with respect to the unknown values of the crystal plasticity parameters. The discrepancy norm to be minimized is defined as (Eq. 25)

$$\text{MSE}\,(x) = \sum_{i=1}^{n} \sum_{i=1}^{m} (u_{ij}^{\text{sim}}\,(x) - u_{ij}^{\text{exp}})^2, \tag{25}$$

where u_{ij}^{sim} $(i, j = 1 \ldots N)$ are the displacements describing the height profile of the simulated indentation topology over the entire pile-up area, the number of sampling points is given by $N \approx 200$. The corresponding experimental values are denoted by u_{ij}^{exp} and $\mathbf{x}$ is the vector of unknown simulation parameters. To minimize Eq. (25) an in-house derivative-free parallelized optimization algorithm (Begau 2008) based on a work about efficient global optimization of expensive black-box function using the Kriging metamodeling approach (Jones et al. 1998) was chosen. The optimization algorithm and strategy is especially designed for optimization of FE problems, requires a small number of usually expensive function calls, i.e. FE simulations of the microindentation, and returned good estimates very close to the global minimum in all of our test cases showing fast convergence behavior. Instead of using starting values that can run into local minima the algorithm samples the entire domain of possible material parameters and successively finds better estimates based on various criteria. Hence, it is in the responsibility of the user to define appropriate boundaries for the crystal plasticity parameters. For the sake of brevity the reader is kindly referred to (Jones et al. 1998) for details of the optimization algorithm.

To obtain the results shown in the following the experimental microindentation was performed into a large grain of known orientation of a coarse grained α-iron material (ARMCO iron). The inverse analysis was applied with knowledge of the experimental boundary conditions and of the residual imprint topography leading to results for the unknown crystal plasticity parameters that describe plastic yielding and hardening behavior of each glide system. Later in this section the microscopic crystal plasticity results are homogenized and verified with respect to uniaxial tensile tests of polycrystalline α-iron.

For the experimental procedure a rather large indentation depth of 3 μm for a sphero-conical indenter of 10 μm tip radius and opening angle of 90° was chosen. The simplified flow rule for the plastic slip rate $\dot{\gamma}^\alpha$ on each slip system α is formulated as Eq. 26

$$\dot{\gamma}^\alpha = \dot{\gamma}_0 \left| \frac{\tau^\alpha}{\tau_c^\alpha} \right|^m \text{sign} \left(\tau^\alpha \right), \tag{26}$$

with the work hardening dependent critical resolved shear stress (Eq. 27)

$$\tau_c^\alpha = \tau_{c,0} + \int_T \dot{\tau}_c^\alpha dt \tag{27}$$

where T is the current time of the FE simulation. The work hardening rate is given by Eq. 28

$$\dot{\tau}_c^\alpha = \sum_{\beta=1}^{N_S} \dot{h}_{\text{init}} \left(1 - \frac{\tau_c^\alpha}{\tau_s^\alpha} \right)^n M_{\alpha\beta} \left| \dot{\gamma}^\beta \right|. \tag{28}$$

This yields the unknown material parameters to be determined by the inverse method, namely the reference shear rate $\dot{\gamma}^0$, the initial critical resolved shear stress $\tau_{c,0}$, the initial hardening rate $\dot{h}_{\text{init}}$, the saturated critical resolved shear stress τ_s^α, the hardening exponent n and the flow rule exponent m.

We considered $N_S = 12$ glide systems. τ^α is the resolved shear stress on glide plane α. The diagonal terms of the Schmidt matrix $M_{\alpha\beta}$ (self-hardening) in all simulations were chosen as 1.0 and for the off-diagonal terms (latent hardening) a value of 1.4 was set, as typical literature values (Roters et al. 2010).

Three-dimensional (3D) simulations similar to the ones described in Sect. 4 have been performed during the inverse analysis. The optimization procedure usually was carried out up to 20 generations, after that no noticeable reduction of the values of the objective function values was observed. The achieved geometrical similarity of simulated and measured imprint topology can be seen in Fig. 15. A more quantitative comparison of the pile-up heights between simulation and experiment is given in Fig. 16, where the height coordinates are plotted against the circumferential angle for two different radii. The identified parameters obtained through the inverse analysis are given in Table 4.

5.2 Homogenization and Verification of Results

The identification and verification of parameters for crystal plasticity models requires vhasvbeen shown above that the inverse analysis of nanoindentation imprints yields

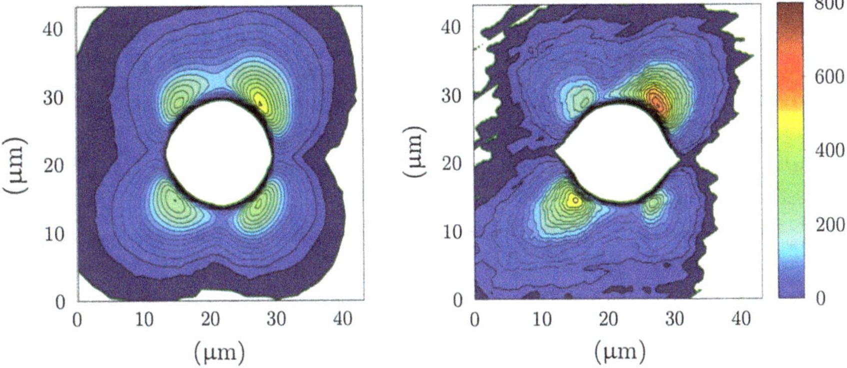

Fig. 15 *Left* crystal plasticity FE simulation using material parameters from inverse analysis (Table 4); *Right* experimental result of microindentation on α-iron (orientation of indent can be found in Table 4). Only positive displacement above the nominal surface is displayed for visualization purposes. The residual imprint depth of the experiment is 3.2 μm, whereas the imprint depth of the inverse analysis is fitted to 3.0 μm

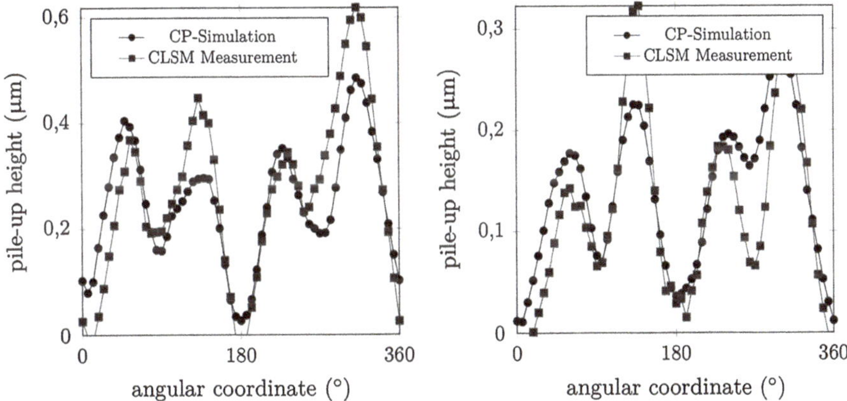

Fig. 16 Circumferential pile-up height from inverse analysis and experiment of α-iron at a radius of 8.7 μm (*left*) and 14.5 μm (*right*) from the center of the imprint

Table 4 Material parameters for α-iron of crystal plasticity constitutive behavior identified by nanoindentation in grain with the given orientation by Bunge Euler angles and inverse analysis of the presented approach; r and Φ are indenter tip-radius and indenter opening-angle respectively

r (μm)	Φ ($^\circ$)	$\dot{h}_{init}$ (MPa/s)	τ_s^α (MPa)	$\tau_{c,0}$ (MPa)	$\dot{\gamma}_0$ (s^{-1})	m	n	Fitness	Euler angle ($^\circ$)
9.9	89.4	961	310	78.3	3.23×10^{-3}	26.7	7.81	16.3	[194,9,164]

such parameters. Here it will be shown in addition that the identified parameters can truly be used for crystal plasticity FE simulations of the respective alloy, by applying a homogenization approach and comparing the homogenized effective polycrystal properties to those obtained from tensile tests. The microstructure of the real material is described in a simplified way by a representative volume element (RVE) consisting of 91 randomly oriented grains. This 3D periodic RVE is created by a Voronoi tessellation (Voronoi 1908). The mechanical behavior of each grain is described with the crystal plasticity model defined above using the identified material parameters from inverse analysis method (Table 4). Neighboring grains have perfect deformation compatibility between them such that no special constitutive relations for grain boundary mechanics are considered here. The geometrical model and the corresponding mesh of the RVE are shown in Fig. 17.

The RVE was subject to a virtual tensile test yielding an effective stress–strain curve for a polycrystal shown in Fig. 18 (left). Comparing the stress–strain behavior to the results obtained by (real) uniaxial straining tests shows good agreement for the yield strength and for the hardening up to the maximum value of accumulated plastic strain of 13%. It is noted that the tensile test reveals a yield point phenomenon that is not described by the crystal plasticity model.

The microindentation load–displacement curve has not been considered in the inverse analyses carried out in this work. However, for completeness it is given in Fig. 18 (right) to demonstrate the quality of the agreement between crystal plasticity FE simulation end experiment.

Fig. 17 *Left* geometrically periodic artificial microstructure achieved through Voronoi-tessellation is used for the virtual tensile tests. The representative volume element consists of 91 randomly oriented grains. *Right* corresponding mesh with 7,857 elements (fully integrated with linear shape functions)

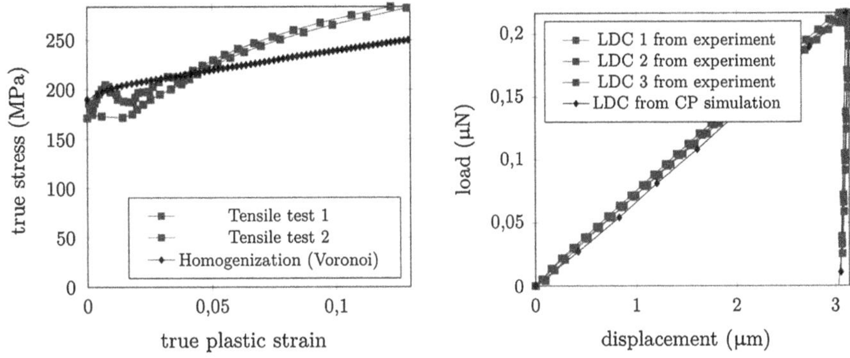

Fig. 18 *Left* homogenized stress–strain behavior through virtual tensile test of a Voronoi poly-crystal compared to results obtained by experimental tensile tests. *Right* comparison of microin-dentation load–displacement curves for a single α-iron grain obtained from FE simulation with parameters from inverse analysis and from experiment

5.3 Highlights

In this section the applicability of a simple local crystal plasticity model for the simulation of microindentation, where indentation size effects do not need to be considered, is demonstrated. The accuracy for the load displacement curve and in particular for the remaining imprint topology is high enough such that a quantitative comparison with experimental data is possible. This allows us to apply inverse methods to determine true material parameters for crystal plasticity models in a unique and robust way. In future work this method will be applied to nanoindentation under consideration of size effects. Thus a direct link between experiment and non-local crystal plasticity models can be established.

6 Concluding Remarks

In this contribution we demonstrated the applicability of simulation methods on different length scales to describe the nanoindentation process from the atomic scale, where dislocation nucleation and evolution of dislocation densities during the initial stages of plastic deformation can be studied, to the macroscale, on which the remaining imprint topology can be described with such an accuracy that inverse methods become applicable to determine realistic material parameters.

Major advancements on the atomic scale have been achieved through recent developments concerning the analysis of atomic structures. With these analysis methods it is not only possible to determine the positions of dislocation segments, but also to assess their Burgers vectors and line directions. From such information important insight can be gained on the evolution of densities of statistically stored and geometrically necessary dislocation densities, respectively. Thus, it could be shown that the deformation mechanisms occurring during atomic simulations are the same as those taking place during experiment, although the length scales of the nanoindenter sizes still differ by one to two orders of magnitude.

From the comparison of the results of atomic simulation and experiment we learn that the microstructural state of the indented material is of utmost importance for its mechanical response. If the indentation occurs into a dislocation-free region, homogeneous dislocation nucleation at a stress level close to the theoretical strength of the material has to take place at the most highly stressed spot below the indenter. By this event plastic deformation is initiated with a strain burst that reveals itself in form of a pop-in. After this initial dislocation nucleation further dislocation generation takes place at much lower stresses and hence in much larger volumes, either by dislocation multiplication or by heterogeneous nucleation at slip lines in the contact area. Mesoscale simulations suggest that strain gradients play a dominant role during the spreading of the plastically deformed zone into a dislocation-free material. If, in contrast, indentation is performed close to preexisting dislocations, these dislocations can be moved at significantly lower stresses. Consequently, mesoscale simulations indicate that under these conditions pop-in behavior is significantly reduced or completely suppressed, which is in agreement with experiment.

On the macroscale, finally, the importance of a reliable description of strain gradients by appropriate evolution laws for densities of statistically stored dislocations and geometrically necessary dislocations, respectively, has been demonstrated. Such dislocation density evolution laws can be consistently formulated within non-local crystal plasticity models. From the comparison of local and non-local crystal plasticity models we predict that the difference in the work hardening behavior due to geometrically necessary dislocation densities reveals itself in significant differences in the remaining imprint topology. This opens a new way for the direct validation of non-local plasticity models, also known as strain-gradient models, with experiment. A major challenge currently faced when applying physically sound non-local plasticity models based on the evolution of dislocation densities is the high numerical effort to solve the underlying equations. Further advancement and a broader applicability of such models require new formulations that increase the numerical efficiency without introducing too many simplifications. Potentially an efficient parallelization strategy for continuum mechanical methods is also a suited way for overcoming current numerical limitations.

While the methods on the atomic scale and the macroscale have reached a certain maturity, the mesoscale methods, where plasticity is described on the basis of dislocation generation and motion, still deserve more attention. The mesoscale is particularly important to fully understand the competition of the deformation mechanisms that are responsible for effects, like pop-in behavior and indentation size effects. Open questions are for example what determines the magnitude of the pop-in and how this information can be related to microstructural quantities. Furthermore, it will be interesting to study the influence of the indentation size effect on the imprint topology, which has a direct impact on inverse methods to determine material parameters from shallow nanoindentations. From such studies further understanding can be expected that will render nanomechanical methods into efficient tools to characterize true material behavior.

Acknowledgments   The authors acknowledge financial support through ThyssenKrupp AG, Bayer MaterialScience AG, Salzgitter Mannesmann Forschung GmbH, Robert Bosch GmbH, Benteler Stahl/Rohr GmbH, Bayer Technology Services GmbH and the state of North-Rhine Westphalia as well as the European Commission in the framework of the European Regional Development Fund (ERDF).

References

Ackland GJ, Jones AP (2006) Applications of local crystal structure measures in experiment and simulation. Phys Rev B 73(5):054104

Alcalá J, Barone AC, Anglada M (2000) The influence of plastic hardening on surface deformation modes around Vickers and spherical indents. Acta Mater 48(13):3451–3464

Arsenlis A, Parks DM (1999) Crystallographic aspects of geometrically-necessary and statistically-stored dislocation density. Acta Mater 47(5):1597–1611

Atkinson M (1995) Further analysis of the size effect in indentation hardness tests of some metals. J Mater Res 10(11):2908–2915

Begau C (2008) Metamodelle zur Optimierung der Pressnahtlage bei Verbundstrangpresssimulationen. Tech. Rep. Algorithm Engineering Reports TR08-2-008, Fakultät für Informatik, TU Dortmund, Germany

Begau C, Hartmaier A, George EP, Pharr GM (2011) Atomistic processes of dislocation generation and plastic deformation during nanoindentation. Acta Mater 59:934–942

Begau C, Hua J, Hartmaier A (2012) A novel approach to study dislocation density tensors and lattice rotation patterns in atomistic simulations. J Mech Phys Solids 60(4):711–722

Bei H, Gao YF, Shim S, George EP, Pharr GM (2008) Strength differences arising from homogeneous versus heterogeneous dislocation nucleation. Phys Rev Lett 77(6):060103

Bei H, George EP, Hay JL, Pharr GM (2005) Influence of indenter tip geometry on elastic deformation during nanoindentation. Phys Rev Lett 95(4):045501

Cheng Y-T, Cheng C-M (1998) Scaling approach to conical indentation in elastic-plastic solids with work hardening. J Appl Phys 84(3):1284–1291

Cordero NM, Gaubert A, Forest S, Busso EP, Gallerneau F, Kruch S (2010) Size effects in generalised continuum crystal plasticity for two-phase laminates. J Mech Phys Solids 58:1963–1994

De Guzman MS, Neubauer G, Flinn P, Nix WD (1993) The Role of Indentation Depth on the Measured Hardness of Materials. In: MRS proceedings, vol 308. Cambridge University Press, 1993

Demir E, Raabe D, Zaafarani N, Zaefferer S (2009) Investigation of the indentation size effect through the measurement of the geometrically necessary dislocations beneath small indents of different depths using EBSD tomography. Acta Mater 57:559–569

Devincre B (1995) Three dimensional stress field expressions for straight dislocation segments. Solid State Commun 93:875–878

Durst K, Backes B, Göken M (2005) Indentation size effect in metallic materials: correcting for the size of the plastic zone. Scripta Mater 52(11):1093–1097

Eidel B, Gruttmann F (2007) Squaring the circle—A curious phenomenon of fcc single crystals in spherical microindentation. Comput Mater Sci 39(1):172–178

Engels P, Ma A, Hartmaier A (2012) Continuum simulation of the evolution of dislocation densities during nanoindentation. Int J Plast 38:159–169

Evans AG, Hutchinson JW (2009) A critical assessment of theories of strain gradient plasticity. Acta Mater 57:1675–1688

Fleck NA, Muller GM, Ashby MF, Hutchinson JW (1994) Strain gradient plasticity: theory and experiment. Acta Metall Mater 42(2):475–487

Forest S, Barbe F, Cailletaud G (2000) Cosserat modelling of size effects in the mechanical behaviour of polycrystals and multiphase materials. Int J Solids Struct 37:7105–7126

Gao H, Huang Y, Nix WD, Hutchinson JW (1999) Mechanism-based strain gradient plasticity—I. theory. J Mech Phys Solid 47(6):1239–1263

Geers MGD, Brekelmans WAM, Bayley CJ (2007) Second-order crystal plasticity: internal stress effects and cyclic loading. Modell Simul Mater Sci Eng 15:133–145

Gottstein G (2004) Physical Foundations of Materials Science. Springer, Berlin

Gurtin ME (2002) A gradient theory of single-crystal viscoplasticity that accounts for geometrically necessary dislocations. J Mech Phys Solids 50:5–32

Han CS, Ma A, Roters F, Raabe D (2007) A Finite Element approach with patch projection for strain gradient plasticity formulations. Int J Plast 23:690–710

Hayashi I, Sato M, Kuroda M (2011) Strain hardening in bent copper foils. J Mech Phys Solids 59:1731–1751

Hirth JP, Lothe J (1982) Theory of dislocations. Wiley, London

Hua J, Hartmaier A (2010) Determining Burgers vectors and geometrically necessary dislocation densities from atomistic data. Modell Simul Mater Sci Eng 18(045007):1–11

Huang Y, Zhang F, Hwang KC, Nix WD, Pharr GM, Feng G (2006) A model of size effects in nano-indentation. J Mech Phys Solid 54:1668–1686

Jones DR, Schonlau M, Welch WJ (1998) Efficient global optimization of expensive black-box functions. J Global Optim 13(4):455–492

Kalidindi SR, Bronkhorst CA, Anand L (1992) Crystallographic texture evolution in bulk deformation processing of FCC metals. J Mech Phys Solid 40(3):537–569

Kelchner CL, Plimpton SJ, Hamilton JC (1998) Dislocation nucleation and defect structure during surface indentation. Phys Rev B 58:11085–11088

Knap J, Ortiz M (2003) Effect of indenter-radius size on Au(001) nanoindentation. Phys Rev Lett 90:226102

Kraft O, Gruber PA, Moenig R, Weygand D (2010) Plasticity in confined dimensions. In: Clarke DR, Ruhle M, Zok F (eds) Annual review of materials research, vol 40, Annual Reviews, pp 293–317

Lee EH (1969) Elastic-Plastic Deformation at Finite Strains. J Appl Mech 36(1):1–6

Liang HY, Woo CH, Huang H, Ngan AHW, Yu TX (2003) Dislocation nucleation in the initial stage during nanoindentation. Phil Mag 83(31–34):3609–3622

Lloyd DJ (1994) Particle reinforced aluminium and magnesium matrix composites. Int Mater Rev 39(1):1–23

Lodes MA, Hartmaier A, Göken M, Durst K (2011) Influence of dislocation density on the pop-in behavior and indentation size effect in CaF_2 single crystals: experiments and molecular dynamics simulations. Acta Mater 59:4264–4273

Ma A, Roters F, Raabe D (2006) A dislocation density based constitutive model for crystal plasticity FEM including geometrically necessary dislocations. Acta Mater 54:2169–2179

Ma Q, Clarke DR (1995) Size dependent hardness of silver single crystals. J Mater Res 10(04):853–863

McElhaney KW, Vlassak JJ, Nix WD (1998) Determination of indenter tip geometry and indentation contact area for depth-sensing indentation experiments. J Mater Res 13(05):1300–1306

Mecking H, Kocks UF (1981) Kinetics of flow and strain hardening. Acta Metall 29:1865–1875

Miller R, Rodney D (2008) On the nonlocal nature of dislocation nucleation during nanoindentation. J Mech Phys Solids 56:1203–1223

Minor A, Asif SAS, Shan Z, Stach E, Cyrankowski E, Wyrobek T, Warren O (2006) A new view of the onset of plasticity during the nanoindentation of aluminium. Nat Mater 5:697–702

Mishin Y, Mehl MJ, Papaconstantopoulos DA, Voter AF, Kress JD (2001) Structural stability and lattice defects in copper: Ab initio, tight-binding, and embedded-atom calculations. Phys Rev B 63(22):224106

Morris JR, Bei H, Pharr GM, George EP (2011) Size effects and stochastic behavior of nanoindentation pop in. Phys Rev Lett 106(16):165502

Mughrabi H (2007) On the current understanding of strain gradient plasticity. Mater Sci Eng, A 387–389:209–213

Nix W, Gao H (1998) Indentation size effects in crystalline materials: a law for strain gradient plasticity. J Mech Phys Solid 46:411–425

Nix WD (1989) Mechanical properties of thin films. Metall Trans A 20:2217–2245

Nye JF (1953) Some geometrical relations in dislocated crystals. Acta Metall 1:153–162

Ohashi T (2005) Crystal plasticity analysis of dislocation emission from micro voids. Int J Plast 21:2071–2088

Oliver WC, Pharr GM (1992) An improved technique for determining hardness and elastic modulus using load and displacement sensing indentation experiments. J Mater Res 7(6):1564–1583

Poole WJ, Ashby MF, Fleck NA (1996) Micro-hardness of annealed and work-hardened copper polycrystals. Scripta Mater 34(4):559–564

Roters F, Eisenlohr P, Hantcherli L, Tjahjanto DD, Bieler TR, Raabe D (2010) Overview of constitutive laws, kinematics, homogenization and multiscale methods in crystal plasticity finite-element modeling: theory, experiments, applications. Acta Mater 58:1152–1211

Roters F, Raabe D, Gottstein G (2000) Work hardening in heterogeneous alloys-a microstructural approach based on three internal state variables. Acta Mater 48(17):4181–4189

Roters F, Wang Y, Kuo JC, Raabe D (2004) Comparison of single crystal simple shear deformation experiments with crystal plasticity finite element simulations. Adv Eng Mater 6(8):653–656

Schmaling B, Hartmaier A (2012) Determination of plastic material properties by analysis of residual imprint geometry of indentation. J Mater Res 27:2167–2177

Shim S, Bei H, Pharr GM, George EP (2008) A different type of indentation size effect. Scripta Mater 59(10):1095–1098

Shrotriya P, Allameh SM, Lou J, Buchheit T, Soboyejo WO (2003) On the measurement of the plasticity length scale parameter in LIGA nickel foils. Mech Mater 35(3–6):233–243

Stadler J, Mikulla R, Trebin HR (1997) IMD: a software package for molecular dynamics studies on parallel computers. Int J Mod Phys C 8(5):1131–1140

Stelmashenko NA, Walls MG, Brown LM, Milman YV (1993) Microindentations on W and Mo oriented single crystals: an STM study. Acta Metall Mater 41(10):2855–2865

Stölken JS, Evans AG (1998) A microbend test method for measuring the plasticity length scale. Acta Mater 46(14):5109–5115

Subramaniyan AK, Sun CT (2008) Continuum interpretation of virial stress in molecular simulations. International J Solids Struct 45(14-15):4340–4346

Suresh S, Nieh TG, Choi BW (1999) Nano-indentation of copper thin films on silicon substrates. Scripta Mater 41:951–957

Suzuki K, Matsuki Y, Masaki K, Sato M, Kuroda M (2009) Tensile and microbend tests of pure aluminum foils with different thicknesses. Mater Sci Eng, A 513–514:77–82

Swadener JG, George EP, Pharr GM (2002) The correlation of the indentation size effect measured with indenters of various shape. J Mech Phys Solids 50:681–694

Tymiak NI, Kramer DE, Bahr DF, Wyrobek TJ, Gerberich WW (2001) Plastic strain and strain gradients at very small indentation depths. Acta Mater 49(6):1021–1034

Uchic MD, Dimiduk DM, Florando JN, Nix WD (2004) Sample dimensions influence strength and crystal plasticity. Science 305:986–989

Vliet KJV, Li J, Zhu T, Yip S, Suresh S (2003) Quantifying the early stages of plasticity through nanoscale experiments and simulations. Phys Rev B 76:104105

Voronoi G (1908) Nouvelles applications des paramètres continus à la théorie des formes quadratiques. Deuxième mémoire. Recherches sur les parallélloèdres primitifs. Journal für die reine und angewandte Mathematik 1908(134):198–287

Weertman J (2002) Anomalous work hardening, non-redundant screw dislocations in a circular bar deformed in torsion, and non-redundant edge dislocations in a bent foil. Acta Mater 50:673–689

Wit RD (1967) Some Relations for Straight Dislocations. Phys Status Solidi (b) 20:567–573

Zagrebelny AV, Lilleodden ET, Gerberich WW, Carter CB (1999) Indentation of silicate-glass films on Al2O3 substrates. J Am Ceram Soc 82(7):1803–1808

Zambaldi C, Yang Y, Bieler TR, Raabe D (2012) Orientation informed nanoindentation of alpha-titanium: indentation pileup in hexagonal metals deforming by prismatic slip. J Mater Res 27(01):356–367

Zhu T, Li J, Van Vliet KJ, Ogata S, Yip S, Suresh S (2004) Predictive modeling of nanoindentation-induced homogeneous dislocation nucleation in copper. J Mech Phys Solids 52(3):691–724

Ziegenhain G, Urbassek H, Hartmaier A (2010) Influence of crystal anisotropy on elastic deformation and onset of plasticity in nanoindentation: a simulational study. J Appl Phys 107:061807

Zimmerman JA, Kelchner CL, Klein PA, Hamilton JC, Foiles SM (2001) Surface step effects on nanoindentation. Phys Rev Lett 87:165507

Zimmerman JA, Webb Iii EB, Hoyt JJ, Jones RE, Klein PA, Bammann DJ (2004) Calculation of stress in atomistic simulation. Modell Simul Mater Sci Eng 12(4):S319–S319

Elastic–Plastic Behaviors of Vertically Aligned Carbon Nanotube Arrays by Large-Displacement Indentation Test

Y. Charles Lu, Johnson Joseph, Qiuhong Zhang, Feng Du and Liming Dai

Abstract This chapter describes the large-displacement indentation test method for examining elastic–plastic behaviors of vertically aligned carbon nanotube arrays (VA-CNTs). The principle of this test is explained by using a cavity expansion model. The experiments have been performed on VA-CNTs synthesized by the chemical vapor deposition (CVD) method. Under a cylindrical, flat indenter, the VA-CNTs exhibit two distinct deformation stages: a short, elastic deformation at small displacement and a plateau-like, plastic deformation at large displacement. The critical indentation stress, a measure of yield stress or collapsing stress of the VA-CNT arrays, has been obtained. The deformation mechanism of the VA-CNTs at large displacement is revealed with scanning electronic microscope (SEM) images of the deformed VA-CNTs and finite element simulations.

1 Introduction

Carbon nanotube (CNT) is a new class of material that is discovered during the late 1990s by Iijima (1991). In its simplest form carbon nanotube may be understood as a molecule with atoms connected by c–c bonds in a hexagonal ring structure pattern. An individual CNT may be visualized as having formed from a graphene sheet

Y. C. Lu (✉) · J. Joseph
Department of Mechanical Engineering, University of Kentucky,
Lexington, KY 40506, USA
e-mail: chlu@engr.uky.edu

Q. Zhang
University of Dayton Research Institute, University of Dayton, Dayton, OH 45469, USA

F. Du · L. Dai
Department of Macromolecular Science and Engineering, Case Western Reverse University,
Cleveland, OH 44106, USA

A. Tiwari (ed.), *Nanomechanical Analysis of High Performance Materials*,
Solid Mechanics and Its Applications 203, DOI: 10.1007/978-94-007-6919-9_16,
© Springer Science+Business Media Dordrecht 2014

with atoms that are interconnected with bonds in hexagonal chains and upon rolling this sheet forms a tube structure. Based on their directions of rolling vectors, the CNTs are classified into three types: (1) arm chair, (2) zigzag, and (3) chiral. Like diamond, carbon nanotube is also allotrope of carbon and thus has exceptional mechanical, electrical, and thermal properties. The CNT is believed to have a Young's modulus up to 1 TPa and can have 100 times strength of steel at one-sixth the weight (Treacy et al. 1996; Krishnan et al. 1998; Wong et al. 1997). The current carrying capacity of the CNT can be as high as 4×10^9 A/cm^2, which is 1,000 times greater than that of copper (Dresselhaus et al. 2001; Hong and Myung 2007). The electron mobility of the CNT is 100,000 cm^2/V/s, as compared to silicon having 1,400 cm^2/V/s, thus has superior electrical conductivity (Hone et al. 1999; Yao et al. 2000). The CNT also has very high thermal conductivity: 3,500 W/m/K, as against 385 W/m/K for copper (Biercuk et al. 2002; Pop et al. 2006).

From practical application point of view, it is highly desirable to produce carbon nanotubes in large scales. This has resulted in a new form of carbon nanotube material: the vertically aligned carbon nanotube arrays (VA-CNTs). VA-CNTs was probably first grown by Terrones et al. (1997). Since then, various techniques have been used to synthesize these materials. Among them, the chemical vapor deposition, CVD, appears to be the most promising method for producing large-scale VA-CNT arrays (Ishigami et al. 2008; Bajpai et al. 2004; Chen et al. 2010). The VA-CNTs so grown can have the size as large as several square centimeters and the height as tall as several millimeters. Since the VA-CNTs can be readily integrated (grown) onto various substrates and devices, they have found a wide range of applications in areas such as the electrical interconnects (Kreupl et al. 2002), thermal interfaces (Cola 2009), energy dissipation devices (Liu et al. 2008), and microelectronic devices (Fan et al. 1999). VA-CNTs can also be grown on non-planar substrates, i.e., the rounded carbon fibers. VA-CNTs on carbon fibers have had significant potentials in aerospace and space applications. They have added multi-functionality to traditional composites (Baur and Silverman 2007; Ci et al. 2008; Zhang et al. 2009), improved the fiber-matrix interface strength (Sager et al. 2009; Patton et al. 2009), and used as flow or pressure sensors on micro air vehicles (MAVs) (Zhang et al. 2010).

The mechanical properties and deformation behaviors of the VA-CNTs have been investigated lately, mostly through the conventional nanoindentation test (Qi et al. 2003; Mesarovic et al. 2007; McCarter et al. 2006; Pathak et al. 2009; Patton et al. 2009). The conventional nanoindentation test is a technique for measuring mechanical properties of materials and structures in small dimensions. The depth of the indentation is typically small (a few nanometers or microns) and therefore the test is primarily used for measuring elastic properties of materials and structures. To measure the elastic response of the VA-CNTs, an indenter of either three-face pyramidal shape (Berkovich indenter) or parabolic shape (spherical indenter) has been used to compress the specimen and then withdrawn from it. The indentation load-depth curves are obtained and then analyzed following the standard Oliver-Pharr method (Oliver and Pharr 1992). The modulus and hardness of the VA-CNT arrays have been obtained.

This chapter reports the use of large-displacement indentation test to measure the elastic–plastic properties of the VA-CNT arrays. With small displacement actuators, most conventional nanoindenters cannot be easily used to conduct tests to large displacements. Conventional nanoindenters have also been limited to the examination of load–displacement responses rather than the real-time observations of deformation during indentation. The present large displacement experiments are conducted with an in situ nanoindenter equipped inside the chamber of a scanning electronic microscope (SEM). The technique can thus reveal both quantitative information (load–displacement) and phenomenological behaviors of the CNT arrays. Cylindrical, flat tip geometry is chosen for the indenter since the stress analysis under a tip of this form has been well established (Sneddon 1946; Barquins and Maugis 1982). Compared to indenters of three-face pyramidal and parabolic shapes, the contact area of a cylindrical flat indenter does not change with displacement, and the extent of the stress field scales with the diameter of the indenter.

2 Principle of Large-Displacement Indentation Test

The large-displacement indentation test can be schematically described as shown in Fig. 1. As an indenter is pushed into the material, a deformation zone is developed surrounding the indenter. The overall process resembles to the opening of a cavity in a solid and the stress required to open such a cavity can be estimated. For blunt indenters (as opposed to sharp indenters), the cavity is typically assumed to be in cylindrical shape.

The cavity starts with an initial radius a_o, and opens to a final radius a, equal to the radius of the indenter. The opening of such a cavity also expands a surrounding plastic zone from an initial radius r_0 to a final radius c. The radical strain in the cylindrical polar coordinates (z, r, θ) is (Eq. 1)

$$\varepsilon_r = \ln\left(\frac{r}{r_o}\right) = \frac{1}{2} \cdot \ln\left(1 - \frac{a^2 - a_0^2}{r^2}\right) \tag{1}$$

Let σ_r, σ_θ be the radial and tangential stresses in the cylindrical polar coordinates. For a general elastic–plastic solid, the constitutive relationship between true stress σ and true strain ε in rectangular coordinates can be described by Eq. 2:

$$\sigma/\sigma_y = f(\varepsilon) \tag{2}$$

where σ_y is the yield stress of the material. So, the stress–strain relation becomes (Eq. 3)

$$\frac{\sigma_\theta - \sigma_r}{\sigma_y} = \frac{2}{\sqrt{3}} \cdot f\left(\frac{2}{\sqrt{3}} |\varepsilon_r|\right) \tag{3}$$

Integrating the condition of the equilibrium, namely (Eq. 4)

$$\frac{d\sigma_r}{dr} = \frac{\sigma_\theta - \sigma_r}{r} \tag{4}$$

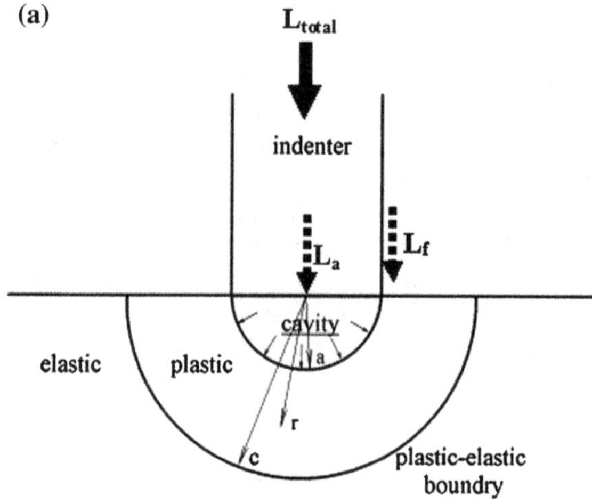

Fig. 1 Schematic diagrams showing the large-displacement indentation test. **a** Front view, **b** Top view

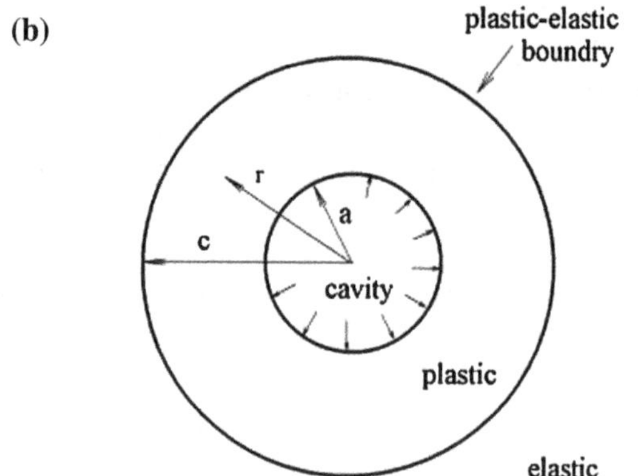

throughout the plastic region from a to r, we obtain for the pressure P on the boundary of the hollow cylinder (the value of $(-\sigma_r)$ at that point) (Eq. 5)

$$\frac{P_m}{\sigma_y} = \frac{2}{\sqrt{3}} \cdot \int_a^r f\left\{-\frac{1}{\sqrt{3}}\ln\left(1 - \frac{a^2 - a_0^2}{r^2}\right)\right\} \frac{dr}{r} \tag{5}$$

The above equation indicates that there exists a cavitation limit P_m as $a/a_0 \rightarrow \infty$. An example solution of Eq. 5 is shown in Fig. 2, as calculated by using a typical yield strain of $\varepsilon_y = 0.1$ and a typical strain harden coefficient n = 1.2 for a general elastic–plastic solid. It is seen that the ratio of P_m/σ_y reaches a constant once the indenter is fully compressed into the material. The magnitude of the ratio is bounded between 1 and 3, varying with the type of the materials.

Clearly, the above equation is similar to the broadly applicable empirical relationship suggested by Tabor (1996) (Eq. 6):

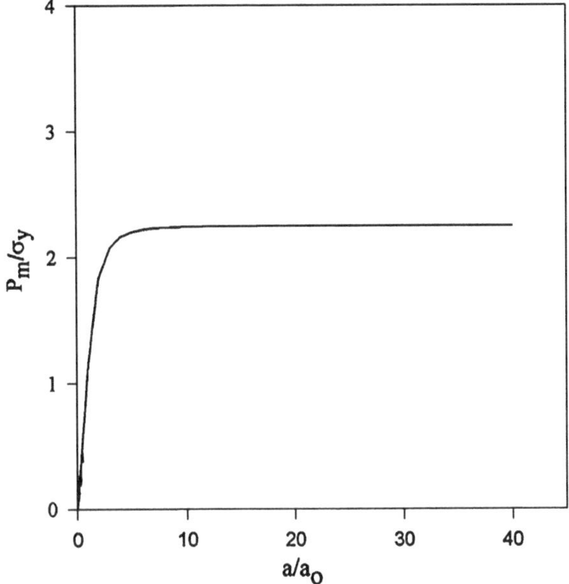

Fig. 2 Prediction of the critical indentation stress from the cavity model. The calculation is obtained by using a typical yield strain of $\varepsilon_y = 0.1$ and a typical strain harden coefficient n = 1.2 for a general elastic–plastic solid

$$\frac{P_m}{\sigma_y} = C \tag{6}$$

where C is called the constraint factor.

Both theoretical analysis (Eq. 5) and empirical analysis (Eq. 6) show that the critical indentation stress (P_m) is proportional to the uniaxial yield stress (σ_y) for an elastic–plastic material. In general, the critical indentation stress beneath an indenter is greater than the uniaxial compressive yield stress of the material because of the confining pressure generated by the surrounding elastically strained material in the indentation stress field. For ductile metals, a value of $C \approx 3$ is generally considered to be appropriate (Tabor 1996; Johnson 1985). For soft polymers, the value of C becomes smaller (Wright et al. 1992; Lu and Shinozaki 1998, 2008; Lu et al. 2008). For foam-like materials, the value of C generally approaches to unity, i.e., $C \approx 1$ (Wilsea et al. 1975; Olurin et al. 2000; Flores-Johnson and Li 2010).

The critical indentation stress (P_m) can be determined experimentally through the large-displacement indentation test. As illustrated in Fig. 1, when a cylindrical indenter of radius a is pressed onto a specimen, the total load (L_{total}) applied to the indenter is (Eq. 7)

$$L_{total} = L_a + L_f \tag{7}$$

where L_a is the axial load acting on the indenter end face and L_f the frictional load acting on the indenter side wall. The mean indentation pressure (P_m) acting on the indenter end is simply expressed as Eq. 8a

$$P_m = L_a / \pi a^2 \tag{8a}$$

The frictional load (L_f) on the indenter side wall is defined by Eq. 8b

$$L_f = 2\pi a h_c \tau \tag{8b}$$

where τ is the frictional shear stress and h_c the contact depth. Assume that the frictional stress is constant on the indenter wall, then the frictional load (L_f) should increase linearly with indentation depth, since the lateral surface area in contact with the material ($2\pi a h_c$) increases almost linearly.

Substituting Eqs. 8a and 8b into Eq. 7 yields Eq. 9:

$$\frac{L_{total}}{\pi a^2} = P_m + \frac{2d}{a}\tau \tag{9}$$

The above equation shows that there exist a linear relationship between indentation stress and normalized displacement at large displacements. The critical indentation stress, P_m, can be determined simply by extrapolating the indentation stress-displacement curve back to zero displacement ($d = 0$), where the frictional load (L_f) vanishes.

The large-displacement indentation test has been a proved method for measuring the plastic properties of materials and structures in small volumes, including metals (Riccardi and Montanari 2004; Lu et al. 2008), isotropic polymers (Wright et al. 1992; Lu and Shinozaki 1998, 2008; Lu et al. 2008), and oriented polymers (Lo et al. 2005; Shinozaki et al. 2008).

3 VA-CNTs Preparations

The present VA-CNTs were synthesized by low pressure chemical vapor deposition of acetylene on planar substrates (SiO_2/Si wafers). A 10 nm thick Al layer was first coated on the wafers before the deposition of 3 nm Fe film in order to enhance the attachment of grown nanotubes on the silicon substrates. The catalyst coated substrate was then inserted into the quartz tube furnace and remained at 750 °C in air for 10 min, followed by pumping the furnace chamber to a pressure less than 10 m Torr. Thereafter, the growth of the CNT arrays was achieved by flowing a mixture gases of 48 % Ar, 28 % H^2, 24 % C^2H^2 at 750 °C under 10 ~ 100 Torr for 10–20 min.

The microstructure of the carbon nanotube samples have been examined using a field emission Hatachi S800 scanning electron microscope, as shown in Fig. 3. Results show that the VA-CNTs are multiwalled (2–3 walls) carbon nanotubes and have a narrow uniform diameter distribution between 10 and 20 nm. By counting the numbers of the carbon nanotubes on the substrate, the areal density of the VA-CNT arrays can be estimated as: $\rho = 10^{10} \sim 10^{11}$ tubes/cm^2. The lengths of the VA-CNTs are in the range of a few hundred to several thousand microns, as achieved by controlling the deposition time and pressure. It is found that the VA-CNT arrays can adhere strongly to the growth silicon substrate, with a measured pull force up to 150 N/cm^2 (Qu et al. 2008).

Fig. 3 SEM images showing the morphology of the side surface of the vertically-aligned carbon nanotube arrays (t ≈ 210 μm). The order of magnification increases from (**a**) to (**c**)

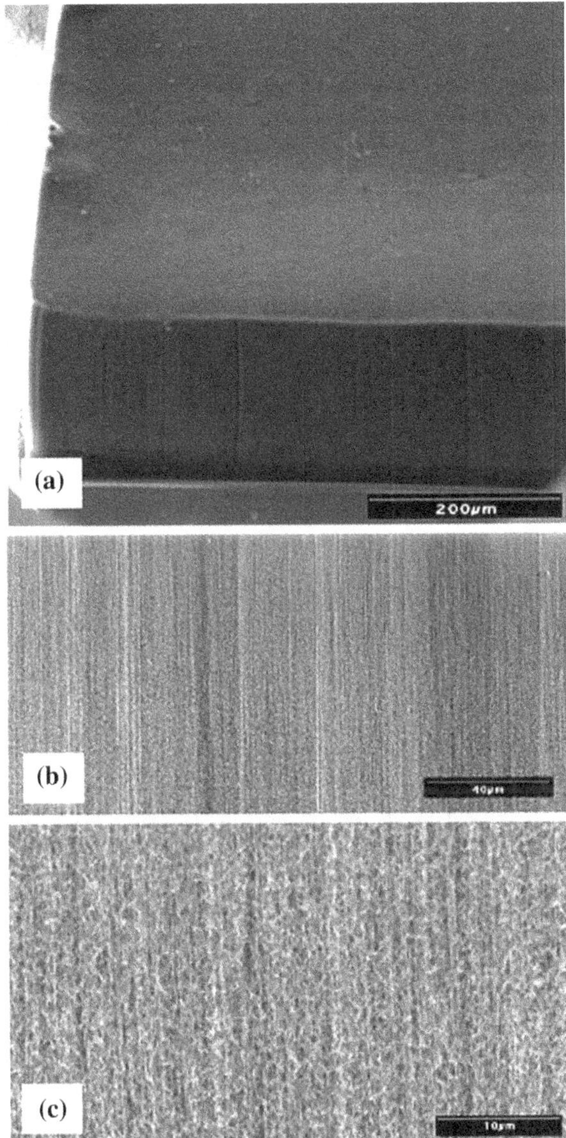

4 Large-Displacement Indentation Test on VA-CNTs

The large-displacement indentation tests were conducted with a custom designed in situ nanoindenter equipped inside the SEM (FEI Sirion). The indenter used was a 100 μm diameter flat-faced cylinder, with a polished contact face. The cylindrical indenter was attached to a strain-gage based load cell, which was connected in series to a piezoelectric actuator. The piezoelectric actuator provided displacement control with sub-nanometer resolution. Resultant forces were measured through the load cell. The VA-CNT array samples were positioned on a piezoelectric positioning stage, which provided x-y-z movements with nanometer-scale resolution and with zero back-lash. The entire nanoindenter device is measured as 50 mm (width) × 50 mm (height) × 150 mm (length) and thus fits well inside the SEM chamber without disturbing the SEM's function. Instrumentation control and data acquisition were achieved by using the Labview software from National Instrument (NI).

During the test, specimens were incrementally loaded at a rate of 100 nm/sec and high resolution SEM images were acquired between displacement intervals. Load and displacement data were recorded and used to compute the indentation stress and strain. The scan images can be analyzed individually and further stitched together to produce videos and synchronized to correlate load and displacement data to the observed deformation phenomena.

5 Results and Discussion

5.1 Indentation Stress-Displacement Curves of CNT Arrays

The large displacement deformation of the VA-CNTs has been recently examined through the axial, micro-compression test (Cao et al. 2005; Hutchens et al. 2010; Maschmann et al. 2011). The CNT arrays are found to behave as an open-cell foam-like material. The stress–strain curve displays three distinct stages: a short elastic region, followed by a prolonged plateau region, and finally a densification region. Under compression, the CNT arrays folded themselves in wavelike pattern. The wavelike folding has been observed to initiate mostly at the bottom of the CNT arrays and then to propagate towards the top.

Micro-compression test often requires special sample preparation technique. For example, the photolithography method has been used to prepare small, cylindrical CNT pillars to be fitted at the testing platens (Hutchens et al. 2010). Here we examine the deformation behaviors of the VA-CNT arrays by conducting large-displacement indentation test directly on the as-grown samples. For this type of test, the size (heights) of the specimens has to be chosen carefully to avoid the influence of rigid substrates from underneath. The problem of a flat punch indenting an isotropic elastic half-space has been solved by Sneddon (Sneddon 1946; Barquins and Maugis 1982). It is found that the stress field beneath a cylindrical, flat indenter is typically confined within one indenter diameter. Given the size of current indenter ($2a = 100$ μm), VA-CNT specimens with heights greater than 100 μm were used for the experiments.

Figures 4 and 5 show the indentation stress-displacement curves of two vertically aligned carbon nanotube arrays, with height equal to approximately 210 and

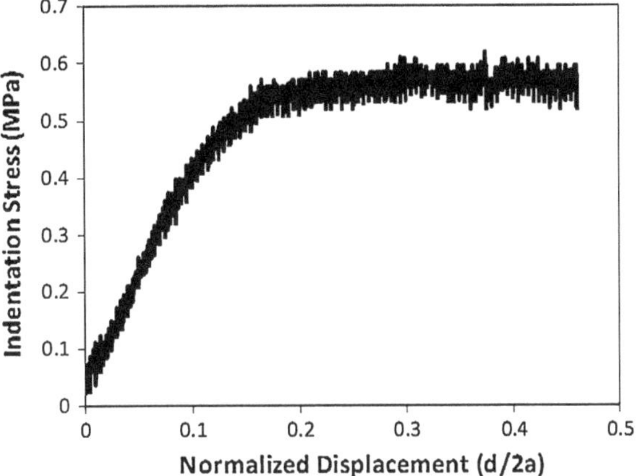

Fig. 4 Indentation stress-displacement curves of a vertically aligned carbon nanotube arrays (height $\approx 210\ \mu$m) with cylindrical, flat-faced indenter

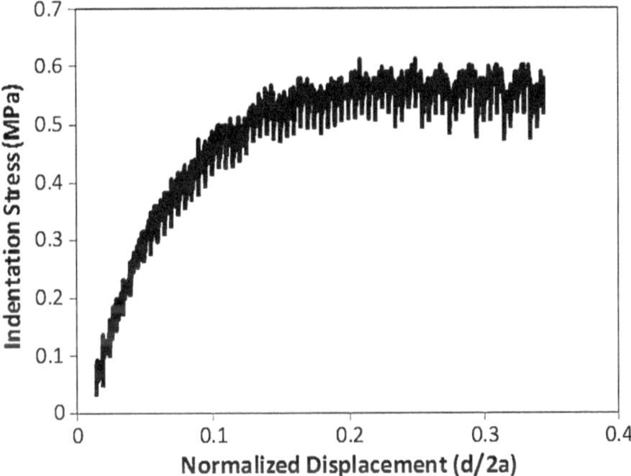

Fig. 5 Indentation stress-displacement curves of a vertically aligned carbon nanotube arrays (height $\approx 1{,}100\ \mu$m) with cylindrical, flat-faced indenter

1,100 μm, respectively. Results reveal that the material initially deforms elastically with the applied load on the indenter, and yields at some point as the applied load is increased. The plastic deformation field and consequently the stress field progressively change with displacement, until some steady state is achieved. The indentation of elastic–plastic solids has gained considerable attention recently, with the purpose of determining the plastic characteristics of the materials such as yield strength, work hardening rate, etc. Most of the work involves the uses of indenters of parabolic shapes, i.e., spherical and conical indenters (Mesarovic and Fleck 1987; Park and Pharr 2004). The present test has chosen a cylindrical, flat indenter. The chief advantage of this type of indenter is that the contact area remains constant during indentation; therefore the applied stress measured by the indenter at the steady state is constant. This allows the measurement of steady state deformation under the indenter, as indicated by the linear stress-displacement response at large displacements (Figs. 4 and 5).

The plateau region indicates the plastic collapses of carbon nanotubes beneath the indenter face. Such collapse allows the strain increase while the stress stays approximately constant. A series of "load-drops" in the plateau regions is observed, which corresponds to the folding of additional carbon nanotubes. If further penetration is permitted until all folding is completed, a third region would appear: the densification region. In that region, the folding of all nanotubes under the indenter face has been completed and the compression of the folded/collapsed materials has started. As a result, the stress would start to rise sharply. Alternatively, the densification response can be observed by indenting a shorter specimen. Overall, the stress-displacement response of the VA-CNT arrays is identical to those reported on open-cell, low-density foams (Wilsea et al. 1975; Olurin et al. 2000; Flores-Johnson and Li 2010).

5.2 Critical Indentation Stress (P_m) of CNT Arrays

Following Eq. 9, the critical indentation stress, P_m, is determined by extrapolating the large-strain indentation stress-displacement curve back to zero displacement ($d = 0$), where P_m is the intercept. The magnitude of P_m so obtained for the present CNT arrays is approximately 0.56 MPa, from Figs. 4 and 5.

The same VA-CNT arrays have been tested under uniaxial compression using the same in situ testing apparatus and the yield strength has been determined: $\sigma_y \approx 0.50$ MPa (Lu et al. 2012). A comparison with the indentation test shows that the critical value of the indentation stress (P_m) is very close to the uniaxial yield stress (σ_y) for the VA-CNT arrays: $P_m/\sigma_y \approx 1.12$. This is because the foam-like CNT arrays has a nearly zero plastic Poisson's ratio (the ratio of transverse to longitudinal plastic strain under compression). Therefore, the large indentation has resulted in very little lateral spreading of the CNT fibers under the indenter and the constraint factor becomes unity.

For the present CNT arrays, the slopes of stress-displacement at large displacements are almost zero (Figs. 4 and 5), indicating that the friction shear stress (τ) acting on the indenter wall due to the elastic compression from surrounding nanotubes is negligible. Therefore, the interfacial friction between the CNT and the indenter side-wall is very small, which is constant with the finding reported by Tu et al. (2003, 2004).

5.3 Large Displacement Deformation Phenomenology

The indentation test is performed on an in situ nanoindenter that is equipped inside the chamber of a SEM, and thus allows for real-time observation and video recording of the deformation process while the CNT arrays are compressed. To view of the deformation process, the in situ nanoindentation was performed at the edge of the CNT array specimen. Figure 6 shows the large displacement phenomenology of a CNT arrays ($t \approx 1,100$ μm) at various indentation stages.

Fig. 6 SEM images showing the development of plastic deformation in the vertically aligned carbon nanotube arrays (height $\approx 1,100$ μm) under a cylindrical flat indenter

The early stage of penetration is dominated by the elastic deformation, as reveled by larger slopes in the load–displacement curves (Figs. 4, 5). Larger slopes

indicate that the CNT materials have greater stiffness initially. Further penetration of the indenter results in the plastic collapse of the carbon nanotubes beneath the indenter head (Fig. 6). The measured stiffness thus decreases with increasing depth of indentation. Observations show that the plastic collapse of the nanotube arrays is limited in extent to the zone directly underneath the indenter face where the shear stress is large. The size of this collapsing zone is much smaller as compared to the larger, hemispherical shaped plastic zones occurred on dense, solid materials, such as polycarbonate (Wright et al. 1992) and polyethylene (Lu and Shinozaki 1998). The nanotubes outside the collapsing zone are seen to exhibit no fracture or tearing.

The series of load-drops in the stress-displacement curves are results of continuous collapsing of nanotubes as the indenter tip moves. The force required for crushing additional nanotubes is relatively small (because its volume is a small fraction of the material under load), so the measurement of the stress associated with the buckling movement are small. Therefore, the total stress at the large strain region has stayed relatively constant.

To better understand the deformation mechanism, the stress/strain fields of the VA-CNTs during indentation were analyzed by using the finite element method. Commercial nonlinear finite element (FE) code ABAQUS was used (ABAQUS 2010). The CNT specimen was modeled with second order, 8-node axisymmetric elements to handle the potential large distortion of the elements occurred during large indentation. The indenter was assumed undeformable and thus modeled with rigid surfaces. The contact between specimen and indenter was treated as frictionless. The CNT arrays were treated as open-cell, foam-like materials and the crushable foam plasticity model was used (Deshpande and Fleck 2000). For comparative purpose, the indentation process of a dense, solid polymer was also modeled. The solid polymer was treated as a power-law work-hardening, elastic–plastic solid, as described in detail elsewhere (Lu and Shinozaki 2008).

Figures 7 and 8 show the contours of the 1st principle stress (σ_1) for foam-like VA-CNTs and dense polymer, respectively. σ_1 is defined by

$$\sigma_1 = \frac{\sigma_r + \sigma_z}{2} + \left[\left(\frac{\sigma_r - \sigma_z}{2}\right)^2 + \tau_{rz}^2\right]^{1/2}$$ and σ_r, σ_z, and τ_{rz} are the radial, normal and shear stresses in the cylindrical polar coordinates. Figures 9 and 10 show the contours of the equivalent plastic strain (ε^{eq}) for foam-like VA-CNTs and dense polymer, respectively. ε^{eq} is defined by $\varepsilon^{eq} = \sqrt{\frac{2}{3}\left(\varepsilon_1^2 + \varepsilon_2^2 + \varepsilon_3^2\right)}$ and ε_1, ε_2, ε_3, are the principal strains. For $\varepsilon^{eq} > 0$, the material has plastically deformed.

It is observed that the stress field and deformation process of VA-CNT arrays under compression are distinctly different from those of solid polymers. For a dense, solid polymer, the distribution of the stress (σ_1) under the flat indenter is in a hemispherical shape. The size (elastic–plastic boundary) of the stress field approximates the diameter of the indenter (2a), as illustrated by the cavity model described earlier. In contrast, the stress field (σ_1) for the foam-like VA-CNT arrays under the flat indenter is much smaller. The stress is primarily concentrated right beneath the indenter face and does not get extended to far field.

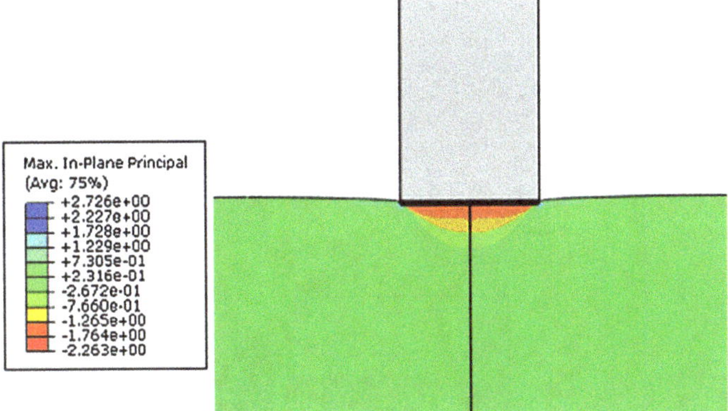

Fig. 7 Contour of 1st principle stress in the vertically aligned carbon nanotube arrays under a flat indenter. The material is treated as an open-cell foam-like material

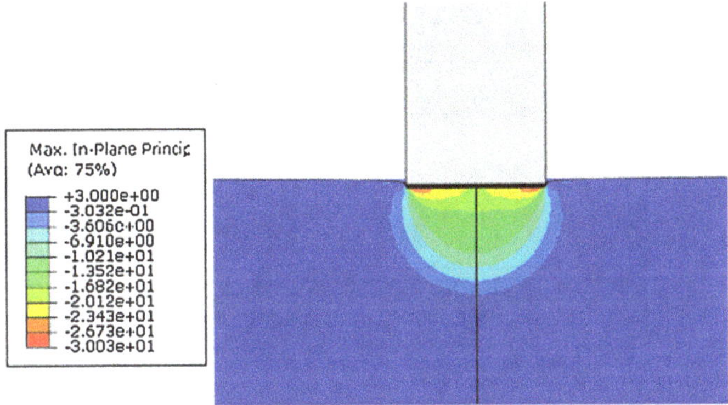

Fig. 8 Contour of 1st principle stress in a dense, solid material under a flat indenter. The material is treated as a power-law work hardening, elastic–plastic solid

The large-displacement indentation process can be understood by the progressive developments of equivalent plastic strain (ε^{eq}). For a dense, solid polymer, the initial inelastic deformation starts near the corners of the indenter. As the depth of indentation increases, the deformed zone increases in size. After a depth of approximately half to one indenter diameter, the deformation zone becomes fully developed surrounding the indenter and then remains relatively constant in size. The diameter of the deformed zone is about twice the diameter of the indenter, again in consistence with the cavitation model. It is also seen that a conical zone directly ahead of the flat indenter tip shows little deformation. However, for the foam-like VA-CNT arrays, the equivalent plastic strain (ε^{eq}) is distributed right beneath the indenter face. The shape of this plastic zone is much narrower, as opposed to a larger, hemispherical zone occurred in the dense, solid polymers. The

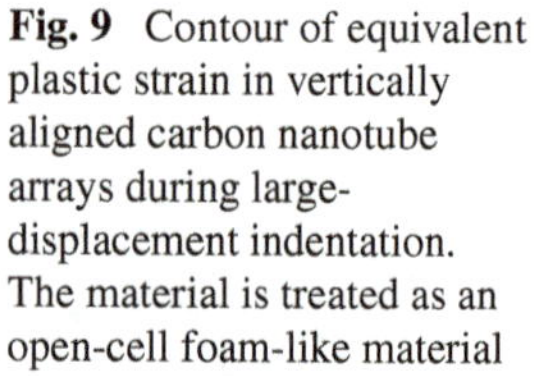

Fig. 9 Contour of equivalent plastic strain in vertically aligned carbon nanotube arrays during large-displacement indentation. The material is treated as an open-cell foam-like material

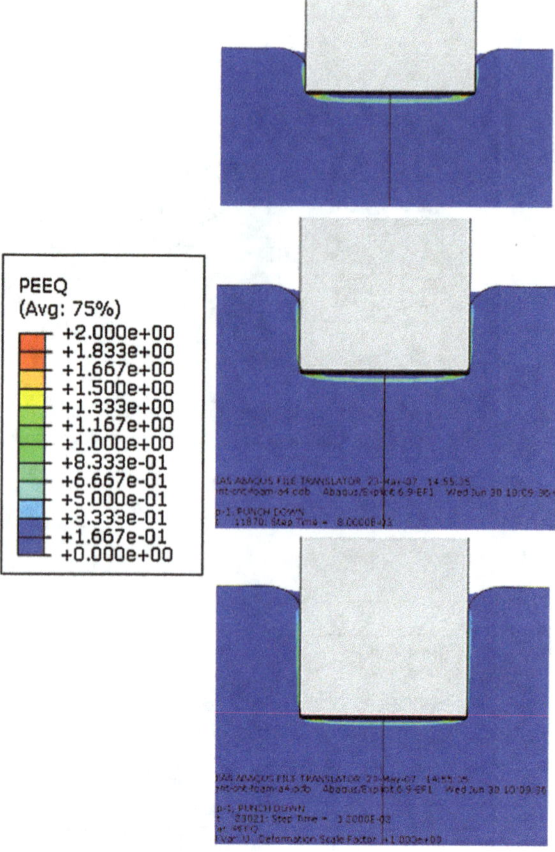

simulated deformation is consistent with the experimental observations (Fig. 6). Overall results confirm that the VA-CNTs behave like low-density foams and the crushable foam plasticity model is appropriate for modeling such materials.

6 Concluding Remarks

We have demonstrated that the large-displacement indentation test is an effective tool for measuring the elastic–plastic properties of the vertically aligned carbon nanotube arrays (VA-CNTs). The large displacement indentation under a cylindrical, flat indenter has been analyzed by modeling the opening of a cavity in a solid, from which a critical indentation stress (P_m) can be predicted. The experiments have been conducted on an in-situ indentation apparatus inside a SEM. Results show that the VA-CNTs exhibit a transient elastic deformation at small

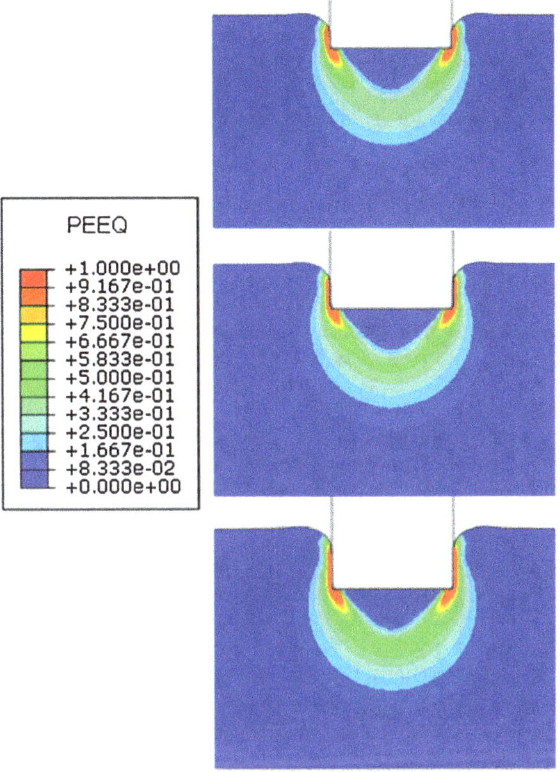

Fig. 10 Contour of equivalent plastic strain in a dense, solid polymer during large-displacement indentation. The material is treated as a power-law work-hardening, elastic-plastic solid

displacement and then steady sate plastic deformation at large displacement. The critical indentation stress (P_m) can be extrapolated from the indentation stress-displacement curves. The magnitude of P_m is a measure of the yield stress or collapsing stress of CNT arrays. The large-displacement indentation has also revealed the deformation mechanism of the VA-CNTs. Under the cylindrical, flat indenter, the nanotube cells collapsed plastically immediately beneath the indenter, a region of the highest stress/strain. Both experiment results and finite element simulations have shown that the sizes of stress/strain zones are much smaller in foam-like VA-CNTs, as opposed to much larger, hemispherical stress/strain zones observed in the dense, solid polymers.

References

ABAQUS (2010) ABAQUS Theory' manual. Simulia Inc., Pawtucket

Bajpai V, Dai L, Ohashi T (2004) Large-scale synthesis of perpendicularly aligned helical carbon nanotubes. J Am Chem Soc 126:5070–5071

Barquins M, Maugis D (1982) Adhesive contact of axisymmetric punches on an elastic half-space-the modified Hertz-Hubers stress tensor for contacting spheres. J Mec Theo Appl 1:331–357

Baur J, Silverman E (2007) Challenges and opportunities in multifunctional nanocomposite structures. MRS Bull 32:328–332

Biercuk MJ, Llaguno MC, Radosavljevic M, Hyun JK, Johnson AT (2002) Carbon nanotube composites for thermal management. Appl Phys Lett 80(15):2767–2769

Cao A, Dickrell PL, Sawyer WG, Ghasemi-Nejhad MN, Ajayan PM (2005) Super-compressible foamlike carbon nanotube films. Science 310(5752):1307–1313

Chen H, Roy A, Baek JB, Zhu L, Qu J, Dai L (2010) Controlled growth and modification of vertically-aligned carbon nanotubes for multifunctional applications. Mater Sci Eng Rep 70:63–91

Ci L, Suhr J, Pushparaj V, Zhang X, Ajayan PM (2008) Continuous carbon nanotube reinforced composites. Nano Lett 8(9):2762–2766

Cola BA (2009) Contact mechanics and thermal conductance of carbon nanotube array interfaces. Int J Heat Mass Transf 52(15–16):3490–3503

Deshpande VS, Fleck NA (2000) Isotropic constitutive model for metallic foams. J Mech Phys Solids 48:1253–1276

Dresselhaus MS, Dresselhaus G, Avouris P (2001) Carbon nanotubes. Springer, Berlin

Fan S, Chapline MG, Franklin NR, Tombler TW, Cassell AM, Dai H (1999) Self-oriented regular arrays of carbon nanotubes and their field emission properties. Science 283:512

Flores-Johnson EA, Li QM (2010) Indentation into polymeric foams. Int J Solids Struct 47:1987–1995

Hone J, Whitney M, Zettl A (1999) Thermal conductivity of single-walled carbon nanotubes. Synth Met 103(1–3):2498–2499

Hong S, Myung S (2007) Nanotube electronics: a flexible approach to mobility. Nat Nanotechnol 2(4):207–208

Hutchens SB, Hall LJ, Greer JR (2010) In situ mechanical testing reveals periodic buckle nucleation and propagation in carbon nanotube bundles. Adv Funct Mater 20(14):2338–2346

Iijima S (1991) Helical microtubules of graphitic carbon. Nature 354:56–58

Ishigami N, Ago H, Imamoto K, Tsuji M, Iakoubovskii K, Minami N (2008) Crystal plane dependent growth of aligned single-walled carbon nanotubes on sapphire. J Am Chem Soc 130(30):9918–9924

Johnson KL (1985) Contact mechanics. Cambridge University Press, Cambridge

Kreupl FGAP, Duesberg GS, Steinhogl W, Liebau M, Unger E, Honlein W (2002) Carbon nanotubes in interconnect applications. Microelectron Eng 64:399–408

Krishnan A, Dujardin E, Ebbesen TW, Yianilos PN, Treacy MMJ (1998) Young's modulus of single-walled nanotubes. Phys Rev B B(58):14013–14019

Liu Y, Qian WZ, Zhang Q, Cao AY, Li ZF, Zhou WP, Ma Y, Wei F (2008) Hierarchical agglomerates of carbon nanotubes as high-pressure cushions. Nano Lett 8:1323

Lo JCW, Lu YC, Shinozaki DM (2005) Kink band formation during microindentation of oriented polyethylene. Mater Sci Eng A 396(15):77–86

Lu YC, Shinozaki DM (1998) Deep penetration microindentation testing of high density polyethylene. Mater Sci Eng A 249:134–144

Lu YC, Shinozaki DM (2008) Characterization and modeling of large displacement micro-/nano-indentation of polymeric solids. ASME J Eng Mater Technol 130:041001

Lu YC, Kurapati S, Yang F (2008) Finite element analysis of cylindrical indentation for determining plastic properties of materials in small volumes. J Phys D Appl Phys 41:115415

Lu YC, Joseph J, Zhang Q, Dai L, Baur J (2012) Large-displacement indentation of vertically aligned carbon nanotube arrays. Exp Mech 52(9):1551–1554

Maschmann MR, Zhang Q, Wheeler R, Du F, Dai L, Baur J (2011) In situ SEM observation of column-like and foam-like CNT array nanoindentation. ACS Appl Mater Interfaces 3:648–653

McCarter CM, Richards RF, Mesarovic SDJ, Richards CD, Bahr DF, McClain D, Jiao J (2006) Mechanical compliance of photolithographically defined vertically aligned carbon nanotube turf. J Mater Sci 41:7872–7878

Mesarovic SD, Fleck NA (1987) Spherical indentation of elastic-plastic solids. Proc R Soc Lond 455:2707–2728

Mesarovic SD, McCarter CM, Bahr DF, Radhakrishnan H, Richards RF, Richards CD, McClain D, Jiao J (2007) Mechanical behavior of a carbon nanotube turf. Scripta Mater 56:157–160

Oliver WC, Pharr GM (1992) An improved technique for determining hardness and elastic modulus using load and displacement sensing indentation experiments. J Mater Res 7:1564

Olurin OB, Fleck NA, Ashby MF (2000) Indentation resistance of an aluminum foam. Script Mater 43:983–989

Park YJ, Pharr GM (2004) Nanoindentation with spherical indenters: finite element studies of deformation in the elastic-plastic transition regime. Thin Solid Films 447–448:246–250

Pathak S, Cambaz ZG, Kalidindi SR, Swadener JG, Gogotsi Y (2009) Viscoelasticity and high buckling stress of dense carbon nanotube brushes. Carbon 47(8):1969–1976

Patton ST, Zhang Q, Qu L, Dai L, Voevodin AA, Baur J (2009) Electromechanical characterization of carbon nanotube grown on carbon fibers. J Appl Phys 106:104313

Pop E, Mann D, Wang Q, Goodson K, Dai HJ (2006) Thermal conductance of an individual single-wall carbon nanotube above room temperature. Nano Lett 6:96–100

Qi HJ, Teo KBK, Lau KKS, Boyce MC, Milne WI, Robertson J, Gleason KK (2003) Determination of mechanical properties of carbon nanotubes and vertically aligned carbon nanotube forests using nanoindentation. J Mech Phys Solids 51:2213–2237

Qu L, Dai L, Stone M, Xia Z, Wang ZL (2008) Carbon nanotube arrays with strong shear binding-on and easy normal lifting-off. Science 322:238

Riccardi B, Montanari R (2004) Indentation of metals by a flat-ended cylindrical punch. Mater Sci Eng A 381(1–2):281–291

Sager RJ, Klein PJ, Lagoudas DC, Zhang Q, Liu J, Dai L, Baur JW (2009) Effect of carbon nanotubes on the interfacial shear strength of T650 carbon fiber in an epoxy matrix. Compos Sci Technol 69(7–8):898–904

Shinozaki DM, Lo JCW, Lu YC (2008) Depth-dependnet displacement modulated indentation in oriented polypropylene. Mater Sci Eng A 491:182–191

Sneddon IN (1946) Boussinesq's problem for a flat-ended cylinder. Proc Cambridge Philos Soc 42:29

Tabor D (1996) Indentation hardness: fifty years on a personal view. Philos Mag A 74:1207

Terrones M, Grobert N, Olivares J, Zhang JP, Terrones H, Kordatos K, Hsu WK, Hare JP, Townsend PD, Prassides K, Cheetham AK, Kroto HW, Walton DRM (1997) Controlled production of aligned-nanotube bundles. Nature 388:52

Treacy MMJ, Ebbesen TW, Gibson JM (1996) Exceptionally high Young's modulus observed for individual carbon nanotubes. Nature 381:678–680

Tu JP, Zhu LP, Hou K, Guo SY (2003) Synthesis and frictional properties of array film of amorphous carbon nanofibers on anodic aluminum oxide. Carbon 41:1257–1263

Tu JP, Jiang CX, Guo SY, Fu MF (2004) Micro-friction characteristics of aligned carbon nanotube film on an anodic aluminum oxide template. Mater Lett 58:1646–1649

Wilsea M, Johnson KL, Ashby MF (1975) Indentation of foamed plastics. Int J Mech Sci 17:457–460

Wong EW, Sheehan PE, Lieber CM (1997) Nanobeam mechanics: elasticity strength and toughness of nanorods and nanotubes. Science 277:1971–1975

Wright SC, Huang Y, Fleck NA (1992) Deep penetration of polycarbonate by a cylindrical punch. Mech Mater 13:277

Yao Z, Kane CL, Dekker C (2000) High-field electrical transport in single-wall carbon nanotubes. Phys Rev Lett 84:2941–2944

Zhang Q, Liu J, Sager R, Dai L, Baur J (2009) Hierarchical composites of carbon nanotubes on carbon fiber: influence of growth condition on fiber tensile properties. Compos Sci Technol 69(5):594–601

Zhang Q, Lu YC, Du F, Dai L, Baur J, Foster DC (2010) Viscoelastic creep of vertically aligned carbon nanotubes. J Phys D: Appl Phys 43:315401

Index

A. Tiwari (ed.), *Nanomechanical Analysis of High Performance Materials*,
Solid Mechanics and Its Applications 203, DOI: 10.1007/978-94-007-6919-9,
© Springer Science+Business Media Dordrecht 2014